Phosphorus Chemistry in Everyday Living

SECOND EDITION

Arthur D. F. Toy

and

Edward N. Walsh

AMERICAN CHEMICAL SOCIETY

Washington, DC 1987

Library of Congress Cataloging-in-Publication Data

Toy, Arthur D. F. (Arthur Dock Fon), 1916–
Phosphorus chemistry in everyday living.

Bibliography: p.
Includes index.

1. Phosphorus. 2. Phosphates. 3. Organophosphorous compounds.

I. Walsh, Edward N., 1925–   . II. Title.

QD181.P1T69   1987   546′.712   86–32240
ISBN 0–8412–1002–0

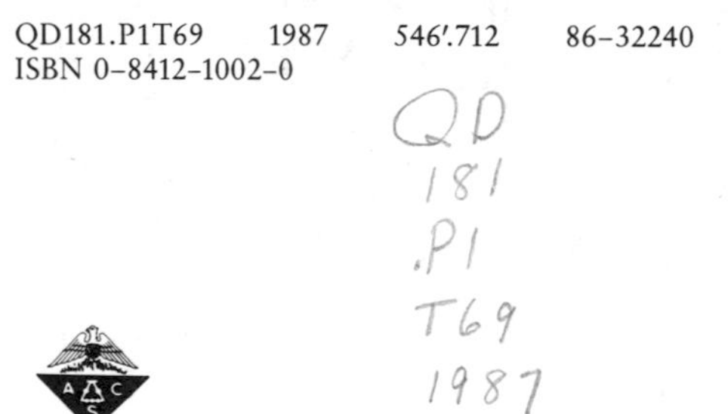

PRINTED IN THE UNITED STATES OF AMERICA

To
Hui-I Toy
&
Mary C. Walsh

About the Authors

Arthur D. F. Toy, presently a consultant, was formerly the Director of Research of Stauffer Chemical Co. He has actively pursued research on inorganic and organic phosphorus chemistry for more than 40 years. Many of the new chemical reactions, especially those involving the formation of carbon to phosphorus bonds that he and his co-workers have discovered, are used to manufacture industrial phosphorus compounds and intermediates. Practical applications of his research include the invention of the insecticide Aspon for the control of chinch bugs in lawns and turfs; processes for synthesizing intermediates for insecticides used to control pests in cotton, rice, and other crops; intermediates for making stabilizers used in nylon; and processes for making organic phosphorus polymers and flame-retardant compounds used in plastics. Dr. Toy approaches his research by carrying out basic studies on the nature of chemical reactions and then uses his findings to develop industrial chemical processes and compounds as well as their practical applications. He is the author of more than 25 scientific publications and holds more than 80 U.S. patents on phosphorus chemistry. He has lectured widely on various aspects of phosphorus chemistry in the United States and abroad.

Dr. Toy received his Ph.D. in 1942 from the University of Illinois. He has worked both as a research scientist and as a research administrator. His previous positions in research at Stauffer Chemical Company include that of Director Eastern Research Center, Chief Scientist, Senior Scientist, Manager of Specialties Department, and Director of Research, Victor Chemical Workers Division.

Dr. Toy belongs to numerous professional and technical societies and has been active in the American Chemical Society (ACS). He served as Chairman of the New York Section (1974–1976), as Chairman of the Westchester Chemical Society (1970–1971), and as a member of several ACS standing committees.

Edward N. Walsh has over 35 years of experience in phosphorus chemistry. His research has been directed primarily to the industrial, agricultural, and food applications of inorganic and organic phosphorus chemistry. Dr. Walsh has written 12 scientific papers and holds more than 60 U.S. patents related to the synthesis and use of phosphorus compounds. He previously served Stauffer Chemical Co. in several managerial and scientific roles, including Senior Scientist, Manager of the Chemical Product Development Department, and Manager of the Chemicals Department. He is currently consulting and serving as an Assistant Professor in the Chemistry Department of St. Peter's College, Jersey City, NJ.

Dr. Walsh received his Ph.D. in 1965 from Illinois Institute of Technology. He is a member of a number of professional and technical societies, including the American Chemical Society (ACS), where he served as Chairman of the New York Section in 1982 and as a member of several ACS standing committees.

Contents

Preface

Since the publication of the first edition, the role of phosphorus compounds in everyday living continues to grow. In updating the book, I am fortunate to have the collaboration of my friend and former colleague, Edward Walsh, as the coauthor. Walsh has worked in the area of industrial phosphorus chemistry for more than 30 years both as a research scientist and as a research administrator.

In the first edition, in many of the chapters I included anecdotes from my personal experiences. The personal pronoun "I" was liberally used. In the chapters updated by me in this edition, where personal anecdotes were included, I also use the personal pronoun "I". So in the text, wherever "I" is found, it refers to me.

The salts of phosphoric acids are ionic. The structural formulas in the text are drawn without showing the charge signs on the cations and anions so that the relationship of these salts to H_3PO_4 is more readily apparent. For the sake of simplicity, the charge signs are also omitted in some of the formulas for the salts of organic acids.

Besides expanding and updating the various chapters, we have also added two new chapters: Organic Phosphorus Polymers and Organic Phosphorus Plant Growth Control Agents, Herbicides, and Fungicides. At the end of each chapter, we have also added some pertinent references. We feel that such references may be of interest to those readers who want to delve more deeply into the subject.

The aim of this edition remains the same as that of the first edition. This book is intended primarily for practicing chemists, chemical executives, and chemical sales people who have a need for knowledge on phosphorus chemistry. The second edition is also intended for college students and their professors to provide them with a better understanding of the importance of the role phosphorus chemistry plays in their everyday living.

ARTHUR D. F. TOY

Acknowledgment

We thank our friends in the chemical industry for supplying us with specific items of information. Special thanks are given to J. E. Blanch, A. R. Petersen, C. Y. Shen, Edith Flanagan, W. F. Gentit, A. M. Aaronson, H. Harnisch, C. Kezerian, G. Fesman, C. A. Ertell, R. M. Lauck, and M. A. Kuck.

We are also indebted to S. F. Adler, E. P. Doane, R. I. Freudenthal, and E. D. Weil for review of the manuscript. Their comments and suggestions are appreciated. We also thank V. P. Moore for typing the manuscript.

ARTHUR D. F. TOY
Stamford, CT

EDWARD N. WALSH
Stauffer Chemical Company
Eastern Research Center
Dobbs Ferry, NY 10522

Introduction

Phosphorus is an unusual chemical element. Although the element can burst into flame spontaneously, some of its compounds impart nonflammability to a host of materials. In fact, they participate in some of the best flame-retarding agents known.

From a biological standpoint, too, phosphorus shows a peculiar ambivalence. Tiny quantities are essential for the proper nutrition of plants, animals, and man. Life's reproductive processes depend on the phosphorus-containing genetic code carriers DNA and RNA. Also, life cannot go on without energy exchanges mediated by vital phosphates such as adenosine triphosphate—called ATP by biochemists. Yet some phosphorus compounds, such as the nerve gases, are so poisonous that small amounts are lethal.

My own introduction to this element came with a high-school chemistry demonstration. Our teacher added a few drops of a solution to a piece of filter paper, set it aside, and went on with his lecture. Within a few minutes, much to our surprise, we noticed that the paper ignited spontaneously. Our teacher explained that he had dissolved a little phosphorus in the solvent, carbon disulfide (CS_2), which he added to the paper. When the volatile carbon disulfide evaporated, it left behind the elemental phosphorus, which quickly revealed its fiery nature by combining with oxygen in the air to produce the hot flame that ignited the filter paper.

This incident stayed with me vividly because when the teacher stepped out of the room, one of the students took a piece of filter paper and added quite a bit of the carbon disulfide–phosphorus solution to it. He wrapped this carefully with more paper and put it in his hip pocket to take home. Despite these precautions, the carbon disulfide must have evaporated because the phosphorus paper caught fire in his pocket, and he could not sit comfortably for quite a spell.

Within a single class of phosphorus compounds, some uses are quite similar although others vary greatly. Take the case of the family of calcium phosphate compounds. Monocalcium phosphate (CaH_2PO_4) is used as a leavening acid in baking to make tender biscuits. Dicalcium phosphate ($CaHPO_4 \cdot 2H_2O$) is used as a polishing agent in toothpaste. Tricalcium phosphate [$Ca_5(PO_4)_3OH$] is the conditioning agent in salt that keeps it flowing freely during humid weather.

Even a single phosphorus compound can have an amazing range of applications—for example, sodium tripolyphosphate ($Na_5P_3O_{10}$). This is an important ingredient in household detergents. In fact, it constitutes up to 45% by weight of most detergents displayed in supermarkets. Sodium tripolyphosphate is also used in applications quite unrelated to detergents. For example, one of the major ingredients used to pickle hams is sodium tripolyphosphate.

The tart taste of most carbonated cola beverages and root beer comes from their phosphoric acid content. Rustproofing of steel also involves treatment with phosphoric acid. A final example is organic phosphates that are used to make plastics pliable and workable and are also added to gasoline to make auto engines run more smoothly.

In short, phosphorus and its compounds show remarkable versatility. As shown in the main body of the book, phosphorus and its compounds are very important to our everyday living.

Chapter One

Phosphorus the Element

MORE THAN 300 YEARS AGO, in 1669, Hennig Brandt, a Hamburg alchemist, like most chemists of his day, was trying to make gold. He let urine stand for days in a tub until it putrified. Then he boiled it down to a paste, heated this paste to a high temperature, and drew the vapors into water where they could condense—to gold. To his surprise and disappointment, however, he obtained instead a white, waxy substance that glowed in the dark.

Brandt had discovered phosphorus, the first element isolated other than the metals and nonmetals, such as gold, lead, and sulfur, that were known in ancient civilizations. The word phosphorus comes from the Greek and means light bearer.

Although news of the discovery spread quickly throughout Germany, Brandt remained secretive. He sold his method first to Dr. Johann Daniel Krafft of Dresden and later to others. Krafft, in turn, tried to sell it and traveled through Europe and eventually to America exhibiting and demonstrating this luminous substance. A few other chemists learned how to make phosphorus after making deals with Brandt or Krafft or by working out their own method from hints of the original one. Robert Boyle, the illustrious English scientist, is said to have discovered the process independently. However, not until 1737, when the French Government studied the process and published its report, did this carefully guarded secret became public knowledge (*1*).

Pure phosphorus is a white, transparent, crystalline, waxy solid that glows in the dark and ignites in air just above room temperature (30 °C). Most of the commercially produced element, however, is not pure; it is yellowish and semitransparent. Hence, the terms yellow and white phosphorus are used for the same material. Phosphorus melts at 44.2 °C and can easily be cut with a knife. Because phosphorus ignites readily when exposed to air, it is usually stored under

1002–0/87/0001$06.00/1

water or in a closed container under a layer of inert gas such as carbon dioxide.

Alchemist Brandt made his phosphorus, which he called "cold fire", in a single, small pot, but today, phosphorus manufacture is a huge industry. In 1984, the United States alone produced more than 750 million pounds of the element. Today phosphorus is made in huge electric furnaces rather than in a small pot over a charcoal fire. In spite of this difference of scale, the chemistry of the two processes is remarkably similar. Brandt's process was based on a high-temperature reaction of inorganic phosphates with carbon. Inorganic phosphates, such as calcium phosphates and magnesium ammonium phosphates, are present in urine, and carbon is present in the form of organic compounds. The modern process for making phosphorus has improved Brandt's method only to the extent of adding silica—sand—to the reaction mixture.

Mining

For some time, calcium phosphate from animal bones was used as the source of inorganic phosphate, but eventually this supply became inadequate. By the early 1800s, the need for bones in England became so great that European battlefields were being combed for them. Present sources of calcium phosphate are minerals, commonly known as phosphorite or phosphate rocks. These minerals are mined in enormous, open pits with dragline buckets so large that an automobile can turn around in them. The phosphate rocks have an approximate composition of the mineral fluorapatite [$Ca_5(PO_4)_3F$], a calcium phosphate containing 4% fluorine. Phosphate rocks are contaminated with iron and aluminum oxides, carbonates, silicates, and organic matter. The largest known American deposits of phosphate rocks lie in Florida, Tennessee, North and South Carolina, Kentucky, Virginia, and the western states of Utah, Idaho, and Montana. Florida is the largest producer of phosphate rock in the United States. In 1983, Florida produced 80% of this country's rock and nearly one-third of the world's supply. Florida's phosphate industry is a $5 billion per year business. In other parts of the world, large deposits are found in the Kola Peninsula, near Kirovsk in the Soviet Arctic part of Scandinavia. Important deposits are also located in Algeria, Tunisia, Egypt, and Morocco. The origin of these phosphate rocks is believed to be the small quantities of phosphate normally present in Earth's granite rocks. Over eons, through weathering and leaching, the phosphates eventually found their way into the sea where marine animals absorbed and concentrated them into their shells, bones, and tissues. The remains of these animals accumulated on the bottom of the sea, and over the years, through geological changes, some of the phosphatic sediments appeared as deposits in dry land.

Electric furnaces for producing phosphorus are usually located near phosphate deposits and a source of cheap electric power. In the United States, most phosphorus furnaces are in Florida, Tennessee, Idaho, and Montana.

Manufacture

Mined phosphate rocks are actually granular particles. Before being charged into the electric furnace, the rocks are first washed to remove most of the clay impurities so that the resultant rock contains at least 28–30% P_4O_{10}. The rocks are then fused in a high-temperature kiln into lumps. These lumps are then crushed and classified into nodules about 1 in. in diameter; the pieces of silica and coke coreactants are matched in size. Coke is the source of carbon for the reaction.

The fusion of the granular phosphate particles into nodules is an important step in the process. Omitting this step results in the formation of fine phosphate dust that carries over to the condenser with the phosphorus vapor. When the phosphorus is condensed as a liquid under water, the presence of fine dust creates a phosphorus–dust emulsion known as phosphorus sludge. Recovery of the phosphorus from this sludge requires a separate and difficult step. Even with the fusion of phosphate particles into nodules, sufficient dust forms to create a sludge. However, such a sludge contains only 5–10% of the phosphorus produced.

After the phosphate nodules, silica, and coke are loaded into the furnace (*see* Figure 1.1), a high-voltage electric arc is struck in the furnace to produce temperatures of 1200–1450 °C. The exact nature of the chemical reaction is still not clear; chemists have speculated on it for years. Most textbooks show the reaction in one step:

$$2Ca_3(PO_4)_2 + 6SiO_2 + 10C \longrightarrow 6CaSiO_3 + 10CO + P_4$$

Another view is that the reaction takes place in two steps. When the mixture is heated to a high temperature, silica unites with the calcium oxide portion of the calcium phosphate to liberate phosphorus(V) oxide. When the temperature reaches 1450 °C in the electric furnace, the phosphorus oxide vaporizes. As this vapor meets carbon, the vapor is reduced to elemental phosphorus and the carbon is oxidized to carbon monoxide. In the two reactions shown here, calcium phos-

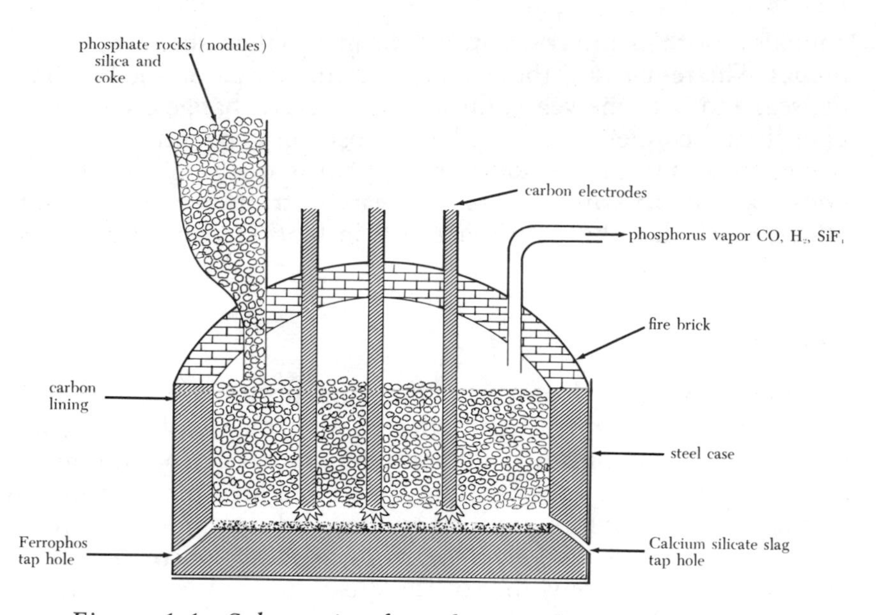

Figure 1.1. Schematic of an electric phosphorus furnace.

phate is written in the alternative form of two oxides (*2*):

$$2(3CaO{\cdot}P_2O_5) + 6SiO_2 \longrightarrow P_4O_{10} + 6CaSiO_3$$

$$P_4O_{10} + 10C \longrightarrow P_4 + 10CO$$

$$2(3CaO{\cdot}P_2O_5) + 6SiO_2 + 10C \longrightarrow P_4 + 6CaSiO_3 + 10CO$$

A second hypothesis visualizes three reactions. First, the carbon reacts with calcium phosphate to form carbon monoxide (CO) and calcium phosphide (Ca_3P_2) (*3*). The calcium phosphide then reacts with more calcium phosphate to form phosphorus and byproduct calcium oxide. The calcium oxide reacts immediately with silica (a base reacting with an acid) to form the salt, calcium silicate:

$$Ca_3(PO_4)_2 + 8C \longrightarrow Ca_3P_2 + 8CO$$

$$Ca_3P_2 + Ca_3(PO_4)_2 + 2C \longrightarrow P_4 + 2CO + 6CaO$$

$$6CaO + 6SiO_2 \longrightarrow 6CaSiO_3$$

$$2Ca_3(PO_4)_2 + 10C + 6SiO_2 \longrightarrow P_4 + 10CO + 6CaSiO_3$$

The summation of these equations, and of those in the two-step hypothesis, is identical with that of the one-step equation.

A third hypothesis is that the reduction of the calcium phosphate proceeds in stages and the formation of calcium phosphite is the first stage (*4*):

$$Ca_3(PO_4)_2 + \text{reductant} \longrightarrow Ca_3(PO_3)_2 + \text{oxide of reductant}$$

The phosphorus is formed as a result of the disproportionation of the calcium phosphite (*5*):

$$10Ca_3(PO_3)_2 \longrightarrow 12CaO + 6Ca_3(PO_4)_2 + 2P_4$$

Recent studies support this hypothesis (*6*).

The reaction could still take place in other ways, and chemists are still actively studying the mechanism of this phosphorus-producing reaction. With a better understanding of the nature of the reaction, possibly the process can be improved and a higher phosphorus yield obtained. Reference 5 gives a useful summary of studies conducted to understand the mechanism.

Silica in the furnace mixture is vital to large-scale production of phosphorus. Silica removes the calcium oxide that is formed so that the reaction can continue. Calcium oxide requires a very high temperature to melt. At the 1450 °C inside the electric furnace, calcium oxide is still a solid and would be difficult to remove in order to recharge the furnace with more raw material. Silica reacts with calcium oxide to form calcium silicate, which melts at the furnace temperature and collects at the bottom where it can be easily drained in a spectacular ribbon of fire.

Other chemicals are formed in the modern phosphorus furnace because phosphate rock also contains iron oxide, calcium fluoride, and clay. Even though the rock is washed to remove most of the clay, the rock is still quite impure.

Silica in the furnace mixture reacts with the calcium fluoride impurity to form silicon tetrafluoride, a toxic compound that is a gas at furnace temperature. So that the atmosphere is not polluted with this gas, silicon tetrafluoride is removed from the furnace vapors by scrubbing with a lime solution.

The iron oxide impurity is taken care of easily. High-temperature carbon reduces the iron impurity to molten iron, which in turn reacts with phosphorus to form iron phosphide, commonly called ferrophos. Ferrophos is a heavy liquid at electric furnace temperatures. As it forms, ferrophos sinks to the bottom of the furnace beneath the molten calcium silicate and can be drained away. The molten calcium silicate is removed after the removal of ferrophos.

To the phosphorus producer, formation of byproduct ferrophos is a costly nuisance. Every 100 lb of ferrophos formed takes about 25 lb of the valuable phosphorus. This amount is a major loss, but an unavoidable one because the iron oxide is present in phosphate ore and cannot be removed beforehand economically. Today, the few uses of ferrophos depend primarily on its high density. As a fine powder, ferrophos is mixed with dynamite for blasting. Its high density also makes ferrophos useful as a filler in high-density concrete used as radiation shields in nuclear reactor installations. Recently, ferrophos has been found useful as an additive for zinc-based anticorrosion paints. These paints are used to coat ship hulls. In this case, ferrophos is reported to also act as a flux should welding be required.

Phosphorus producers sell the calcium silicate slag byproduct as ballast for roadbeds in highway and railway construction.

Some efforts have also been made to recover fluorine from the byproduct silicon tetrafluoride. Reaction with water and soda ash converts silicon tetrafluoride to sodium hexafluorosilicate (Na_2SiF_6), which is used for the fluoridation of water.

The carbon monoxide byproduct from the phosphorus-producing reaction is also of value. When it burns in air to form carbon dioxide, carbon monoxide gives off heat, so it is used as fuel to supplement the heat needed to fuse the calcium phosphate mineral into nodules needed for the furnace reaction.

As a result of all these assorted manipulations, losses, and gains, the phosphorus produced in the United States sold for about $1800/ton, or 90¢/lb, in 1984.

In 1980, about 37 million tons of phosphate rock was consumed in the United States. Approximately 7% was used to produce elemental phosphorus. The remainder was converted to wet-process phosphoric acid and fertilizer products.

Literature Cited

1. Weeks, M. E. *Discovery of the Elements*, 4th ed.; Mack Printing: Easton, PA, 1939.
2. Jacob, K. D.; Reynolds, D. S. *Ind. Eng. Chem.* **1928,** *20*, 1204.
3. Franck, H.; Fuldner, H. *Z. Anorg. Allg. Chem.* **1932,** *204*, 97.
4. Postnikov, N. N. *Izv. Akad. Nauk. SSSR, Ser. Khim.* **1965,** 67–72.
5. Kriklivyi, D. I. *Z. Prikladnoi Khim.* **1984,** *11*, 2409–2417.
6. Ershov, V. A. *Tr. Leningr. Nauchno-Issled. Proektn. Inst. Osnovn. Khim.* **1972,** *5*, 57–65.

Chapter Two

Phosphorus in Matches and in Warfare

EARLY CHEMISTS MUST HAVE BEEN INTRIGUED by the fact that phosphorus ignites easily in air, especially because fire had to be started by nursing flame from a spark produced by striking steel and flint together. Although the idea of starting a fire with phosphorus occurred to many people, modern matches, which we take so much for granted, were not developed overnight. More than 200 years passed before the familiar book matches of today evolved.

The early phosphorus matches were crude. In the 1780s, the expensive "ethereal match" was fashionable. This match consisted of a slip of paper tipped with white phosphorus and sealed in a glass tube. When a flame was needed, the glass tube was broken, air entered and oxidized the phosphorus, and the paper ignited. A modification of this device was the "phosphorus box", which consisted of sulfur-tipped wood splints sold with a small bottle of impure phosphorus that, because of the impurities, is a liquid. So that a fire could be started, the splint was dipped in the phosphorus and then held in air. Air ignited the phosphorus, which ignited the sulfur, which ignited the splint.

The next improvement was the "strike-anywhere" match. Research showed that white phosphorus does not ignite as readily when coated with glue or starch. White phosphorus will ignite, however, by heat generated from friction such as by striking. This discovery led to the development of today's wooden matches. The early strike-anywhere match consisted of a head of white phosphorus as the igniting agent and underneath a body of potassium chlorate ($KClO_3$) as the oxidizing agent. These substances were held together by animal glue and starch around the sulfur-coated tip of a wood splinter. This invention took advantage of a special property of potassium chlorate,

1002–0/87/0007$06.00/1

which loses all its oxygen when heated or rubbed with phosphorus, by oxidizing the phosphorus to phosphoric oxide (P_4O_{10}). The reaction generates large amounts of heat. When a phosphorus-tipped chlorate match is struck, frictional heat causes the phosphorus to ignite in air. At the same time, the phosphorus and organic binders are rapidly oxidized by potassium chlorate. The heat of the reaction quickly ignites the sulfur, which transmits the flame to the wooden splint. Because this rapid reaction can be explosive, inert materials, such as diatomaceous earth or ground glass, are added to retard combustion.

We face two major problems in using white phosphorus for matches. First, white phosphorus is extremely poisonous and can cause "phossy jaw", an illness that exacted a fearful toll among early matchmakers. The disease is so named because it is caused by phosphorus fumes that are inhaled or absorbed through cavities in the teeth and that then attack and destroy bones, particularly the jaw bone. Phossy jaw is usually fatal and was very prevalent before automatic matchmaking machines were invented. Because matches were first made by hand in the homes of the poor, many of these people suffered this horrible death. Other innocent persons, including babies, were poisoned by accidentally eating or chewing these white phosphorus matches. The fatal dose of phosphorus to humans is about 0.1 g, less than 0.01 oz. Phossy jaw is a constant threat even today to people who work with elemental phosphorus. The danger, however, is minimized by avoiding phosphorus vapors and by strict adherence to good dental care—absolutely no unfilled cavities. Each phosphorus-producing facility in the United States has a resident dentist or dental technican to check workers for cavities and to fill them.

The second problem encountered with white phosphorus matches is that they ignite so easily. Accidental fires can be readily started, sometimes even by the gnawing of rats and mice. Obviously, white phosphorus had to be replaced with a material that would be safer—from both the health- and fire-hazard standpoints.

This material turned out to be a noncrystalline (amorphous) form of phosphorus, called red phosphorus. Red phosphorus is produced by heating white crystalline phosphorus to about 400 °C under an atmosphere of an oxygen-free inert gas, such as nitrogen or argon. During the heating, the white phosphorus (P_4), which has a tetrahedral molecular structure containing four phosphorus atoms

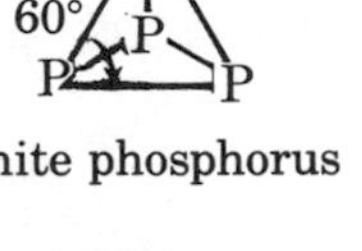

P
2.21 Å
60°
P
P P

white phosphorus

is believed to rearrange by the cleavage of one or more of the P–P bonds. The cleaved bonds of many P_4 were first thought to join to form a polymer of the chain structure

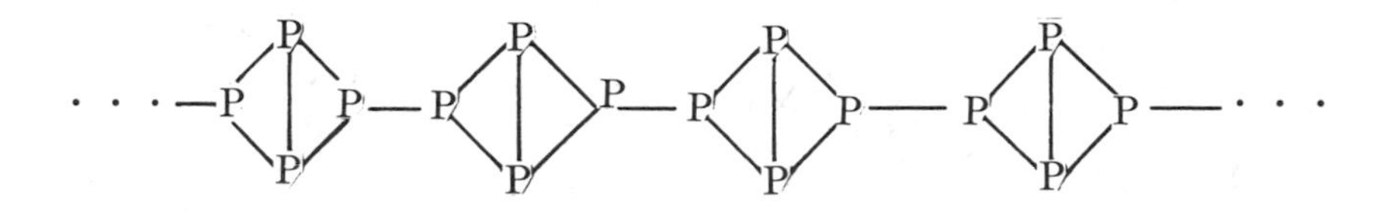

Although the structure of red phosphorus is still the topic of considerable discussion (*1–4*), recent reports indicate the basic structural unit of amorphous red phosphorus is a pentagonal tube of phosphorus atoms. Therefore, this structure may be considered as an intermediate that subsequently rearranges to the more stable pentagonal tube structure. Similar pentagonal tubes of phosphorus atoms are known to be the structural units of Hittorf's phosphorus, a crystalline, monoclinic violet form of phosphorus. Examples of the structure of these pentagonal tubes in Hittorf's phosphorus are shown in Figure 2.1. Each corner in the pentagonal tube is occupied by a phosphorus atom.

The amorphous form of red phosphorus, incidentally, can be converted to crystalline forms by heating to temperatures below the melting point. Various crystal forms have been reported, depending on the heat treatment used during this conversion. For example, at 540 °C, amorphous red phosphorus is converted to a tetragonal form; at 600 °C, a cubic form is obtained; and after red phosphorus is annealed at 550 °C, a triclinic form can be isolated.

Amorphous red phosphorus is not regarded as poisonous in comparison to white phosphorus. Also, red phosphorus has an ignition temperature significantly higher than that of the white form (260 versus 30 °C). These properties led to the development of red phosphorus as a replacement for white phosphorus in matches. This discovery paved the way for safety matches, because despite its higher ignition temperature, amorphous red phosphorus still undergoes the rapid oxidative action with potassium chlorate ($KClO_3$), which leads to flame formation.

Matches and Warfare

These safety matches (wooden as well as book matches) had as basic ingredients a head of potassium chlorate and sulfur bound, with glue, around the tip of a splint that had been dipped in paraffin. The

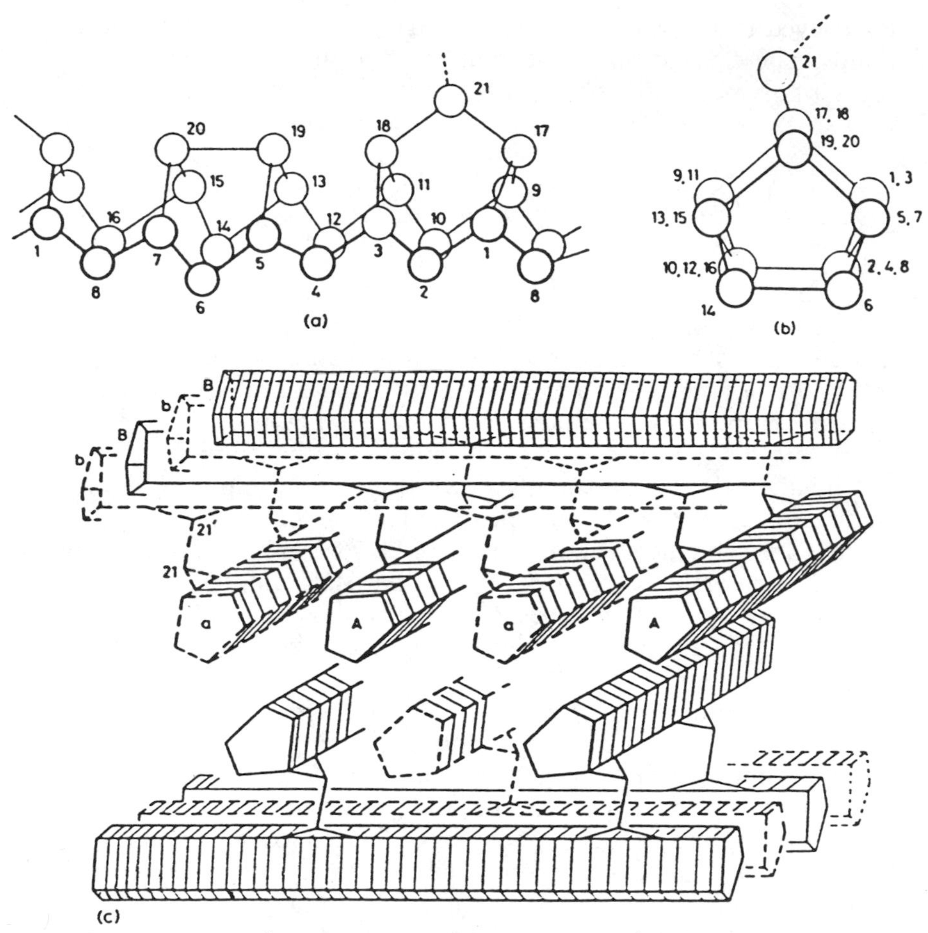

Figure 2.1. The structure of Hittorf's phosphorus. (a) Side view of a single pentagonal "tube", (b) end view of a tube, and (c) assembly of tubes in crystal structure. (*Reproduced from reference 5. Copyright 1974 Elsevier.*)

separate friction striking surface had two basic ingredients—mostly red phosphorus as the igniting agent and glass powder to provide the friction. These components are bound by glue. Zinc oxide and calcium carbonate stabilizers are added to protect the red phosphorus against degradation on exposure to air and moisture. Without such stabilizers, the striking surface would turn black and soggy and would not light the match when struck.

The wooden or paper splint of the match is treated first with a solution of monoammonium phosphate ($NH_4H_2PO_4$), a good flame retardant, to prevent afterglow when the flame is extinguished. The no-afterglow match, for obvious reasons, was called the drunkard's match. The no-afterglow treatment has prevented many accidental fires from careless users who discard a still smouldering match. Sulfur was once used with the red phosphorus, in both strike-anywhere and safety matches. However, at the end of the 19th century, phosphorus sesquisulfide (P_4S_3), in which phosphorus and sulfur are chemically combined in a four P to three S ratio, was introduced as the igniting agent. This compound is even less toxic than red phosphorus and is less affected when exposed to the atmosphere. Phosphorus sesquisulfide does not react with water at ordinary temperatures. Above 100 °C, phosphorus sesquisulfide burns in air. This compound is produced by the reaction of white phosphorus with molten sulfur at 320–380 °C. Phosphorus sesquisulfide is then separated from other phosphorus sulfides formed during the reaction, such as those having P-to-S ratios of 4:10 and 4:7, by washing with water or dilute sodium bicarbonate solution and drying at 40–50 °C.

The ingredients of our modern strike-anywhere match are thus a tip of phosphorus sesquisulfide as igniting agent, zinc oxide as the stabilizer, and potassium chlorate as the oxidizer, held by glue with powdered glass. The bulb contains less phosphorus sesquisulfide and more potassium chlorate to assist the initial oxidation whereby the match splint is raised above its flame point.

The next time you strike a match, consider all the chemical ingenuity that has gone into that small, instant flame. Although the invention and development of matches have solved many of the problems of obtaining an instant flame, the quest for improved products is never ending. Phosphorus-containing matches must be thrown away after each use. What a challenge the development of a reusable permanent match other than the cigarette lighter would be.

Use of Phosphorus for Incendiaries

White phosphorus burns rapidly in air to form phosphoric oxide (P_4O_{10}). In the atmosphere, phosphoric oxide appears as a dense white smoke. During World War I, the military forces discovered that the obscuring power of this smoke per unit weight of phosphorus was greater than that of any smoke-generating chemicals then known. Taking advantage of this property, munitions makers produced many phosphorus shells for artillery use. The opaque cloud of phosphoric oxide smoke hides an army from the enemy yet is sufficiently innocuous for soldiers to penetrate it with little discomfort. Phosphorus-filled shells and grenades continued to be used in World War II. In application on the battlefield, these shells were also effective weapons. A barrage of phosphorus shells rained small particles of burning phosphorus that stuck tenaciously to clothing and skin. The result was often burning the enemy and subjecting them to injury or death or forcing them out of their foxholes and exposing them to gunfire.

Phosphorus shell manufacture was refined in World War II. Normally, the shells are filled with molten phosphorus, which solidifies into a large chunk. When the shell explodes, solid phosphorus is broken up and ignited, but a nonuniform distribution of white phosphorus smoke for screening is produced. If small particles of the solid phosphorus are uniformly mixed with a rubber-gasoline gel, the smoke screen obtained is more homogeneous.

Self-igniting phosphorus, when combined with benzine (gasoline), is called a Molotov cocktail. During World War II, right after the evacuation from Dunkirk, France, beer and milk bottles were used by the British to prepare millions of Molotov cocktails for the defense of England. Storage of these bottles gave the local citizenry rather exciting experiences. For safety, the bottles were usually submerged in a nearby stream. Occasionally, however, boxes of 1000 or so bottles broke loose and floated away; considerable concern and several spectacular fireworks displays resulted.

The sensitivity of phosphorus to ignition is also used for amusement. Caps for toy pistols are made with a mixture of two separate pastes—one of potassium chlorate and the other of red phosphorus, sulfur, and calcium carbonate. A water slurry of the pastes containing a gum binder is dabbed on paper to make caps. The impact of metal on the mixture causes the explosive ignition heard when the trigger is pulled.

Literature Cited

1. Krebs, H.; Gruber, H. U. *Z. Naturforsch., A* **1967,** *22,* 96.
2. von Thurn, H.; Krebs, H. *Acta Crystallogr., Sect. B* **1969,** *41,* 327.
3. Beyeler, H. U.; Veprek, S. *Philos. Mag.,* [*Part*] *B* **1980,** *41,* 327.
4. Olego, D. J.; Baumann, J. A.; Kuck, M. A.; Schachter, R.; Michel, C. G.; Raccah, P. M. *Solid State Commun.* **1984,** *52* (3), 311.
5. Corbridge, D. E. *The Structural Chemistry of Phosphorus*; Elsevier: New York, 1974.
6. Brophy, L. P.; Fisher, G. J. B. *The Chemical Warfare Service: Organizing for War*; Department of Army: Washington, DC, 1959.

Chapter Three

Phosphoric Acids

A HOT DAY AND VIGOROUS EXERCISE will make most of us reach for a refreshing cola drink. We never think about the fact that we are actually drinking a flavored, carbonated, sweetened, dilute solution of phosphoric acid. We give even less thought to the fact that the bottle cap is made from sheet metal that has been phosphatized in a phosphoric acid solution so that the cola will not rust through the enamel coating or that the bottle is cleaned in an alkaline bath containing sodium phosphates. Without realizing it, we use phosphoric acid and its derivatives every day of our lives. Although the phosphoric acid family has many members, only two fairly pure forms can be isolated in large quantities: orthophosphoric and pyrophosphoric acid. Orthophosphoric acid—or simply phosphoric acid—is manufactured by two processes, thermal and wet.

Thermal Phosphoric Acid

Thermal phosphoric acid is the general term applied to phosphoric acid prepared by the reaction of water with phosphoric anhydride [also called phosphorus(V) oxide and often written as P_4O_{10}] obtained by burning elemental phosphorus in air. The reactions are

$$\underset{\text{phosphorus}}{P_4} + \underset{\text{oxygen}}{5O_2} \longrightarrow \underset{\text{phosphoric anhydride}}{P_4O_{10}}$$

$$P_4O_{10} + 6H_2O \longrightarrow \underset{\text{phosphoric acid}}{4H_3PO_4}$$

Although this process involves two steps, in practice the reactions are carried out consecutively in a single system of reactors. The phosphoric anhydride (P_4O_{10}) formed by burning phosphorus is hydrated

1002–0/87/0013$06.00/1

immediately with water. Both reactions generate considerable heat; hence, the term "thermal acid".

The heat generated is removed by cool water; a great cloud of steam is produced from the tops of the reactors. On a cold day, the cloud looks like the P_4O_{10} smoke screen from a phosphorus bomb. At a production plant where I was once working, a pollution control officer thought we were allowing P_4O_{10} to escape into the atmosphere. Our engineer had to climb with him to the top of the reactors, condense some of the steam, and then drink it to prove that the vapor was harmless.

The concentration of industrially produced phosphoric acid is usually 80–90%. This phosphoric acid contains some arsenic as arsenic oxide or arsenic acid. The source of arsenic is calcium arsenate impurity in the phosphate rock reduced in the electric furnace to elemental arsenic. Arsenic codistills with phosphorus and oxidizes along with it during burning. Because a lot of thermal phosphoric acid is used in food, toxic arsenic must be removed. The usual technique is to add gaseous hydrogen sulfide, which smells like rotten eggs. Arsenic is precipitated as the insoluble arsenic sulfides and is removed by filtration. Any residual hydrogen sulfide is then removed by blowing air through the acid.

Next, the treated phosphoric acid is diluted to 85, 80, or 75% strength—the common commercial concentrations. Treated phosphoric acid can also be concentrated by evaporating the water at high temperature to give the equivalent of 105% H_3PO_4. This compound is "superphosphoric acid", and its P_4O_{10} content is 76.5%. Finally, superphosphoric acid can be concentrated to 82–84% P_4O_{10}, the strength of commercial polyphosphoric acid.

Wet-Process Phosphoric Acid

Unlike thermal phosphoric acid, wet-process phosphoric acid is made directly from calcium phosphate rock. Sulfuric acid reacts with calcium phosphate to form water-insoluble calcium sulfate ($CaSO_4 \cdot 2H_2O$, gypsum) and a water solution of phosphoric acid. The gypsum precipitate is removed by filtration, and the dilute water solution of phosphoric acid is then concentrated by boiling off water until the desired phosphoric acid concentration is obtained. When the mineral fluorapatite, $Ca_5(PO_4)_3F$, is used as the calcium phosphate source, the reaction is

$$Ca_5(PO_4)_3F + 5H_2SO_4 + 10H_2O \longrightarrow \underset{\text{gypsum}}{5CaSO_4 \cdot 2H_2O} + \underset{\text{phosphoric acid}}{3H_3PO_4} + \underset{\text{hydrogen fluoride gas}}{HF}$$

Today two basic wet processes exist for phosphoric acid: the dihydrate process and the hemihydrate process. The names indicate the type of calcium sulfate formed during the phosphate rock digestion. The dihydrate process is operated at 170–180 °F, a temperature range at which calcium sulfate dihydrate ($CaSO_4 \cdot 2H_2O$) or gypsum is formed. The desired temperature range is maintained by removing the excess heat from the reaction by boiling off some water under reduced pressure. This form of calcium sulfate has a crystal size and shape that are easy to filter; however, the process produces a more dilute solution of phosphoric acid containing 25–28% P_4O_{10}. This dilute solution requires more thermal energy to concentrate it to the desired P_4O_{10} concentration of around 54%, which corresponds to 75% H_3PO_4. The hemihydrate process (*1*) is operated at higher temperatures, 190–200 °F, where the calcium sulfate precipitates as $CaSO_4 \cdot 0.5H_2O$. This process has the advantage of producing a more concentrated solution of phosphoric acid right from the start, containing 42–50% P_4O_{10}. However, the crystals of $CaSO_4 \cdot 0.5H_2O$ obtained are of such shape and fine size that they are difficult to filter. Also, in this process, an anhydrous calcium sulfate ($CaSO_4$) is readily formed as a scale or coating on the reactor and heat exchanger walls; thus, the efficiency of heat transfer is reduced. Because of the higher temperatures used, greater corrosion of the equipment occurs. Today, worldwide, approximately 80–85% of the wet-process phosphoric acid plants use the dihydrate process.

Commercial production of wet-process phosphoric acid dates back to about 1850. The original commercial production of elemental phosphorus (on a very small scale) used wet-process phosphoric acid as raw material. Most earlier wet-process acid was used to manufacture monocalcium phosphate for use in baking powder. By the 1870s, the acid was being used in Germany to make fertilizers.

Originally, animal bones were the source of calcium phosphate. As described in Chapter 1, at one time the shortage of bones was so acute in England that old battlefields in Europe were turned to for their supply of bones of dead soldiers. In many areas, bones were the raw material used into the 20th century.

August Kochs, the founder of Victor Chemical Works, later acquired by the Stauffer Chemical Co., was one of the pioneers in the manufacture of phosphoric acid and phosphates. Kochs started his business near Chicago in the early 1900s. Every morning he would drive his horse-drawn wagon to the Chicago stockyard to buy animal bones. In the afternoon, he went out to sell his products. For 50 years, the phosphate industry grew, and so did his business. Animal bones were eventually replaced by phosphate minerals, and wet-process phosphoric acid grade was superseded by thermal phosphoric acid. By the time Kochs died, the business he founded was sold for over $100 million.

The greatest difficulty in manufacturing wet-process phosphoric acid is to control the reaction conditions so that precipitated gypsum is easy to filter and wash free of residual phosphoric acid. Originally, only a 15% concentration of phosphoric acid could be made directly. Today, through process modifications, as indicated earlier, concentrations of 40–50% before evaporation are possible. Even though these concentrations are good enough for many applications, impurities such as fluorosilicate, iron, aluminum, and magnesium must be removed to produce the industrial-grade sodium phosphates used in detergents and cleaning products (but not pure enough for use in food).

Applications. Most wet-process phosphoric acid is used in fertilizer. Fertilizer triple superphosphate is prepared by the action of wet phosphoric acid on finely ground phosphate rock. The wet-process acid—when concentrated to superphosphoric acid—is used to make high-analysis liquid fertilizer.

Today, liquid fertilizers are often prepared from ammonia and wet-process phosphoric acid. These solutions are usually unstable and must be used within a few days of mixing. These fertilizer solutions are unstable in the sense that they are supersaturated with many of the salts originally present in wet-process phosphoric acid as impurities. Because many of these salts, such as those of zinc, copper, manganese, and boron, are also micronutrients for the plants, preferably, they should not be allowed to precipitate. Also, if the salts do precipitate, the precipitated sludge can make dispensing the fertilizer more difficult. Therefore, these fertilizer solutions should be used soon after they are made. Materials such as urea are sometimes added in place of part of the ammonia. Urea too provides usable nitrogen to the plant, but it also extends the time the fertilizer solution can be stored before metal salts precipitate.

Industrial-grade sodium phosphates are made after removing some impurities from wet-process acid. First, the acid solution is adjusted with sodium carbonate (Na_2CO_3, soda ash) to a pH of about 2. At this point, most of the fluoride impurity precipitates out as insoluble sodium fluorosilicate (Na_2SiF_6) and is filtered off. (Incidentally, pure sodium fluorosilicate is used to fluoridate water.)

Next, the pH of the solution is raised to around 5. At this point, most of the iron and aluminum impurities, along with some residual sodium fluorosilicate, come out. The mixture is filtered again, and the resulting solution is essentially monosodium phosphate. This compound can be converted to other sodium phosphates by reaction with more soda ash or sodium hydroxide or by other treatments.

These phosphate salts still contain impurities that make them unsuitable for food applications, but they can be used in detergents and other cleaning products.

The American production of phosphoric acid for 1984 was equivalent to more than 11,000,000 tons of 100% P_4O_{10}.

A New Process. Wet-process phosphoric acid is made with sulfuric acid rather than hydrochloric acid so that the byproduct, gypsum, which is insoluble in water, can be removed by filtration. If hydrochloric acid were used, the byproduct would be water-soluble calcium chloride ($CaCl_2$):

$$Ca_5(PO_4)_3F + 10HCl \longrightarrow 3H_3PO_4 + 5CaCl_2 + HF$$

Separation of water-soluble calcium chloride from a water solution of phosphoric acid is a sticky problem, but it was solved neatly by Israel Mining Industries (2). Phosphate rock is digested with hydrochloric acid to give a water solution of phosphoric acid and calcium chloride. The acid is then selectively extracted by using an organic solvent. Phosphoric acid is then stripped from the organic solvent with water. Both butyl and amyl alcohol are water-insoluble and can dissolve phosphoric acid but not calcium chloride. Other solvents are also suitable. In 1983, many plants used versions of this process. Some of these plants are located in Spain, Mexico, Japan, Great Britain (3), and France. Together, these plants have a total capacity of over 120,000 tons of purified wet-process phosphoric acid.

Even crude phosphoric acid obtained by the sulfuric acid wet-process can be purified by the solvent extraction process described previously. After each extraction stage, more and more impurities are left in the discarded water layer. The number of extractions used depends on the purity needed for the specific end use. The higher the number of extractions used, the purer is the acid. Although many problems remain, many of the solvent extraction processes that are currently being operated commercially are producing a high-quality grade of phosphoric acid, suitable for use in manufacturing food-grade phosphates.

Classes of Phosphoric Acids

Many members are included in the phosphoric acid family. If the members are examined individually, their relationship to each other will be clear. (Superphosphoric acid is discussed in Chapter 6 on fertilizers.)

The first member is orthophosphoric acid; it is common phosphoric acid and usually the prefix ortho is omitted. Its formula is H_3PO_4.

If two molecules of phosphoric acid are heated to remove one molecule of water, the result is pyrophosphoric acid (also called diphosphoric acid) containing two phosphorus atoms. Its P_4O_{10} content is 79.8%. The reaction is

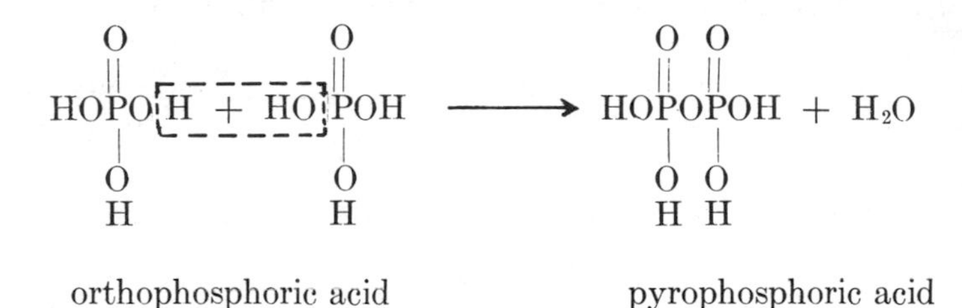

orthophosphoric acid pyrophosphoric acid

If three molecules of phosphoric acid are heated to eliminate two molecules of water, tripolyphosphoric acid is formed:

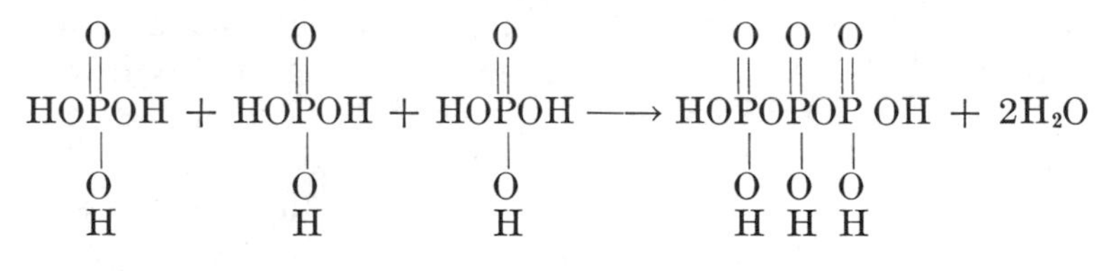

orthophosphoric acid tripolyphosphoric acid

Tripolyphosphoric acid has a P_4O_{10} content of 82.6%. By continuing in this way, theoretically, phosphoric acids of various chain lengths containing more and more phosphorus atoms can be made. An acid with four phosphorus atoms is called tetrapolyphosphoric acid, one with five is called pentapolyphosphoric acid, and so on.

Three molecules of orthophosphoric acid can also be condensed into one molecule by eliminating three instead of two molecules of water. Theoretically, in this case, these molecules could form a cyclic trimetaphosphoric acid:

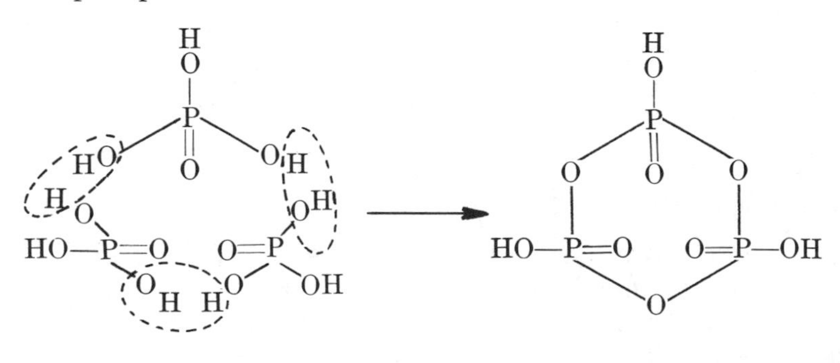

orthophosphoric acid cyclic trimetaphosphoric acid

Similarly, four molecules of phosphoric acid theoretically could condense into a cyclic tetrametaphosphoric acid by eliminating four molecules of water.

Although these cyclic acids exist, they are not made by this water-elimination method. These acids are only mentioned to give a complete picture of phosphoric acids and polyphosphoric acids and their relationship to each other.

Properties of Phosphoric Acids

Except for orthophosphoric and pyrophosphoric acid, large quantities of the individual polyphosphoric acids are difficult to isolate. As indicated earlier, when two molecules of phosphoric acid are heated to remove one molecule of water, the residue contains 79.8% P_4O_{10}. Because this amount is the exact theoretical percentage of P_4O_{10} for pyrophosphoric acid, we would expect this residue to be pure pyrophosphoric acid. However, analysis by paper chromatography shows that the residue contains only about 43% pyrophosphoric acid; the remainder is about 17% ortho-, 23% tripoly-, 11% tetrapoly-, a lesser amount of pentapoly-, and "hypolyphosphoric acid".

[Paper chromatography is a special technique for separating and detecting compounds. When compounds of the same class are treated with a proper solvent on absorbent paper such as filter paper, the smallest (lowest molecular weight) moves the fastest and the farthest away from the starting point. The largest (highest molecular weight) moves the slowest and travels the shortest distance. The in-between members travel varying distances in inverse relation to their molecular weights. Thus, when a mixture of phosphoric acid of various chain lengths or molecular weights is put as a spot on a filter paper and washed with solvent, after some time, orthophosphoric acid will have moved the farthest from the origin, followed by pyrophosphoric acid, then tripolyphosphoric acid, tetrapolyphosphoric acid, pentapolyphosphoric acid, and so on. Because the acids are colorless, these migration spots cannot be seen. However, the acids can be made visible if they are all converted to the orthophosphate by partial hydrolysis and then permitted to react with special reagents of molybdate and stannous chloride to produce blue spots. Present techniques can separate polyphosphoric acids as large as those with nine phosphorus atoms. Polyphosphoric acids above nine are lumped together as "hypolyphosphoric acid".]

When three molecules of phosphoric acid are condensed by eliminating two molecules of water, the residue has a P_4O_{10} content of 82.6%, the theoretical value for pure tripolyphosphoric acid. Paper chromatographic analysis shows a content of only about 17% tripolyphosphoric acid, the total being various amounts of acids of different chain lengths ranging from the ortho with one phosphorus atom to acids with two, three, four, five, six, seven, eight, and nine phosphorus atoms on up.

Syrupy condensed phosphoric acid with a specific P_4O_{10} content is thus not a pure compound. It is a mixture of polyphosphoric acids in equilibrium with each other. The P_4O_{10} content controls the equilibrium composition. The equilibrium composition of the acid is the same whether the acid is made by removing water from orthophosphoric acid or by adding P_4O_{10}.

Orthophosphoric acid has a theoretical P_4O_{10} content of 72.4% (*see* Table 3.I). According to some investigators, a liquid phosphoric acid with this P_4O_{10} content contains less than 90% orthophosphoric acid; the remainder is pyrophosphoric acid and some unreacted water. However, when orthophosphoric acid is crystallized from a liquid of 72.4% P_4O_{10} content, the equilibrium shifts toward the pure H_3PO_4 and eventually all of the material is converted to pure, solid crystalline

Table 3.I. Composition of Strong Phosphoric Acids

Composition (wt % P_4O_{10})	*Content (wt %)*									
	Ortho	*Pyro*	*Tri*	*Tetra*	*Penta*	*Hexa*	*Hepta*	*Octa*	*Nona*	*Hypoly*
68.80	100.00	—[a]								
69.81	97.85	2.15								
70.62	95.22	4.78								
72.04	89.91	10.09								
72.44	87.28	12.72								
73.43	76.69	23.31								
74.26	67.78	29.54	2.67							
75.14	55.81	38.88	5.31							
75.97	48.93	41.76	8.23	1.08						
77.12	39.86	46.70	11.16	2.28						
78.02	26.91	49.30	16.85	5.33	1.60					
78.52	24.43	48.29	18.27	6.75	2.26					
79.45	16.73	43.29	22.09	10.69	4.48	1.92	0.80			
80.51	13.46	35.00	24.98	13.99	6.58	3.14	2.84			
81.60	8.06	27.01	22.28	16.99	11.00	5.78	3.72	2.31	1.55	1.28
82.57	5.10	19.91	16.43	16.01	12.64	8.89	6.41	4.11	3.51	6.99
83.48	4.95	16.94	15.82	15.91	12.46	9.71	6.77	5.04	2.99	9.42
84.20	3.63	10.60	11.63	13.05	12.17	9.75	8.19	5.92	4.91	20.16
84.95	2.32	6.97	7.74	11.00	10.45	9.62	8.62	7.85	6.03	29.41
86.26	1.54	2.97	3.31	5.16	5.32	5.54	3.51	3.30	3.30	66.03

SOURCE: Reproduced with permission from reference 4. Copyright 1956 National Research Council of Canada.

[a]Trace.

orthophosphoric acid with a melting point of 42.35 °C. Similarly, the pure pyrophosphoric acid with a melting point of 54.3 °C can be crystallized in the pure form from a liquid mixture of phosphoric acids having a theoretical average P_4O_{10} content of 79.8%. Orthophosphoric and pyrophosphoric are the only acids that have been isolated in a pure form in this manner.

Literature Cited

1. *Chem. Eng. News* **1983,** *Nov. 7*, 37.
2. Blumberg, R. *Solvent Extr. Rev.* **1971,** *1*, 93–104.
3. *Phosphorus Potassium* **1974,** *71*, 35–37.
4. Huhtu, A.-L.; Gartaganis, P. A. *Can. J. Chem.* **1956,** *34*, 790.

Wet-process phosphoric acid plant in Pierce, FL.

Chapter Four

Food Uses of Phosphates

IF A TOOTH IS ALLOWED TO STAND in a glass of carbonated cola drink overnight, some people say the tooth will completely dissolve. When one of my sons was about 5 years old, I decided to test this theory. My wife and I had tried to discourage him from drinking too much soda pop by telling him that the dilute acid solution was not good for his teeth. But that explanation did not impress him. Hoping to demonstrate our point graphically, we put a baby tooth he had just lost into a glass of well-known cola overnight, predicting that the tooth would dissolve by morning. When I awoke first and found the tooth still in the glass, I fished it out. When our son saw that the tooth was gone, he was astonished at this evidence of cola corrosion, but as honest parents, we confessed the little trick we had played on him. However, to show that we were not entirely wrong, we urged him to feel the tooth. It had become quite soft from the etching action of the phosphoric acid.

Tartness in Carbonated Beverages

The tartness that makes drinks so refreshing comes from added acids, such as citric (citrus fruit acid), tartaric (grape acid), malic (apple acid), and phosphoric. Fruit acids are used in fruit-flavored drinks. Cola drinks, root beer, sarsaparilla, and similar soft drinks generally contain phosphoric acid in amounts ranging from 0.013% to 0.084% of 75% phosphoric acid. At this writing, the soft drink market is in excess of 9.2 billion gallons per year. The food-grade phosphoric acid used in these drinks must be very pure, and quality specifications are the most stringent in the business. In analyzing for phosphoric acid content, the cola manufacturer requires three different methods, the results of which must cross-check.

1002–0/87/0023$11.00/1

The acidity of cola is measured by determining its pH, that is, the negative logarithm of its hydrogen ion concentration. The higher the pH, the lower the acidity; a pH of 7 is neutral. Cola contains from 0.057 to 0.084% of 75% phosphoric acid. Colas are strongly acidic, with a pH of around 2.3–2.5. Root beer, which contains only about 0.013% of 75% phosphoric acid, has a pH close to 5. Do not be shocked at such strong acidity in these drinks because the pH of the human stomach normally runs around 2.5 and during digestion can go as low as 0.9. Each acid, depending on its chemical nature, gives a different hydrogen ion concentration when the same amount is dissolved in a fixed volume of water. This situation occurs because each acid dissociates, or ionizes, to a different degree. This ionization results in the liberation of hydrogen ion, which is the measure of acidity. In other words, different weights of the different acids are needed to obtain solutions of the same acid strength.

Acid strength is reflected in the sourness or tartness of a soft drink. The stronger the acid, the more sour the drink. Acid strength can also be influenced by other ingredients—for example, buffering agents. The only true test for sourness is the taste test. In a sample test, on a 100% basis, 1 lb of phosphoric acid was equal in sourness to 4.25 lb of citric acid or 5.5 lb of tartaric acid. Because, in 1984, citric acid cost about 81¢/lb and 75% phosphoric acid cost about 27¢/lb, considerable economic advantage is gained in using phosphoric acid if sourness alone is the controlling factor.

However, the more expensive fruit acids enhance the special flavor of certain fruit drinks. Also, if economic considerations alone dictated acid choice, the cheapest acid is sulfuric or hydrochloric. Neither is used to any extent, however, because their addition to food is regarded as questionable by some nutritionists. On the other hand, phosphoric acid is low cost, and phosphates are not considered harmful; as shown later, they can actually be beneficial to health.

Nutrient Supplements

As a boy in Southern China, I often accompanied my parents when they took gifts to friends with newborn babies. To supplement their diet, the new mothers always ate a stew of pigs' feet that had cooked for many hours in vinegar and sugar. This stew was kept cooking and ready all day on the stove, and as visitors we were always offered some. I remember how delicious it tasted.

My recollection of these visits was not stirred until many years later when I suddenly realized that those Chinese mothers were supplementing their diet with calcium and phosphorus—two important constituents in bones, teeth, and milk. The acetic acid in the vinegar converted the calcium phosphate in the bones of the pigs' feet into a soluble and assimilable form. American mothers and pregnant women supplement their diets also—not with a delicious stew of pigs' feet but with tablets of tasteless calcium phosphates.

Calcium and phosphorus are important not only to humans but also to animals and plants as well. Plants obtain these elements from the soil, and animals eat the plants. Humans eat both plants and animals and metabolize the calcium and phosphorus to build bones, teeth, muscle tissues, and nerve cells. The noncellular bone structure of an average adult consists of 60% of some form of tricalcium phosphates; the teeth, 70%. An average person, therefore, carries in his body around 7–8 lb of tricalcium phosphate. As calcium phosphates are used in various bodily functions, these phosphates are replenished by a continuous cycle. Used phosphorus is carried by the blood to the kidneys and excreted in urine, mainly as soluble ammonium phosphate salts.

An average adult eliminates the equivalent of about 3–4 g of phosphoric acid (H_3PO_4) a day into the sewage system. Some of the sewage eventually returns to the soil where plants absorb it and begin the cycle again. However, some excreted phosphorus finds its way into lakes, streams, and waterways through sewage effluents and causes pollution. This problem is discussed in more detail in Chapter 8.

Nature's cycle of moving calcium and phosphorus from soil to plants and animals and humans and back to soil again is too slow to satisfy humans. Thus, for human and animal consumption, calcium and phosphorus compounds are added as mineral supplements and to stock feeds; for plants, these compounds are added as fertilizers. Before synthetic fertilizers were available, Indians buried a fish in each hill of corn to supply the plants with the calcium and phosphorus from the bones and other parts of the fish; nitrogen was supplied from the degradation of the proteins in the rotting flesh.

Calcium Phosphates for Humans. Calcium phosphates suitable for human consumption in foods are prepared by adding lime to very pure phosphoric acid solution. In this reaction, calcium from the lime (CaO) replaces the hydrogen in phosphoric acid. Because hydrogen ion is monovalent and calcium ion is divalent, each calcium ion can replace two hydrogen ions. For calcium orthophosphates, the particular calcium phosphate compound obtained depends largely on the proportion of lime added to the phosphoric acid. So that a compound in its highest purity can be obtained, the ratio of the reacting compounds and reaction conditions must be adjusted to favor formation of that compound. Table 4.1 shows the common calcium

Table 4.I. Common Calcium Phosphates

Structure	Formula	Name
(HO)₃P=O	H_3PO_4	orthophosphoric acid
Ca[O–P(=O)(OH)₂]₂·H₂O	$Ca(H_2PO_4)_2 \cdot H_2O$	monocalcium phosphate monohydrate[a]
Ca[O–P(=O)(OH)₂]₂	$Ca(H_2PO_4)_2$	anhydrous monocalcium phosphate[a]
Ca(O)₂P(=O)(OH)·2H₂O	$CaHPO_4 \cdot 2H_2O$	dicalcium phosphate dihydrate[a]
Ca(O)₂P(=O)(OH)	$CaHPO_4$	anhydrous dicalcium phosphate[a]
Ca, Ca, HOCa—O, Ca, Ca linked through O to three P=O	$Ca_5(PO_4)_3(OH)$	tricalcium phosphate[a] (hydroxyapatite)
Ca(O)₂P(=O)–O–P(=O)(O)₂Ca	$Ca_2P_2O_7$	calcium pyrophosphate

[a] The prefixes mono, di, and tri used with calcium phosphates indicate that one, two or three of the hydrogens in phosphoric acid are replaced by calcium.

phosphates, their formulas, and their relationships to orthophosphoric acid.

Calcium Phosphates for Animals. STOCK-FOOD-GRADE CALCIUM PHOSPHATE. Quality standards for calcium phosphates used in baking acid, mineral supplements, or dentifrices are high. The products must meet all the requirements of state and federal food laws. Because animal life is not valued as highly as human life, the specifications for calcium phosphates to be used as mineral supplements for animals are not nearly as strict. Accordingly, the methods for making calcium phosphate for animal consumption are not as stringent. Stock food-grade dicalcium phosphate can be made by hydrating hot lime (CaO) with water and allowing the pasty product, calcium hydroxide [$Ca(OH)_2$], to react with phosphoric acid of 75–80% concentration. The reaction gives off heat; the temperature is about 100 °C. Some water evaporates as steam, and the reaction mixture is dried and ground in a mill. The phosphorus content of this product is almost 21%, close to the theoretical value of 22.8% for anhydrous dicalcium phosphate. The phosphorus content consists of monocalcium phosphate, tricalcium phosphate, unreacted lime, and anhydrous dicalcium phosphate as the main component.

Commercially, feed formulators prefer a product with a phosphorus content of 18.5%. So that this requirement is met, enough ground calcium carbonate (limestone) is added to the original product to lower the phosphorus content of the final composition to the desired level.

Another method for making stock-feed dicalcium phosphate involves the reaction of hydrated lime and phosphoric acid produced by the wet process. Dicalcium phosphate dihydrate containing 18.5% phosphorus is obtained directly. Because dicalcium phosphate is prepared from the impure wet-process acid, it is contaminated with the usual metal salts such as iron, aluminum, and magnesium present in the acid. These impurities do not seem to be harmful to animals.

In another manufacturing method, finely ground limestone reacts directly with phosphoric acid of 67% P_4O_{10} concentration to form the dicalcium phosphate dihydrate. The use of limestone eliminates the cost of converting the limestone to lime. However, because limestone is not as reactive as lime, the mixture must be stored for about 24 h to complete the reaction. This product also has a phosphorus content close to 18.5%. In 1980, the United States produced 1.5 billion pounds of dicalcium phosphate with 18.5% phosphorus content.

Farmers and cattlemen buy the stock-food-grade dicalcium phosphate and add it along with other mineral supplements and vitamins to the feed for their animals as dietary nutrients.

DEFLUORINATED PHOSPHATE ROCKS. Phosphate rock is a cheap source of calcium and phosphorus. Unfortunately, phosphate rock cannot be used directly as a stock-food supplement because most of the naturally occurring phosphate rocks have a composition close to that of the mineral fluorapatite [$Ca_5(PO_4)_3F$]. These rocks contain from 2 to as much as 4.5% fluorine, which is regarded as unhealthy for animals. Also, fluorapatite is relatively inert, and cattle, sheep, and hogs cannot convert its calcium and phosphorus into an assimilable form. One solution to this problem is to extract fluoride from the phosphate rock. The resulting product is defluorinated phosphate rock.

Several commercial methods are used to defluorinate phosphate rock. Most methods involve the heating of finely ground phosphate rock to 1200–1400 °C. Heating is done in the presence of water vapor with added phosphoric acid, sodium compounds, and silica. The fluoride is evolved as the volatile silicon tetrafluoride (SiF_4) and hydrogen fluoride (HF). By general agreement among producers, the finished product must have less than 1 part in 10,000 of fluoride for every 1% of phosphorus in the product. The phosphorus content of the defluorinated rock is adjusted to 18.5% by adding phosphoric acid during the defluorination process.

Various methods can be used to measure the value of the calcium and phosphorus to the animals. The American Association of Agricultural Chemists specifies determination of the solubility of the phosphate (calculated as the percent of the total P_4O_{10}) that is dissolved from a 1-g sample in 100 cm^3 of neutral ammonium citrate solution. Probably, what is soluble in an ammonium citrate solution should also be soluble in the digestive system of the animal. The most reliable method, however, is the so-called bioassay. Chickens are fed food that is fortified with the phosphate in question. After a specified time, the bones of these chickens are analyzed for total calcium and phosphorus content. If the calcium phosphate added is in an available form, it shows positively in higher than normal bone content. The test must be carried out under carefully controlled conditions with dozens of chickens in each test group. All the chickens must be fed "off-litter", that is, the chickens are not allowed to eat their droppings. (Chickens have poor digestive systems and eat their droppings over and over again to get the nutritive value from their feed.)

PHOSPHORIC ACID. Phosphoric acid is also used as a nutrient supplement in animal feeds in combination with molasses. In addition to supplying added phosphorus nutrient value, phosphoric acid also speeds up the molasses, that proverbial slow poke. "Slow as molasses in January" is not an exaggeration. We have seen crude molasses that is slow even in July, but when phosphoric acid is added, the acid reduces the viscosity and stickiness of the molasses; thus, both storage and handling are easier. This mixture makes a very good animal-feed supplement.

Phosphate Leavening Acids

History of Chemical Leavening. The use of leavening agents like yeast to make spongelike baked goods dates back to the early Egyptians. During its fermentation reaction with the simple sugars in the dough, yeast generates alcohol and carbon dioxide gas (CO_2). These gas bubbles permeate the soft, pliable dough and make it swell—a process that usually takes several hours. During the exodus from Egypt, the Hebrews did not have time to wait for their bread dough to rise; they only had time to bake unleavened dough; the result was matzos.

Yeast is still important in today's baking industry. Most bread, rolls, and coffee cakes are leavened with it. However, the use of chemical agents for leavening dough has grown to such an extent that by 1980 the volume had reached 100 million pounds a year in the United States alone.

Chemical leavening involves the action of an acid on sodium bicarbonate (baking soda, $NaHCO_3$) to release carbon dioxide gas. Baked goods are prepared mostly from wheat flours that contain the protein, gluten. Gluten, when hydrated with water, is fairly tough and rubbery and can be stretched into films that have a high capacity to retain gas. In preparing dough for baking, the kneading or mixing process disperses the gluten in thin films throughout the system. These films later hold the small nuclei of carbon dioxide bubbles generated by the reaction of chemical leavening agents (or by yeast action and by the air incorporated in the mixing process).

When dough is baked, the gas cells expand as the temperature increases. More gas diffuses into these cells as further chemical reaction takes place in the leavening system; this action causes the dough to rise further. Finally, the gluten sets at around 160–170 °F (71–77 °C) and becomes more rigid, and the starch in the flour gels. The baked goods do not expand much during the remaining 10 min or so in the oven.

The small holes found in breads, cakes, pancakes, or biscuits are therefore made primarily by carbon dioxide gas. The walls of those holes are cooked gluten films. The spaces between those holes contain gelled starch and other baking ingredients. These holes make the baked goods light and tender, and the taste and other properties that characterize cakes or biscuits are the result of the other ingredients.

During mixing and in the early stage of baking, the gas-retaining gluten cell walls are relatively weak. These walls do not become rigid until set by heat near the end of the baking process. Thus, during the first few minutes of baking, a cake can collapse from excess vibration. This problem is the reason why cooks do not like to have children playing in the kitchen when they have a cake baking in the oven.

The first synthetic chemical system for leavening was introduced in the 1850s. The system used cream of tartar (potassium acid tartrate, $KHC_4H_4O_6$, obtained from leftover sediment in the manufacture of wine) as the leavening acid to be used with sodium bicarbonate. Soon, monocalcium phosphate was introduced as a leavening acid. The new composition of monocalcium phosphate–sodium bicarbonate baking powder was invented by a Harvard professor of chemistry, E. N. Horsford. His monocalcium phosphate was crude because his process involved the reaction of partially charred bones with sulfuric acid to form phosphoric acid with calcium sulfate ($CaSO_4$) as the byproduct. To concentrate the material, Horsford boiled the liquid phosphoric acid that he had separated from the solid calcium sulfate by filtration. A precise quantity of bone ash (crude tricalcium phosphate) was then added to form crude monocalcium phosphate. The resulting moist substance was dried with flour or starch. After further drying for a few weeks, the substance was ground into dry acid phosphate granules to be used with sodium bicarbonate as baking powder. Horsford's process now seems both ingenious and somewhat quaint. To commercialize his patents, Horsford and a partner founded a company in Rhode Island. Horsford was dismissed from Harvard's faculty because of the shamefulness of engaging in trade (*1*, *2*).

Monocalcium Phosphate Monohydrate. Refined-grade monocalcium phosphate monohydrate, $Ca(H_2PO_4)_2 \cdot H_2O$, is a versatile chemical. Cooks use it in biscuits, adults use it in effervescent headache tablets, and children sometimes use it to shoot off toy rockets.

This phosphate monohydrate is now prepared industrially by adding hot lime (a base) to 75% phosphoric acid in a controlled volume of water. In this order of addition, excess acid is always present because the object is only to neutralize one of the three hydrogens in the phosphoric acid with 1 equiv of calcium. The reaction temperature is usually kept between 75 and 110 °C and always below 140 °C. Above 140 °C, anhydrous monocalcium phosphate begins to form.

The reaction for the formation of monocalcium phosphate monohydrate from lime and phosphoric acid in water is

$$2H_3PO_4 + CaO \xrightarrow{H_2O} Ca(H_2PO_4)_2 \cdot H_2O$$

Monocalcium phosphate prepared this way is about 90% pure; contamination comes mainly from dicalcium phosphate. This impurity is present because, during the reaction, a small portion of the monocalcium phosphate formed undergoes a decomposition reaction to form dicalcium phosphate and phosphoric acid:

$$Ca(H_2PO_4)_2 \cdot H_2O \longrightarrow CaHPO_4 + H_3PO_4 + H_2O$$

The phosphoric acid formed, of course, reacts with lime to form more monocalcium phosphate. This reaction also occurs when monocalcium phosphate monohydrate is dissolved in water. In other words, the compound does not dissolve in water to give a solution of pure monocalcium phosphate. This phenomenon is called incongruent solubility. The presence of up to 10% dicalcium phosphate seems to have no harmful effect on any of the known applications for monocalcium phosphate monohydrate.

Because monocalcium phosphate is actually phosphoric acid that has only one of its hydrogen ions neutralized with 1 equiv of calcium, the compound is still an acid although an easy-to-handle, edible, solid acid. Most of its uses are based on this acidic property. The most important use for monocalcium phosphate monohydrate is as a leavening acid in baking.

WHAT IS STRAIGHT BAKING POWDER? Although several kinds of baking powder are on the market today, very few still use cream of tartar as the acid. Some baking powders used today are prepared from monocalcium phosphate monohydrate and sodium bicarbonate mixed in with 37–40% starch (the starch keeps the acid separate from the sodium bicarbonate until the powder is used in the dough). This mixture is called straight phosphate baking powder.

Because of the fast reactivity of monocalcium phosphate monohydrate, phosphate baking powder liberates two-thirds of the available carbon dioxide gas during dough mixing. Some of the gas stays in the dough, but a large part of it is lost to the atmosphere and is unavailable for leavening. As a consequence, the straight phosphate baking powder has only a minor share of today's baking business.

When monocalcium phosphate monohydrate in straight phosphate baking powder reacts with sodium bicarbonate in a dough, the inorganic chemists like to assume that these compounds react as if they were in a pure water solution; these chemists write

$$3CaH_4(PO_4)_2{\cdot}H_2O + 8NaHCO_3 \longrightarrow 8CO_2 + Ca_3(PO_4)_2 + 4Na_2HPO_4 + 11H_2O$$

However, chemists who work with flour know that the reaction is not quite this simple. All these chemists will concede is that carbon dioxide gas is generated because they know that many side reactions also occur between the acid and the baking ingredients and between the acidic impurities in the flour and sodium bicarbonate.

WHAT IS COMBINATION BAKING POWDER? Baking powder prepared from monocalcium phosphate monohydrate in combination with a slow-acting acid is called combination baking powder. For household use, the slow acid is generally sodium aluminum sulfate, $NaAl(SO_4)_2$. In 1983, about 85% of all household baking powders were of this type. Sodium aluminum sulfate, commonly called SAS, is believed to react with sodium bicarbonate:

$$NaAl(SO_4)_2 + 3NaHCO_3 \xrightarrow{H_2O} 3CO_2 + Al(OH)_3 + 2Na_2SO_4$$

In a dough system, this reaction does not begin until heat is applied.

A typical combination powder contains 28% sodium bicarbonate, 10.7% monocalcium phosphate monohydrate, 21.4% SAS, and 39.9% starch. When this powder is used in dough, about one-third of the carbon dioxide (CO_2) is liberated during mixing. CO_2 is generated by the action of the monocalcium phosphate monohydrate on some of the sodium bicarbonate and creates the gas nuclei for later expansion. SAS then takes over the leavening action during baking.

However, SAS has its shortcomings; it reacts too slowly, and it continues to react with sodium bicarbonate; carbon dioxide is liberated, even after the starch and gluten have gelled and set. When this happens in a biscuit, the side walls split open. SAS can also impart a bitter taste to the baked goods. As manufactured today, SAS contains iron as an impurity; this impurity prevents the use of SAS in self-rising flours and prepared mixes because iron is a catalyst for the rapid oxidation of fats, which results in the development of rancidity.

SELF-RISING FLOUR AND ALL-PURPOSE FLOUR. Self-rising flour was introduced around 1873 by premixing the baking powder components with flour. Because the original monocalcium phosphate monohydrate varied widely in its composition and quality, this early premixture did not find wide acceptance. However, as its manufacture improved, the quality of the self-rising flour prepared from the premixture also improved.

High-quality monocalcium phosphate monohydrate as we know it today was introduced into self-rising flour in the early 1930s. Because this acid is so active, the moisture in the flour, as well as that which seeps into the box from the atmosphere, causes this acid in the flour to react slowly with sodium bicarbonate; thus, storage life of self-rising flour is only about 3–4 months.

Because the product is convenient, many cooks prefer it to regular flour and separate baking powder. The all-purpose self-rising flour contains around 1.375 lb of sodium bicarbonate, 1.75 lb of monocalcium phosphate monohydrate, and 2.25 lb of salt per 100 lb of flour. Because of the relatively short shelf life of the product, other better and slower acting acids are being developed to replace monocalcium phosphate monohydrate. Some of these new replacements are discussed later in this chapter.

Monocalcium phosphate monohydrate is also added in small quantities (0.25–0.75%) to all-purpose flour. This practice began in the 1920s for flour used in the popular sour-milk biscuits. The reason for its use is that some batches of sour-milk biscuits tasted like soap. Investigation showed that this alkaline taste occurred when the cook used too much sodium bicarbonate. The acid used for leavening in sour-milk biscuits is the lactic acid in sour milk. However, sour milk varies in acid content from batch to batch. Cooks may not be aware of this, however, and they use the same amount of sodium bicarbonate for the same measured amount of liquid. If the sour milk has a low lactic acid level, soapy-tasting biscuits from unreacted bicarbonate are the result.

Chemists found that when monocalcium phosphate monohydrate was added to flour, the monocalcium phosphate monohydrate compensated for any acid deficiency in the sour milk. Later, the added phosphate was found to have other advantages: it modified the gluten so that a softer and more plastic dough was obtained; it prevented "rope development" in the yeast dough. (Rope is caused by the intrusion of bacteria, specifically *Bacillus mesentericus*, which makes the bread stringy or "ropey". Other compounds such as calcium propionate are used widely as preservatives for this purpose.) Monocalcium phosphate monohydrate also supplies extra calcium and phosphorus, which are valuable nutrient supplements. More than 75% of the all-purpose flour sold today contains added monocalcium phosphate monohydrate. The phosphated flour, therefore, is also present in yeast-leavened baked goods.

BREAD IMPROVER. Monocalcium phosphate monohydrate is also used as the source of calcium and phosphorus for "bread-improver" compositions. A typical composition may contain 7.5% ammonium sulfate, 50.0% monocalcium phosphate, 0.3% potassium bromate, 20.0% salt, and 22.0% starch. These compositions are used extensively by large bakeries to improve yeast-leavened products—that is, they stimulate the growth of yeast. About 0.50–0.75% of the composition is added to the flour.

USE IN EFFERVESCENT TABLETS. Another important use for monocalcium phosphate monohydrate is as the ingredient in effervescent tablets that give off bubbles when added to a glass of water. This bubbling is the result of the quick release of carbon dioxide gas by the action of acidic monocalcium phosphate on sodium bicarbonate—a kind of baking powder in tablet form. This chemical reaction has also provided considerable enjoyment to children. They use the rapidly generated gas to power small rockets. Children with knowledgeable and indulgent parents sometimes raid their parents' baking powder for this purpose. If they forget to put the lid of the container back on tight, moisture seeps into the baking power. If the powder is later used to bake a cake, the cake will rise to only a portion of its normal height.

Coated Anhydrous Monocalcium Phosphate: The Improved Phosphate Leavening Acid. The anhydrous form of monocalcium phosphate, $Ca(H_2PO_4)_2$, has been known for a long time. However, this form was never used as a baking acid because it is hygroscopic; it absorbs moisture on the surface of its crystals and causes them to undergo the decomposition reaction to form dicalcium phosphate and free phosphoric acid. The free phosphoric acid formed is even more hygroscopic; it absorbs more water and causes more decomposition. This decomposition reaction is identical with the incongruent solubility of monocalcium phosphate in water described previously. In a baking powder composition, the phosphoric acid formed would also react with the sodium bicarbonate and result in lowered activity.

In the 1930s, almost a century after the origin of chemical leavening, a breakthrough in research on anhydrous monocalcium phosphate occurred. Julian Schlaeger, a chemist, was working on the anhydrous material in the laboratories of Victor Chemical Works, now a part of Stauffer Chemical Co. He and his colleagues developed a delayed-action anhydrous monocalcium phosphate by forming an extremely thin layer of an almost insoluble glassy phosphate on the surface of the fine phosphate particles. The product was given the trade name V–90, and the discovery was hailed by the baking industry as "the most outstanding development in 100 years of use of chemical leavening agents" (*3*).

I asked Schlaeger, a former colleague, how he came upon this discovery. He told me that slowing the reactivity of monocalcium phosphate monohydrate had been the major problem for years. In early attempts, he tried to coat the tiny particles of the material with a wax or lacquer. Practically speaking, these coatings were useless. If the coating were too thick, the product would not react with the sodium bicarbonate even when mixed with moist dough. Too thin a coating usually resulted in incomplete coverage of the particle, and that situation was as effective as no coating at all.

This research was done in the 1930s, long before encapsulation—a technique that permits a uniform coating of almost any desired thickness to be applied to a particle. Even if such technology had been available, the cost would have been prohibitive in this case. After all, monocalcium phosphate is relatively inexpensive—7–9¢/lb in 1972. The cost of encapsulation was many times that. Even in 1984, with the price of monocalcium phosphate at 46¢/lb, encapsulation remains an economically prohibitive process.

Schlaeger thought that possibly a thin coating of calcium acid pyrophosphate or glassy calcium metaphosphate could be formed by heating the monocalcium phosphate at a high temperature. Formation of calcium acid pyrophosphate would eliminate 1 mol of water from 1 mol of monocalcium phosphate:

$$\mathrm{Ca^{2+}}\left(\mathrm{HO}\underset{\mathrm{O^-}}{\overset{\mathrm{O}}{\mathrm{P}}}\mathrm{-O}\boxed{\mathrm{H + HO}}\mathrm{-}\underset{\mathrm{O^-}}{\overset{\mathrm{O}}{\mathrm{P}}}\mathrm{-OH}\right) \xrightarrow{\Delta} \mathrm{Ca^{2+}}\left(\mathrm{HO-}\underset{\mathrm{O^-}}{\overset{\mathrm{O}}{\mathrm{P}}}\mathrm{-O-}\underset{\mathrm{O^-}}{\overset{\mathrm{O}}{\mathrm{P}}}\mathrm{-OH}\right) + \mathrm{H_2O}$$

monocalcium phosphate

or

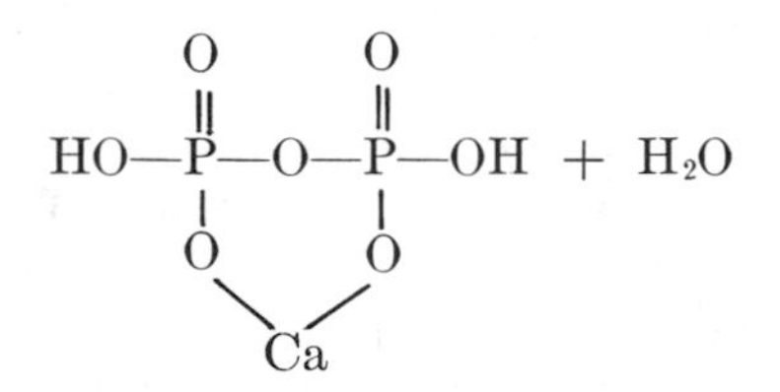

calcium acid pyrophosphate

Formation of calcium metaphosphate would involve the elimination of the same number of moles of water as the number of moles of monocalcium phosphate. Such a product should be a high molecular weight polymeric glass. The overall reaction for the formation of this compound from monocalcium phosphate is

$$(n+1)Ca^{2+}\left(HO{-}\overset{\overset{\displaystyle O}{\|}}{\underset{\underset{\displaystyle O^-}{|}}{P}}{-}OH\right)_2 \xrightarrow{\Delta} (n+1)Ca^{2+}\left[HO{-}\overset{\overset{\displaystyle O}{\|}}{\underset{\underset{\displaystyle O^-}{|}}{P}}\left({-}O{-}\overset{\overset{\displaystyle O}{|}}{\underset{\underset{\displaystyle O}{|}}{P}}{-}\right)_{2n}O{-}\overset{\overset{\displaystyle O}{|}}{\underset{\underset{\displaystyle O^-}{|}}{P}}{-}OH\right]$$

From general knowledge of this type of compound, a temperature of 200 °C would be needed before an appreciable amount of monocalcium phosphate would convert to calcium acid pyrophosphate; for the conversion to calcium metaphosphate, an even higher temperature would be required. Because the desire was to convert only the surface of the particles, this research was exploring new chemical frontiers.

When formed, calcium acid pyrophosphate has two less acidic hydrogens and is less water-soluble than monocalcium phosphate; thus, calcium acid pyrophosphate should have a less tart, acidic taste. The calcium metaphosphate glass should be insoluble in water, and because it does not contain acidic hydrogen ions except for one at each end of the long chain, calcium metaphosphate glass should not have any tart taste. This simple taste test was used as a guide by Schlaeger to follow his coating experiment. In his exploratory work, he heated ordinary monocalcium phosphate monohydrate produced at the Victor plant. Later, we will see that the use of this source of monocalcium phosphate was fortuitous.

The temperature chosen for heating was the arbitrary range 210–220 °C; this range was based solely on the sixth sense that often guides researchers. Schlaeger knew that when monocalcium phosphate monohydrate is heated above 140 °C, it loses its water of hydration, and the anhydrous material forms. At 220 °C, however, he was exploring new chemistry. According to plan, he tasted the product periodically to see if the immediate tart taste had disappeared. After 30 min of heating it had, and this sign was encouraging. This discovery was made at around 4:30 p.m., time to go home. Schlaeger had his assistant determine the neutralizing strength of the heat-treated material. (Neutralizing strength is the measure of the capacity of a baking acid to react with sodium bicarbonate. This property is expressed in terms of parts by weight of the acid required to neutralize exactly 100 parts of sodium bicarbonate.) Self-rising flour was then made with this material. As a control, he also made a self-rising flour containing the regular monocalcium phosphate monohydrate. When he took those two formulations home, made them into biscuit dough, and baked them in his oven, a new baking acid was born.

The manufacturing process developed subsequently for the large-scale production of the new baking acid involves the reaction of lime with a concentrated phosphoric acid at around 140–175 °C. The reaction is done in a batch mixer equipped for efficient mixing. The product is a dry powder of minute crystals of anhydrous monocalcium phosphate. These crystals are then heat-treated at approximately 200–220 °C.

Further development seemed fairly straightforward. Large batches were made, and the products were evaluated in different baking formulations. These baking tests gave excellent results. All the biscuits made under carefully controlled conditions showed a volume increase of approximately one-third over those made with ordinary monocalcium phosphate. This result means that less baking acid and sodium bicarbonate are needed to get the same volume increase. Similarly, self-rising flour compositions lasted much longer on the shelf without deterioration by absorbed moisture.

One day, without warning, batch after batch of the newly produced material failed; the baking acid was no longer superior to ordinary anhydrous monocalcium phosphate. All hands were recruited to investigate this problem. Russell Bell, the microscopist on the project, told me that he spent endless hours peering through his lenses searching for some difference between the minute crystals of the good and bad material. Measurements of the indexes of refraction of the crystals showed values of which all characterize only anhydrous monocalcium phosphate; the good and bad crystals were identical by this test. (The fine crystals were all smaller than 200 mesh; this result meant that the particles had diameters of less than 0.0029 in.)

The first indication of some physical difference came finally when Bell watched the crystals dissolve in water under his microscope. The ordinary anhydrous monocalcium phosphate dissolved quite rapidly, and not much residue remained. This result was also true with the unsatisfactory heat-treated material. However, material from the good heat-treated batches dissolved much more slowly. Even more important, only very fine transparent empty glassy shells having the shape of the original crystals remained on the microscope slide. These glassy shells were then collected and analyzed. Besides phosphorus and calcium, the shells contained large amounts of potassium, aluminum, sodium, and magnesium along with traces of other minor elements.

Apparently, the good heat-treated material was prepared from phosphoric acids containing the four elements as impurities. A check of the phosphoric acids used as raw material showed that that was indeed the case. During that particular period, Victor Chemical Works was making a transition from manufacturing phosphoric acid from phosphorus produced by the blast furnace process to phosphorus from the new electric furnace. The blast furnace acid was far less pure and contained appreciable amounts of the four elements. To check on their idea, Victor chemists made new batches of monocalcium phosphate monohydrate from the pure electric furnace acid but first added small amounts of potassium, aluminum, sodium, and magnesium. When this mixture was heated, these researchers obtained excellent delayed-action baking acid. However, they found that not all of these additives were equally effective. Since then, some of these trace metal impurities were added to phosphoric acid and became part of the new process.

Fortunately, for his first experiment Schlaeger used the monocalcium phosphate monohydrate produced at Victor's own plant and made from the blast furnace acid. If he had used the material from the pure electric furnace acid, V–90 (as the product was subsequently trademarked) might still be undiscovered. Later research showed that the coating is indeed a form of glass with some crystallinity. The major component seems to be a mixed potassium, aluminum, calcium, and magnesium metaphosphate. This composition is formed by the dehydration of the mixed monometal acid phosphates. Later, formation of calcium metaphosphate or calcium acid pyrophosphate as originally envisioned by Schlaeger was shown to require a much higher temperature. Thus, the discovery of coated anhydrous monocalcium phosphate did not turn out as originally theorized; the outcome was much better because of unsuspected impurities. This method of discovery is not unusual in research. A researcher starts work on the basis of a hypothesis and goes on from there. Luck, or serendipity, often plays a large part in the ultimate result.

Speculatively, the following sequence of reactions occurs during manufacturing. When lime reacts with phosphoric acid, pure monocalcium phosphate (being less soluble than, for example, monopotassium phosphate, monosodium phosphate, monomagnesium phosphate, and monoaluminum phosphate) crystallizes out. The impurities are concentrated in the mother liquor, the liquid phase. When the whole product is dried at 140–175 °C, the mother liquor containing the four metal acid phosphates dries uniformly on the surface of the pure anhydrous monocalcium phosphate crystals. Heating converts this metal–acid–phosphate coating into a continuous, glassy, substantially water-insoluble metaphosphate coating. When the anhydrous monocalcium phosphate is dissolved from the interior of the crystal, the hollow shell coating left behind can be seen under a microscope. Crystals of monocalcium phosphate monohydrate must be precipitated originally in such a small size so that after heating they are much smaller than 200 mesh. The particles must be produced in the correct size to begin with for baking purposes. If they are too large, the particles cannot be ground finer since grinding destroys the coating.

The biggest advantage of the coated anhydrous monocalcium phosphate is its resistance to attack by atmospheric moisture. This property makes possible the formulation of self-rising flour with a long shelf life. Food processors use the coated anhydrous monocalcium phosphate in prepared cake, pancake, waffle, and biscuit mix formulations.

In dough, the coating delays the start of the reaction with sodium bicarbonate, and once gas generation does start, the rate is slower. Experiments using V–90 in baking powder to make a biscuit show almost no reaction during the first few minutes of dough mixing. Only about 15% of the carbon dioxide is released during mixing and 35% more during the 10–15 min of waiting at the bench before baking. As we have seen, monocalcium phosphate monohydrate generates about two-thirds of its gas at the dough-mixing stage alone. The delayed action allows the gluten to become saturated with water and forms films to trap the released gas during the bench action period. This action then permits the rest of the leavening gas to be liberated in the early part of the baking. In fact, the coated anhydrous monocalcium phosphate promotes complete leavening action before the gluten and starch set. Each biscuit made with the new phosphates has greater volume, is fluffier, and is more appetizing because of a minimum of splits on the side wall.

Commercial introduction of coated anhydrous monocalcium phosphate in 1939 revolutionized the baking industry. A much greater variety of prepared mixes is now available in the supermarkets, and some people say these mixes make better baked goods.

Dicalcium Phosphate as a Leavening Acid. Besides its use in toothpaste, dicalcium phosphate dihydrate (DPD) (*see* Chapter 5) is also used as a leavening acid. This usage may seem surprising because one reason DPD is used in toothpaste is its inactivity, insolubility, and indifference to the other components in the mixture. The immediate pH of a water slurry of DPD is around 7.4–7.5 or slightly alkaline. When DPD is used as a leavening acid, chemists take advantage of one of its undesirable properties, which had to be corrected before it could be used in a toothpaste. DPD, when unstabilized, dehydrates and decomposes in the presence of water into hydroxyapatite and phosphoric acid (or maybe to the acidic monocalcium phosphate). Heat triggers the reaction. The acid from this process—whether as free phosphoric acid or as acidic monocalcium phosphate—acts as a leavening acid.

Naturally, such a slow-reacting, heat-dependent leavening action would be useless for pancakes or waffles, which require quick leavening, but this action is very useful in cakes. A cake takes about 30 min to bake, and the setting temperature—that is, the temperature at which the cake becomes firm—is high [about 160–170 °F (71–77 °C) inside the cake]. For cakes, most of the leavening action is accomplished by beating air into the batter. In many recipes, sodium bicarbonate is also added. It releases carbon dioxide when heated to 140 °F (60 °C); an alkaline residue and occasionally some unreacted sodium bicarbonate remain. Dicalcium phosphate dihydrate releases its acidity at around 150–160 °F (66–71 °C) just before the cake sets. Actually, the released acidity contributes very little to the leavening of the cake. Its main purpose is to neutralize the soapy alkaline taste from the remaining sodium bicarbonate.

Sodium Acid Pyrophosphate as a Leavening Acid. DOUGHNUTS. Sodium acid pyrophosphate, SAPP, has an acidic property that makes it useful as a baking acid, and its pyrophosphate ion enables it to sequester (tie up) many metal ions such as iron, magnesium, and calcium. These properties are used to make tender doughnuts and to keep boiled potatoes white.

SAPP is prepared by removing 1 mol of water from 2 mol of monosodium phosphate:

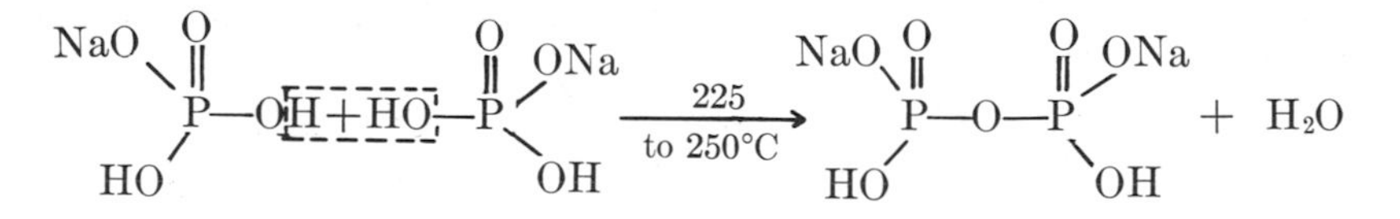

The temperature range of 225–250 °C is critical because at higher temperatures monosodium phosphate is converted to meta- and polyphosphates; these conversions occur at 530–600 and 800–900 °C, respectively.

SAPP with its two acidic hydrogens is the solid acid in the baking powder used by many commercial bakeries. At room temperature, SAPP reacts very slowly with sodium bicarbonate, even after it is mixed in the dough or batter. This slow reactivity at room temperature permits bakers to mix dough or batter in large batches and bake at their leisure. The reaction of SAPP with sodium bicarbonate does not start until the dough is heated in the oven.

The reason for this slow reaction of SAPP with sodium bicarbonate is not well understood. In water, the two compounds react rapidly, as any water-soluble acid is expected to react with sodium bicarbonate. In a batter, other ingredients apparently interfere. Some people believe that this interference is caused by the coating of SAPP particles with insoluble calcium pyrophosphate from the reaction of SAPP with calcium ions from batter ingredients such as milk solids. This speculation is supported partly by the fact that when more SAPP surface area is available (as with finer particles), SAPP is less reactive in the baking system.

In normal chemical reactions, when no surface coating interference is involved, the finer the size of the reactant particles, the faster the reaction. Also, the reaction between SAPP and sodium bicarbonate is quite rapid in a batter that contains no calcium salts.

The reaction between SAPP and sodium bicarbonate—which will not begin until the batter is heated—is ideal for making doughnuts. The preparation of a cake doughnut (in contrast to a bread doughnut, which is raised by a yeast fermentation reaction) is an art that requires a great deal of science. Doughnut batter is not as thick as biscuit dough and is not as thin as cake batter. In doughnut batter, carbon dioxide liberation must be controlled during mixing. If too many gas nuclei form during quick deep-frying, they cause the doughnut to overexpand into a highly porous mass that absorbs too much frying oil; a greasy doughnut results.

What is needed is an acid that reacts with the sodium bicarbonate in the doughnut batter only when heated. Such a temperature-triggered reaction should give a sustained release of carbon dioxide gas for 1.5–2 min. The resulting doughnut is just porous enough to absorb the right amount of fat for good flavor and to have a firm, pleasant-looking shape.

In our laboratories at Stauffer Chemical Co., we have made thousands of doughnuts using an automatic machine, mounted on a sensitive scale, that makes 30–40-dozen doughnuts per hour. We test doughnut formulations containing different baking acids to determine their influence on the absorption of fat by the doughnuts. Because the machine is mounted on a scale, the amount of fat absorbed by a dozen doughnuts can be determined simply by weighing the loss of fat after each dozen of doughnuts is fried.

Still, a good doughnut not only is a function of the baking acid and the other ingredients but also depends on frying conditions. Although our doughnut-making machine is a commercial unit used in specialty shops, it does not quite duplicate the conditions and results of the large machines in doughnut factories that make tens of thousands of doughnuts an hour. The smallest machine we could find to duplicate the results of our customers' big machines was one that makes 100-dozen doughnuts an hour. For each formulation we test, we make several hundred doughnuts, and one of our problems is trying to get rid of all the doughnuts we make. We can only eat so many.

REFRIGERATED BISCUITS. Refrigerated biscuit dough involves ingenious chemistry. The dough prepared from the flour mixture containing SAPP and sodium bicarbonate is rolled into a sheet and then cut into biscuit shapes. This procedure is done at 55–57 °F (13–14 °C) (dough temperature) to slow down the already slow room-temperature reactivity of the system. Cut biscuit doughs are stacked into a fiber (cardboard) can lined inside with a spiral of aluminum sheet in which one section overlaps the next. The ends of the can are then sealed with metal caps.

These canned biscuits are placed in a warm area between 80 and 100 °F (27 and 38 °C) for about 30 min. The reaction between the SAPP and sodium bicarbonate begins, and the dough swells. This swelling squeezes residual air from the can and presses against the overlapping aluminum foil; thus, the can is sealed tight. The can is thus filled with swelled dough under a carbon dioxide atmosphere.

Because the can is under constant pressure, air cannot seep in, and aerobic bacteria fermentation (which depends on oxygen from air) is prevented. Canned biscuits kept at 35–40 °F (2–4 °C)—the normal temperature range of a refrigerator—will keep for 2–3 months, and the cans will withstand pressures up to 90 $lb/in.^2$ (six times atmospheric pressure). A can that has popped open in the refrigerator of a grocery store means that it has not been stored properly.

COMMERCIAL BAKING POWDER. Commercial bakeries use a baking powder of monocalcium phosphate monohydrate mixed with a slow-acting acid. The acid is usually sodium acid pyrophosphate (SAPP) or calcium lactate. The reaction of calcium lactate with sodium bicarbonate is assumed to occur as follows:

$$Ca(C_3H_5O_3)_2 + 2NaHCO_3 \longrightarrow CO_2 + 2NaC_3H_5O_3 + CaCO_3 + H_2O$$

The $CaCO_3$ thus formed may react further with monocalcium phosphate monohydrate to liberate more carbon dioxide gas:

$$Ca(H_2PO_4)_2{\cdot}H_2O + 2CaCO_3 \longrightarrow 2CO_2 + Ca_3(PO_4)_2 + 3H_2O$$

After the CO_2 gas nuclei are formed, further leavening occurs only when heat is applied. Large batches of dough can thus be prepared at one time and baked at leisure.

Currently, low-sodium formulations are attracting interest. Therefore, today we are seeing formulations in which the sodium bicarbonate is replaced by potassium bicarbonate. However, these formulations have not yet attained a significant market share.

Sodium Aluminum Phosphates: A New Family of Baking Acids. Sodium aluminum phosphates are the most recent phosphate leavening acids introduced to the baking industry. The two industrially important members have the formulas $NaH_{14}Al_3(PO_4)_8 \cdot 4H_2O$ and $Na_3H_{15}Al_2(PO_4)_8$. These phosphates are crystalline compounds with individual X-ray diffraction patterns. These compounds are excellent leavening acids and in many respects are quite different from the leavening acids discussed earlier.

In the section on calcium phosphates, we discussed the use of monocalcium phosphate monohydrate as a leavening acid for baking. We also described the invention of coated anhydrous monocalcium phosphate that, through its microscopic crystalline coating, ensures a delayed reaction with sodium bicarbonate. This delayed reaction made possible the formulation of self-rising flour and ready-mixed baking formulations with much longer storage times than previously possible. The reaction also resulted in improved baked goods. In the section on sodium acid pyrophosphate, we discussed the importance of SAPP in many baking systems and noted that when SAPP is used in baking, it provides most of the leavening action only after the batter or dough is heated. This property is important for institutions where a large batch, for example, of pancake or waffle batter is prepared in the morning but must remain useful for the rest of the day.

This summary would imply that all problems concerning leavening acids were solved, but this situation is far from true. The competitive search for an improved leavening acid continued and led to the discovery of sodium aluminum phosphates (SALP). The first sodium aluminum phosphate introduced commercially in the late 1950s was $NaH_{14}Al_3(PO_4)_8 \cdot 4H_2O$. This compound is referred as 1–3–8 SALP with the numbers indicating the ratio of the sodium (Na), aluminum (Al), and phosphorus (P) atoms in the compound. $Na_3H_{15}Al_2(PO_4)_8$, which was introduced later, is called 3–2–8 SALP, with the numbers again indicating the ratio of the three principal atoms. Both compounds can be made by adding the correct ratio of sodium (as sodium carbonate or sodium hydroxide) and alumina (Al_2O_3) to excess, hot concentrated phosphoric acid. When the mixture is cooled, crystals of the respective compounds precipitate out.

Unlike the monocalcium phosphates and SAPP, the structures of 1–3–8 SALP and 3–2–8 SALP are quite complex. Despite this complexity, we know a good deal about their physical and chemical properties and how to use them.

Because we were associated with the inventors of 1–3–8 SALP, we are familiar with their trials and tribulations. We saw the preparation of the compound in a small beaker, its evaluation in baking biscuits to the larger-scale preparation in a pilot plant, and the complete evaluation in many different commercial baking formulations. All this research led finally to the successful commercial production of the compound in a multimillion-pound industrial chemical manufacturing plant.

In the development of 1–3–8 SALP, many of the original obstacles were technical. For example, the pure compound is quite hygroscopic, absorbing moisture from the air and becoming sticky. One can envision bags of it shipped as a free-flowing powder but arriving at the customer's plant as sticky lumps. Also, a leavening acid wetted with absorbed water would be expected to be much more reactive with sodium bicarbonate. Such reactivity would make a self-rising flour inactive almost instantly. Laboratory experiments, however, showed that when 1–3–8 SALP is incorporated in a self-rising flour, the slight moisture the compound picks up is beneficial. The fine flour particles stick to 1–3–8 SALP and protect it from reacting with sodium bicarbonate. The problem of its becoming sticky on storage and shipment, however, had to be solved before 1–3–8 SALP could be marketed. Basic research in inorganic chemistry paved the way. The final answer involved treating the compound with a small amount of a soluble potassium salt. The potassium ions displace some of the hydrogen ions from the surface of the 1–3–8 SALP crystal. This action changes the surface characteristics of the crystal; thus, 1–3–8 SALP's hygroscopicity is reduced to a satisfactory level without affecting the compound's function as a leavening acid. The potassium salt treatment along with some formulation modifications has resulted in a product that remains free-flowing even after prolonged storage.

The discovery of the sodium aluminum phosphate leavening acids coincides with the development of newer emulsifier systems (shortenings) for baking. Gas bubble nuclei are formed in the dough or batter by the mechanical mixing in of air and by the action of the leavening system, which generates some CO_2 gas at that initial stage. Much of the gas usually escapes during mixing. In baking, the gas bubble nuclei, which formed initially in the dough, are expanded by heat and also by the diffusing in of the newly formed CO_2 gas. Thus, a double-acting baking powder was desirable. One action generates the gas bubble nuclei during mixing and bench action, and the other generates gas during baking.

With the advent of the new emulsifiers or shortenings, which are used in the "prepared baking mixes", any bubbles that form during mixing are stabilized by these emulsifiers, and few escape. Thus, the need for a gas-generating system that does its work during mixing is much less. What is needed are gas-generating systems that become active during baking to expand the size of the bubble nuclei already formed and that are stabilized by the new shortening—in other words, a delayed-reaction leavening system. Sodium aluminum phosphate acids provide this delayed reaction.

The reaction of 1–3–8 SALP with sodium bicarbonate in a water system is shown in equation 1. This equation is based only on the reactants used and the amount of CO_2 gas evolved. The structure of the new sodium aluminum phosphate that is formed has not been completely confirmed.

$$2NaAl_3H_{14}(PO_4)_8 \cdot 4H_2O + 23NaHCO_3 \longrightarrow Na_5Al_6(PO_4)_6(OH)_5 \cdot 12H_2O + 10Na_2HPO_4 + 14H_2O + 23CO_2 \quad (1)$$

When 1–3–8 SALP is used as a leavening acid, it releases its acidity to react with sodium bicarbonate very slowly. Only 20–30% of the totally available CO_2 gas is released during mixing and during the period the dough sits on the bench (i.e., during the "bench action"). The remaining 70–80% is released slowly during baking. Experiments have shown that 1–3–8 SALP has a more desirable gas-releasing rate for cakes than does sodium acid pyrophosphate. This property is shown in the CO_2 release rate curves (Figure 4.1).

Sodium aluminum phosphate is the most stable leavening acid available today. Its low reactivity with sodium bicarbonate at room temperature in a batter makes the preparation of pancake and waffle batters that can be delivered in wax cartons by dairies possible. When kept refrigerated, these batters contain sufficient leavening action to make very good pancakes and waffles even after many days of storage. In fact, such batter is usually destroyed by bacterial action before the leavening action is gone. Biscuit doughs made with self-rising flour containing sodium aluminum phosphate lose very little of their leavening action after 24 h in a refrigerator.

These examples show the characteristics of sodium aluminum phosphate as a leavening acid and its superiority over previously known agents. However, SALP is not so perfect as to replace all of the previously known acids. Rather, SALP supplements them, extending and improving their usefulness. SALP has made possible leavening systems heretofore unavailable. In fact, a great portion of the two sodium aluminum phosphates is used today in combination with such leavening acids as the coated anhydrous monocalcium phosphate. A typical example is a formula containing 2 parts of 1–3–8 SALP and 1 part of coated anhydrous monocalcium phosphate.

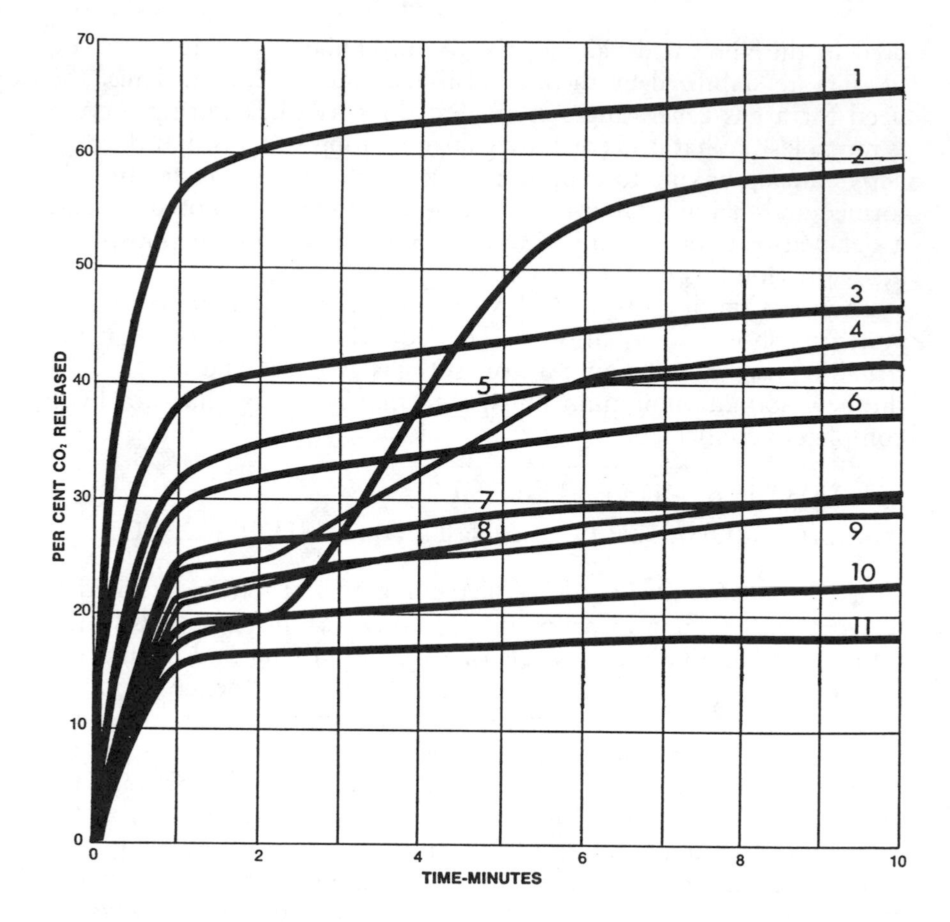

Figure 4.1. Curves show reaction rates of various leavening agents during the mixing and bench period in a typical dough-nut dough system at 27 °C (81 °F): 1, monocalcium phosphate monohydrate; 2, anhydrous monocalcium phosphate; 3, fast-reacting sodium acid pyrophosphate; 4, sodium aluminum phosphate with anhydrous monocalcium phosphate; 5, donut-grade sodium acid pyrophosphate; 6, intermediate-reacting sodium acid pyrophosphate; 7, slow-reacting sodium acid pyrophosphate; 8, sodium aluminum phosphate; 9, sodium aluminum phosphate with anhydrous aluminum sulfate; 10, very slow reacting sodium acid pyrophosphate; 11, dicalcium phosphate dihydrate. (Reproduced with permission from reference 4. Copyright 1971 Stauffer Chemical.)

The SALP combinations are also used in household and commercial baking powders, in self-rising flours, and in prepared biscuit and cake mixes. An example of a baking powder formulated with 1–3–8 SALP is 30.0 parts of sodium bicarbonate (soda), 37.2 parts of 1–3–8 SALP, 32.3 parts of cornstarch, and 5.0 parts of tricalcium phosphate. The cornstarch is added to keep the soda and the 1–3–8 SALP from reacting with each other during storage. The tricalcium phosphate is added to keep the product free-flowing.

From a baker's point of view, sodium aluminum phosphate, when used as a component in the new leavening acid system, produces baked goods with a finer and more even grain and with a tender and resilient crumb. In flavor-sensitive systems, its blandness allows only the desired flavor to predominate. In contrast, SAPP, when used as the leavening acid, usually imparts an astringent aftertaste known as the pyro taste.

Phosphates for Dairy Products

Cheese Emulsifiers. DISODIUM PHOSPHATE DIHYDRATE. Disodium phosphate, as its name implies, is phosphoric acid in which two hydrogen atoms have been replaced by sodium atoms. This compound is prepared by the reaction of 1 mol of sodium carbonate with 1 mol of phosphoric acid. Commercial products are the anhydrous form (Na_2HPO_4) and the dihydrate ($Na_2HPO_4 \cdot 2H_2O$):

$$H_3PO_4 + Na_2CO_3 \longrightarrow Na_2HPO_4 + H_2O + CO_2$$

Of course, the reaction is not as simple as adding the correct amount of sodium carbonate to a water solution of phosphoric acid. In practice, the sodium carbonate used is not basic enough to drive the reaction to completion. About 15% of the sodium in the disodium phosphate must be supplied from the more basic sodium hydroxide. Further, reactant concentration and reaction temperature must be controlled carefully. If too concentrated a solution of reactants is used with too high a temperature, tetrasodium pyrophosphate tends to form as a coproduct. As much as 0.5% of tetrasodium pyrophosphate in the disodium phosphate is detrimental to its use in process cheese, for example.

The largest use for disodium phosphate is as an emulsifier for pasteurized American process cheese—that is, to distribute the butterfat uniformly throughout the cheese. All cheese is the result of controlled fermentation of a precipitated milk curd (milk protein) that contains entrapped butterfat. Blue cheeses, such as Roquefort and Danish blue, are the result of curd fermentation by a specific mold (*Penicillium roqueforti* and *Penicillium glaucum*). Cheddar cheese, the most popular cheese in the United States, is produced when milk curd is ripened with a special culture of bacteria. When such spoilage goes unchecked and the bacterial decay continues, aged Cheddar results. However, not all precipitated milk curds, even when inoculated with the same bacteria strain and aged to the same extent, result in cheese with the same degree of flavor development; several grades, all with varying flavors, are possible.

So that uniform flavor and consistency are ensured, the pasteurized cheese process was developed. In pasteurization, various grades of Cheddar of different properties—including some grades called "stinkers"—are blended and cooked with added steam at around 160 °F (71 °C) into a homogeneous mass. Cooking and pasteurization inactivate the bacteria and prevent further degradation or flavor changes. The melted cheese is poured into a container where it solidifies into a loaf or it is extruded as a ribbon and cut into slices.

Although process cheese was being made in Europe as early as 1895, James L. Kraft, a cheese peddler in the United States, began experimenting with emulsifying salts and pasteurization in 1911. Between 1916 and 1922, Kraft received several U.S. patents covering the processes of heating, blending, and packaging cheese as well as the addition of 5% sodium phosphate to prevent fat separation during heating and sterilization. During the same period, Elmer E. Eldredge, a bacteriologist and chemist, working for the Phenix Cheese Co., showed that he could duplicate a Swiss Gruyere cheese by blending and heating cheese with 2% sodium citrate. Both Phenix Cheese Co. and Kraft had filed patents, and an interference case between claims was settled in the U.S. Patent Office. Although Kraft's patents were issued earlier, starting in 1916, and the Eldredge patent assigned to Phenix Cheese Co. was not issued until 1921, the two companies agreed to share their rights. Thus, the first cheese emulsifiers were born.

The cheese emulsifier is necessary to distribute the butterfat (originally trapped in the milk curd) homogeneously throughout the cheese during the cooking or storage. Although both sodium citrate and sodium phosphate can be used, the most popular emulsifer is disodium phosphate ($Na_2HPO_4 \cdot 2H_2O$) (DSP) because of its ready solubility in the melted cheese mixture and its generally lower cost.

The mechanism of the emulsifying action of DSP is not yet completely understood. We do know that DSP prevents fat globules from separating when cheese is melted. The U.S. Food and Drug Administration permits the use of a maximum of 3% added emulsifier, whether citrate or phosphate or a combination of both. The practical limit for disodium phosphate use is about 1.8%. Above this level, DSP tends to crystallize out in the cheese as the dodecahydrate

($Na_2HPO_4 \cdot 12H_2O$), and these crystals have the appearance of glass splinters. Even though this situation is not harmful, difficulties arise in convincing a cook that he or she is not looking at glass splinters in the cheese. The extent of this effect is related to the amount of water in the cheese and to the age of the cheese. Higher water concentrations and more fully aged cheese lead to greater tolerance to DSP. The more water present, as in the case of cheese spreads or cheese foods, the more DSP can be used. In the United States, process cheese can have no more than 40% water, cheese food no more than 44%, and cheese spread no more than 60%. Also, aged cheese has more carboxylic acid groups, which can convert part of the DSP to monosodium phosphate. Both of these effects reduce the likelihood of crystal formation.

Raw cheeses used for pasteurized process cheese contain about 30–38% moisture. Because federal regulations allow a maximum of 40% water in process cheese, this level is attained by adding steam during processing. The emulsifier is used, therefore, to prevent the separation of not only fat but also water.

Pasteurized process cheese can be made without added emulsifier and still have no separation problems. However, the raw cheese must have the correct natural balance of magnesium citrate and calcium phosphate. Because cows graze on different grasses at different seasons, they produce milk of varying calcium, magnesium, citrate, and phosphate content. Phosphates and other ingredients are added to overcome nature's vagaries.

Disodium phosphate as an emulsifier produces a cheese that melts to a gooey mess when heated. This finding is important when making cheeseburgers. However, if pyrophosphate is inadvertently introduced as an impurity, the cheese will not melt at all when it is heated. Unfortunately, we do not yet have a good scientific answer for this behavior.

As a cheese emulsifier, DSP is also used in combination with insoluble sodium metaphosphate (IMP). Commercial mixtures contain about 40–70% of IMP and 30–60% of DSP (sometimes trisodium phosphate is used in place of DSP). A total of 3% of such a mixture may be used in process cheese. Because IMP is practically insoluble in water, possibly it acts only as a diluent. Test results, however, have confirmed that IMP does contribute to some degree as a cheese emulsifier. IMP was manufactured for use both as a cheese emulsifier and as a polishing agent (abrasive) in some toothpaste formulations that contain fluoride. Recent changes in toothpaste formulations by dentifrice manufacturers have resulted in a change from the use of IMP to other dentifrice polishing agents. The result was a substantial reduction, as much as 90%, in the demand for IMP. Because the dental and cheese IMPs were produced in the same plants, this reduction in demand resulted in a significant increase in the manufacturing cost of IMP. Therefore, the price of IMP rose beyond that needed to be competitive with other cheese emulsifier formulations. This result is an interesting example of the effect of market volume and, therefore, production volume and cost, on determining whether a product, however effective, becomes or remains commerical.

KASAL: A SODIUM ALUMINUM PHOSPHATE. The third commercially important sodium aluminum phosphate, although it has the empirical formula $Na_{15}Al_3(PO_4)_8$, is not a pure crystalline compound. This compound can be made by the reaction of sodium hydroxide, alumina, and phosphoric acid in the ratio indicated by the formula. Drying the reaction mixture yields a free-flowing white powder. Because the powder contains no acidic hydrogen, it is not a leavening acid but an efficient emulsifier for process cheese. Its trade name, Kasal, is derived from the German word Käse for cheese.

Robert Lauck, one of the inventors of Kasal, told me how he happened to discover this compound. In connection with another problem, he was studying the dispersing action of the milk protein, casein, in water. Because he was aware of the action of phosphates on proteins, he had tried various alkaline phosphates as dispersing agents and wanted to try an aluminum phosphate. The 1–3–8 SALP that was available was an acid and not suitable because he needed an alkaline sodium aluminum phosphate. Thus, he neutralized acidic 1–3–8 SALP with sodium carbonate. The resulting product was indeed an effective casein dispersant. He reasoned that because this alkaline sodium aluminum phosphate has some effect on casein, this phosphate might also emulsify pasteurized process cheese.

The important word in the preceding sentence is "might". Chemists use it for their hypotheses all the time, and the probability of converting the term "it might" to "it will" is not very high. In Lauck's case, success was not immediate. After partial successes and partial failures, and with the assistance of a large industrial organization, he finally achieved his goal.

Unlike disodium phosphate dihydrate, which is seldom used at more than a 1.8% level because of the possibility of the glass splinter like crystal formation of disodium phosphate dodecahydrate, Kasal can be used to a maximum level of 3% as permitted by the Food and Drug Administration. Kasal provides a cheese with good melting characteristics, and its blandness permits the cheese flavor to dominate. Thus, a process cheese with improved properties and flavor is obtained.

Disodium Phosphate in Evaporated Milk. Evaporated milk is made by heating raw milk to approximately 240 °F (116 °C) to concentrate it. If it is stored, the milk gels slowly into a mushy solid. If about 0.1% anhydrous disodium phosphate is added before heating, gelation is prevented and the milk remains liquid. Here again phosphate addition helps to maintain the correct balance of calcium and phosphate. Phosphate is needed especially in milk produced in the springtime when phosphate content is normally low. Although many hypotheses exist as to why a correct calcium–phosphate balance prevents gelation, a definitive answer is still needed.

Tetrasodium Pyrophosphate (TSPP) in Buttermilk, Chocolate Milk, and Puddings. The ability of TSPP (Chapter 7) to disperse solids accounts for its use in buttermilk. When whole milk is churned to coagulate the cream or butterfat into butter, the residual liquor after fermentation is buttermilk. Lactic acid formed during fermentation is responsible for buttermilk's sour taste. If the buttermilk is to taste good and look appealing, it must have the right consistency. TSPP not only prevents curd agglomeration but also, if the curds do agglomerate too much and buttermilk becomes too thick, disperses them.

When tetrasodium pyrophosphate is ionized in water, it dissociates into as many as four positively charged sodium ions called cations. TSPP also could produce one pyrophosphate ion with four negative charges called an anion:

$$Na_4P_2O_7 \xrightarrow{H_2O} 4Na^+ + \underset{\text{pyrophosphate anion}}{P_2O_7^{4-}}$$

The ability to disperse solids is common to anions having multiple negative charges, such as the pyrophosphate anion. One theory to explain this phenomenon proposes that the solid particles absorb the negatively charged pyrophosphate anions on their surfaces. That is, the pyrophosphate anions, each with the four negative charges, adhere to the solid particles and in effect give all the particles a negative charge. Because negatively charged particles repel each other, the particles disperse.

Thick chocolate milk also uses TSPP. The thickening of milk products is just the reverse of the dispersing action described for buttermilk. In chocolate milk, calcium ions react with the pyrophosphate to form a weak calcium pyrophosphate gel. This gel interacts with milk protein to give the gel more body. Because milk is often drunk through a straw, the thickness must be limited by adding only about 0.1% tetrasodium pyrophosphate. Unexpectedly, chocolate milk thickened with TSPP also has a richer color and a superior flavor. However, no good theory has been developed to account for these advantages.

Because TSPP reacts with calcium ions and protein in milk to form a gel, it is also used in instant puddings that do not require cooking. These formulations usually contain milk, precooked starch, sugar, salt, flavoring, TSPP, and some added soluble calcium salt such as calcium acetate to supply extra calcium ions. The extra calcium reacts with TSPP to produce more calcium pyrophosphate gel, which then reacts with the protein in milk to give a firm gel (pudding). (This situation contrasts with chocolate milk where only thickening is desired, in which case the calcium in the milk itself is sufficient.)

Phosphates in Meat Products

Sodium Phosphates in Ham Curing. Both disodium phosphate and sodium tripolyphosphate are used to cure hams. The flavor of ham, besides that imparted by smoking, is the result of controlled bacterial decay. The bacteria come from the environment, and the taste of the decayed meat is what we recognize and enjoy as ham flavor.

Normally ham is fermented or cured in a pickling (salt) solution at around pH 6. We were told that disodium phosphate was originally added to increase the pH to about 6.5–6.8; this situation favors bacterial growth, just enough to improve the ham flavor. About 5% disodium phosphate was added to a brine solution containing about 15–20% salt and other pickling ingredients. This solution was injected into the ham through the pig's arteries until the equivalent of 10% of the weight of the ham was added. The ham was cured with the pickling solution for several days and then smoked.

Unexpectedly, hams treated with this solution were more tender and more juicy than those cured without added phosphate. In normal curing, ham usually loses about 10% of its weight. If water is added, the ham becomes soggy. Ham cured with added phosphate loses almost no weight and is juicy but not soggy.

A few years ago, a consumer group was agitating for the discontinuance of the phosphate pickle process. This group complained that producers were selling added water to the public as part of a ham. However, because of the quality improvement in ham when water is retained by the added phosphate, the producers won their case.

The favorable results obtained with sodium tripolyphosphate on ham were also not expected. Originally, sodium tripolyphosphate was added to kill bacteria in sausages and ham. Cured meat products, exposed to air, often develop an unappetizing greenish color; this color is the result of myoglobin, the pigment responsible for the red color of the meat. Myoglobin reacts with the hydrogen peroxide formed by bacteria that grow on meat. This reaction can be demonstrated by cutting a slice of bologna and pouring hydrogen peroxide on it. A greenish color develops almost immediately. Although sodium tripolyphosphate was added originally to kill the bacteria that generate hydrogen peroxide, sodium tripolyphosphate not only improved the color of sausage but also retained the moisture and fat that impart a desired plumpness and texture to the sausage. Added to ham, sodium tripolyphosphate gave a more juicy, tender product with improved flavor and nutritive properties.

If we compare the original concept for using disodium phosphate to cure ham with that of using sodium tripolyphosphate, interestingly, in one case the intention was to grow bacteria to improve the flavor and in the other the intention was to kill bacteria to improve the color. Both approaches are successful in unexpected ways. As a mater of fact, TSPP and tetrapotassium pyrophosphate are also effective for pickling ham. Commercial operations sometimes use a mixture of phosphates. The present hypothesis is that added phosphates somehow increase the water-binding capacity of proteins. This effect is believed to be related to the ability of polyphosphate ions to separate the main proteins in meat into myosin and actin. This separation results in the release of proteins that can bind more water. The elucidation of the precise mechanism for this phenomenon, however, requires more detailed scientific study.

Potassium Polymetaphosphate in Sausage. Potassium polymetaphosphates are quite different from the sodium polymetaphosphates. The commercial water softener Calgon is a water-soluble sodium polymetaphosphate with 15–20 mol of phosphorus/mol of compound. No counterpart to Calgon exists in the potassium metaphosphates. The potassium polymetaphosphates are generally quite insoluble in water. These phosphates are made by heating monopotassium phosphate to 330–400 °C. If a very high molecular weight potassium polymetaphosphate is desired—average molecular weight of over a million—a temperature of about 500 °C is needed.

Physically, potassium polymetaphosphate looks like white asbestos with a fiberlike appearance. Potassium polymetaphosphate can be soaked in distilled water for a long time at room temperature without turning gummy. In water that contains metal ions, potassium polymetaphosphate behaves quite differently. For example, it dissolves to give a very viscous solution in water when a little sodium chloride is added.

Solid potassium polymetaphosphate can be visualized as a series of long chains compacted together tightly in a crystalline lattice. This lattice is so tightly packed that the water molecules cannot get in between the long chains. However, when a soluble sodium salt (or other monovalent ion such as lithium) is added, the sodium ions replace some of the potassium ions. Because the sodium atom is smaller than the potassium atom, the neat, tight symmetry of the chain is disturbed. Now water molecules can enter and pry the chains loose from their crystalline lattice. After leaving the lattice, the chains disperse in water like a gel, which resembles a solution of natural gum, or starch. If, at this point, a bivalent ion such as calcium is added, each calcium ion can replace two of the sodium or potassium ions from two separate chains. When many chains are tied together with many calcium ions, the process is called cross-linking.

Such cross-linked material is no longer soluble in water. Because the material is a network of many long chains with small holes in between, the holes can hold some water. The finished product is a rubbery gum that can be stretched into a film or squeezed into a round bouncing ball. Such properties are unusual indeed for a completely inorganic compound. Unfortunately, no way has yet been developed to hold the water permanently. The water evaporates with time, and the product becomes a solid powder.

Potassium metaphosphate, dispersed in water as a clear and viscous gel, is used in sausage manufacture. The meat mixture contains at least 30% fat, sometimes as much as 60%, and added water. The best ingredient for binding, absorbing, and holding the fat and moisture together during smoking or cooking is hot bull meat. A more correct term would be warm bull meat, that is, meat from a slaughtered bull. The meat is added to the mixture before the meat cells become cold (before rigor mortis sets in). The exact mechanism for the action of hot bull meat in sausage is not clearly understood. Meat cells undergo a physicochemical change on cooling. Apparently, the cell wall hardens and prevents the full release of proteins, albumin, and hemoglobin as well as naturally occurring organic phosphates and other proteins. When bull meat is dumped into the cutter in the sausage manufacturing machinery, the almost living cells somehow help the hemoglobin, albumin, natural phosphate, and other proteins in the meat to act as a binder for fat and moisture. When the resulting sausage is boiled or cooked, it remains juicy, and the fat stays evenly distributed.

However, bull meat is becoming scarce. Many potential bulls are slaughtered while young and sold for veal. To replace the bull meat, new formulations containing potassium polymetaphosphate along with tetrapotassium pyrophosphate and tetrasodium pyrophosphates are being patented. However, potassium polymetaphosphate was not yet approved for food use by the Food and Drug Administration in the United States at the time this book was written. Potassium polymetaphosphate is used in sausages, however, in many European countries.

Phosphates in Seafood Products

Phosphates also improve the quality of freshly processed fish as well as canned seafoods. For example, when a fish is fileted, it usually exudes a slimy liquid, more commonly known as drip, which is a solution of soluble protein. The amount of exudate is increased when the frozen filets are thawed. Prior to freezing of the filets, if the filets are dipped in a solution of about 12.5% sodium tripolyphosphate and 4% salt, the drip loss—and thus the protein loss—is significantly decreased. This development has led to improved quality in frozen fish filets as well as improved nutritive value.

Many other seafood products such as lobsters, shrimp, crab meat, haddock, cod, and salmon are preserved by canning. Over a period of storage time, crystals of magnesium ammonium phosphate ($MgNH_4PO_4$), known as struvite, form. Although harmless both physically and nutritionally, the crystals do look like sharp pieces of glass and have resulted in consumer rejection. Addition of 0.25–1.50% of sodium acid pyrophosphate or sodium polyphosphate based on the total moisture content of the canned seafood prevents struvite formation. The effectiveness of these polyphosphates is based on their ability to sequester magnesium so that it is no longer available to form $MgNH_4PO_4$. One advantage of these phosphates is that they are generally recognized as safe in all food products.

Phosphates in Cereal and Potato Products

Cold Water Gel Starch. DISODIUM PHOSPHATE. Disodium phosphate alone or combined with monosodium phosphate is used to make starch for instant puddings and pie fillings. The starch is slurried with a solution of mono- and disodium phosphate at pH 6.1–6.5. After filtration, the starch is heated in a vacuum oven at approximately 60 °C for a few hours. This treated starch will now form a gel when cold water is added; normally starch only gels on cooking. The cold-water gel from phosphate-modified starch does not become thin on aging or cooking and can be used in desserts as well as for industrial sizing of textiles and papers (filling in the pores in the surface). Even though we find phosphate-modified starch useful, we still do not really know how the starch molecule is modified. The phosphate content in this starch is very low—a fraction of 1% by weight of the starch.

CYCLIC SODIUM TRIMETAPHOSPHATE. Cyclic sodium metaphosphate is an interesting compound, but few industrial applications exist for it. One application is for treating cornstarch to make a product that forms a stable jelly with cold water. The chemistry involves the phosphorylation of the starch, in an alkaline medium, by the cyclic sodium trimetaphosphate to form a starch phosphate (5). The phosphorylation is believed to take place on the terminal hydroxy group of the anhydroglucose groups. The anhydroglucose groups are the basic building block of the starch molecule.

The first step of this reaction is

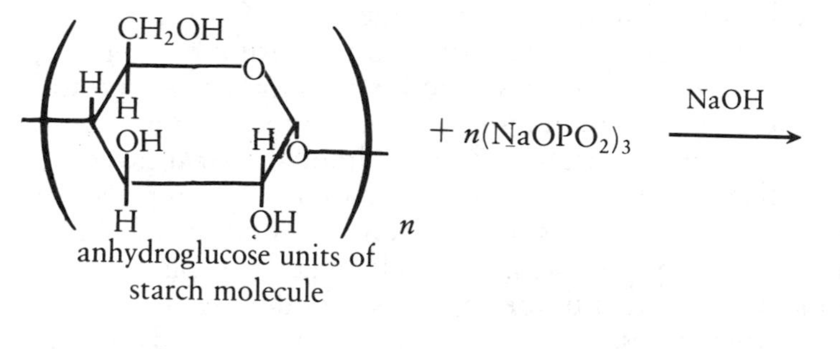

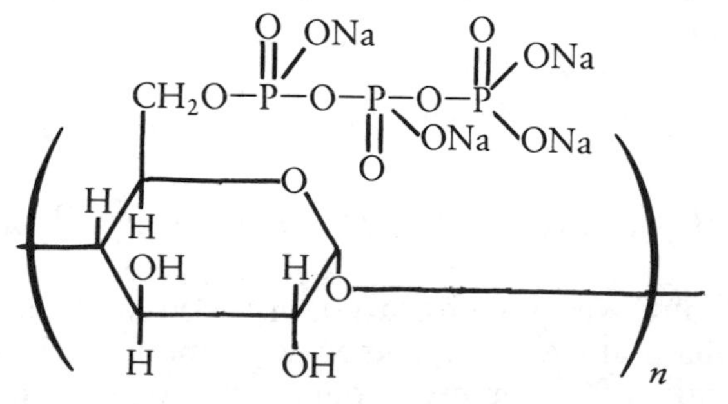

The reaction is much more complex than indicated. The degree of phosphorylation is usually much less than one phosphate per anhydroglucose group, and the phosphorus chain is further hydrolyzed during the process so that the phosphate is primarily present as a

monophosphate ester, for example

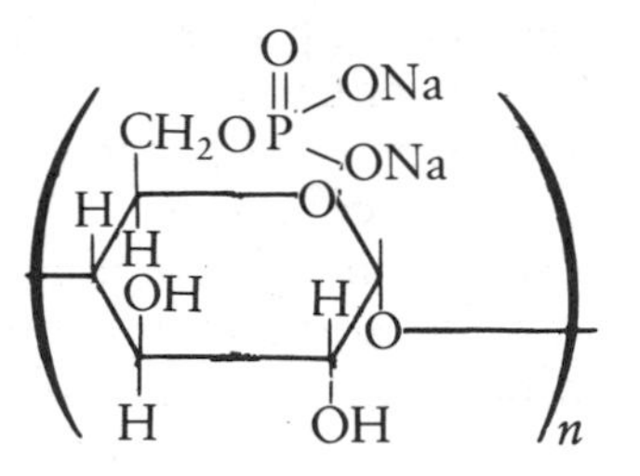

Starch, when phosphorylated in this manner, is not unlike that produced by the disodium phosphate treatment discussed earlier. Starch becomes resistant to hydrolysis and degradation. Resistance to degradation is a very desirable property in those cases where the starch is to be used under stressful conditions. Examples of such applications include the sizing of paper, the formulation of salad dressing, and the thickening of cooked foods. For the sizing of paper, a starch is desired that is resistant to degradation during heating, because this property allows the starch granules to swell without bursting. This characteristic results in better retention of the starch size. For the manufacture of salad dressing, starch granules are desired that can pass through the homogenization without bursting. Also, phosphorylated starch, when used as a thickening agent in cooked foods, will form a thick paste during the cooking, whereas unmodified starch becomes thin and quite fluid.

A second example of using cyclic sodium trimetaphosphate to phosphorylate an organic material is phosphorylation of single-cell protein. Single-cell protein holds great promise as a source of low-cost, food-grade protein. This protein is derived from the growth of single-cell organisms such as yeast. Low-cost organic materials, such as organic waste or petroleum hydrocarbons, are used as the growth medium or feed for the microorganism.

An example of single-cell protein is the protein present in baker's yeast (*Saccharomyces cerevisiae*). One problem with such protein is it is contaminated with nucleic acids, particularly ribonucleic acid (RNA) (*see* Chapter 16 for more details regarding RNA). Nucleic acids are rich in purine moieties, and excessive amounts of purines in the diet can lead to the formation of uric acid, a material that, when deposited in the joints of humans, results in gouty arthritis. Therefore, RNA should be separated from the protein. The separation is difficult to do without denaturing the protein and seriously reducing its functionality. The value of protein in foods is related not only to its nutrient qualities but also to how it functions in the food formulation.

Examples of such functionality are emulsifying properties, whipping properties, and heat-setting properties. As an example, to make an angel food cake, all three of these functions are required to produce a batter that is well emulsified and, when baked, fluffs to a light density and heat-sets without collapsing.

When the yeast is disintegrated by milling and then treated with an alkaline solution of cyclic sodium trimetaphosphate, the protein is phosphorylated and becomes more soluble, although the nucleic acid remains unaffected. When the pH of this extract is adjusted to the isoelectric point of the phosphorylated protein, a pH of 3.5–4.0, the protein precipitates from solution. The protein can then be isolated with a recovery yield of more than 90%, yet the RNA content is reduced from an original 27.0% to 3.5%. Also, the functionality, as measured by its emulsifying capacity and emulsion stability, is significantly improved. The discoverers of this process, scientists (6) from National Taiwan University, Taiwan, believe the cyclic sodium trimetaphosphate functions by phosphorylating the hydroxy groups of the serine and the amino groups of the lysine moieties present in protein.

Cyclic sodium trimetaphosphate is prepared by heating monosodium phosphate to about 530–600 °C. Three moles of monosodium phosphate joins head to tail and loses 3 mol of water to form a ring.

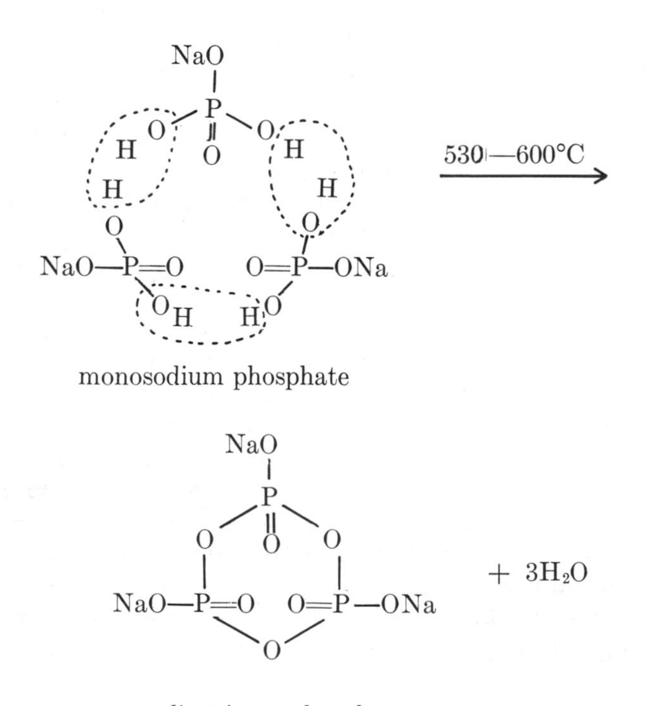

The product is a white crystalline compound. The compound is called cyclic trimetaphosphate; this name means that three metaphosphate ($NaOPO_2$) units are joined in a cyclic ring. Sodium trimetaphosphate can also be prepared from glassy sodium polymetaphosphate (Chapter 12); the glassy material is heated or tempered at 520 °C. The material devitrifies, and the long chains break into three metaphosphate unit chains and join into rings.

Instant Cooking Cereal: Disodium Phosphate. Some instant hot cereals also contain disodium phosphate. Years ago, many hours were required to cook breakfast cereal, especially products such as farina, to a soft, smooth-textured mush. Addition of about 1% disodium phosphate to the farina lowers the cooking time to a few minutes. Cereals require less cooking when the pH is brought slightly above 7, and disodium phosphate does that. (If too much disodium phosphate is added, resulting in too high a pH, the cereal disintegrates. A strongly alkaline solution can actually dissolve starch in the cereal.) Of course, the pH can be adjusted with such bases as sodium carbonate, but this base would impart a soapy taste to the cereal. Disodium phosphate is added in advance to instant cereal. One patented composition consists of a mixture of disodium phosphate and calcium and iron phosphates as mineral supplements. These ingredients are agglomerated into small particles similar to farina. Young cooks who have never had to wait for the old-fashioned farina to cook probably think that instant cooking is something that we have always had.

Potato Processing: Sodium Acid Pyrophosphate. Among its other uses, sodium acid pyrophosphate is also used to treat potatoes by taking advantage of the sequestering property of the pyrophosphate portion of the compound. When a white potato is boiled at home, its end often turns grayish black. The same thing can happen with homemade French fried potatoes. Yet precooked (or frozen) French fries, diced potatoes, and canned boiled potatoes are all bright and white. Commercial potato products have either been dipped or blanched with a 1–2% solution of sodium acid pyrophosphate to keep their pearly color.

Food chemists believe that after-cooking darkening of potatoes is caused by iron compounds. The potato plant absorbs iron from the soil and deposits most of it as a colorless, ferrous (Fe^{2+}) organic complex compound in the stem end of the tuber. During cooking, the iron is freed from the organic complex, and the iron can combine with tannin-like compounds in the potato. On exposure to air, this iron–tannin complex oxidizes to deeply colored ferric (Fe^{3+}) compounds. The pyrophosphate ion sequesters the iron as a colorless iron pyrophosphate complex—that is, a compound held together by both strong ionic bonds and electrostatic forces; the pyrophosphate ion thus prevents the iron from reacting with the tannin-like material later on to form dark compounds.

SAPP prevents cooked sweet potatoes from darkening by the same complexing reaction. For processed sweet potato products (frozen, dehydrated, and pureed), the natural color can be maintained without darkening by adding 0.3–0.4% sodium acid pyrophosphate, alone or in combination with tetrasodium pyrophosphate, to the potato during processing. For obvious reasons, a trade name for the sodium acid pyrophosphate used for potato treatment is Taterfos.

Tricalcium Phosphate as an Anticaking Agent

When we shake salt onto our food, the salt usually flows freely, even in humid weather. Sodium chloride, the common table salt obtained from mines or from the sea, even after many purification steps still contains a small amount of magnesium chloride as an impurity. Magnesium chloride, which is hygroscopic, absorbs moisture and causes the salt to agglomerate. Tricalcium phosphate, $Ca_5(PO_4)_3(OH)$, is used to prevent this agglomeration. This phosphate consists of fine particles, about 1–2 μm in size (1 μm is 1×10^{-6} m), that coat the larger salt crystals and prevent sticking.

Tricalcium phosphate is used also to impart free-flowing properties to other powders—for example, granulated sugar, baking powders, and even fertilizers. Usually, the addition of 1% is sufficient, and seldom more than 2–3% is needed.

Most people use the formula $Ca_3(PO_4)_2$ for tricalcium phosphate. Actually, the industrial tricalcium phosphate that is used for conditioning and other applications has a composition that corresponds to hydroxyapatite: $Ca_5(PO_4)_3(OH)$. Hydroxyapatite is prepared commercially by adding phosphoric acid to a slurry of hydrated lime. This order of addition of reactants is just the reverse of that for preparing the monocalcium and dicalcium phosphates. In this case, the aim is to neutralize all the hydrogen ions in the phosphoric acid, and with this order of addition, excess lime is always present. The hydroxyapatite formed is extremely insoluble and precipitates immediately as fine particles, from less than a micrometer to just a few micrometers in diameter. The precipitate is centrifuged from the mother liquor, dried, and milled to pass through a 325-mesh (less than 0.0018-in. diameter) screen.

Literature Cited

1. Horsford, E. N. U.S. Patent 14 722, 1856.

2. Van Wazer, J. R. *Phosphorus and Its Compounds*; Interscience: New York, 1961; Vol. II, p 1601.
3. Schlaeger, J. R. U.S. Patent 2 160 232, May 30, 1939 (to Victor Chemical Works).
4. *Modern Leavening with Sodium Aluminum Phosphates*; Stauffer Chemical: Westport, CT, 1971.
5. Kerr, R. W.; Cleveland, F. C. U.S. Patent 2 801 252, 1957.
6. Sung, H.-Y.; Cheng, H.-J.; Chuan, S.-J. *Proc. Natl. Sci. Counc., Repub. China, Part 1* **1983,** *3*, 181.

Chapter Five

Dentifrices and Pharmaceutical Tableting

Dicalcium Phosphates

Dicalcium Phosphate Dihydrate ($CaHPO_4 \cdot 2H_2O$). Many toothpastes contain a phosphate as a polishing agent, usually dicalcium phosphate dihydrate. Silica is also used as a polishing agent. Of the 60–70 million pounds of crystalline dicalcium phosphate dihydrate produced annually in the United States, about two-thirds winds up in toothpaste.

Dicalcium phosphate dihydrate is prepared industrially by adding a dilute slurry of hydrated lime [$Ca(OH)_2$], a base, to a 30–40% water solution of phosphoric acid. In this reaction, two hydrogen atoms in phosphoric acid are replaced by one calcium atom. The reaction is controlled by addition of the correct ratio of lime to phosphoric acid; so that only the dihydrate (and no anhydrous material) is formed, the temperature is kept at 40 °C or lower. The reaction is

$$H_3PO_4 + Ca(OH)_2 \longrightarrow CaHPO_4 \cdot 2H_2O$$

Dicalcium phosphate dihydrate precipitates and is separated from the mother liquor by centrifuging. The centrifuged material is dried and milled to such fine particles that 99.5% of it will pass through a 325-mesh screen (325-mesh particles have a diameter of less than 0.0018 in. or 44 μm).

PREVENTION OF DEHYDRATION. Before dicalcium phosphate dihydrate can be used in a dentifrice, it must first be kept from dehydrating to the anhydrous form after it is formulated into toothpaste. Dehy-

1002–0/87/0063$06.00/1

dration results in the growth of relatively large crystals of anhydrous dicalcium phosphate, which makes the paste gritty or, worse, may cause it to harden like cement in the tube. A pyrophosphate ion is added until its concentration in the finished product is about 1% to stabilize the formulation. At present, we have no good explanation as to how this process works.

The oldest method of stabilization is to add 2–3% trimagnesium phosphate to the finished product. Again, we have no clear-cut explanation as to why trimagnesium phosphate acts as a stabilizer. Minor modifications in additives and stabilizers are necessary to fulfill special requirements desired by particular toothpaste manufacturers. For example, dicalcium phosphate dihydrate stabilized with trimagnesium phosphate is not generally used in toothpastes containing the surfactant (i.e., a cleaning agent like soap) and cavity preventative sodium lauryl sarcosinate. The magnesium ion in trimagnesium phosphate would react with sodium lauryl sarcosinate to precipitate crystalline magnesium lauryl sarcosinate, which would give the toothpaste a gritty consistency.

DEHYDRATION AND DECOMPOSITION. Dehydration in the case of dicalcium phosphate dihydrate (DPD) would mean the loss of 2 mol of water to form the anhydrous product:

$$CaHPO_4 \cdot 2H_2O \longrightarrow CaHPO_4 + 2H_2O$$

Actually, the chemistry of this particular dehydration reaction is very complex and not well understood. We do know that the reaction is catalyzed by moisture, acid, and unknown impurities; heat also helps. Thus, chemists are surprised that DPD dehydrates faster in a moist atmosphere than when it is kept dry. In other words, to keep DPD wet (hydrated), you must keep it dry.

At lower temperatures, the dehydration reaction yields anhydrous dicalcium phosphate. At higher temperatures, and especially in the presence of hot or boiling water, the reaction proceeds as follows. First, dicalcium phosphate dihydrate hydrolyzes to the more basic octacalcium phosphate, $Ca_8H_2(PO_4)_6 \cdot 5H_2O$, and free phosphoric acid (*1*, *2*):

$$8CaHPO_4 \cdot 2H_2O \longrightarrow Ca_8H_2(PO_4)_6 \cdot 5H_2O + 2H_3PO_4 + 11H_2O$$

As more heat is added, octacalcium phosphate hydrolyzes to the still more basic hydroxyapatite [tricalcium phosphate, $Ca_5(PO_4)_3OH$] and more free phosphoric acid:

$$5Ca_8H_2(PO_4)_6 \cdot 5H_2O \longrightarrow 8Ca_5(PO_4)_3OH + 6H_3PO_4 + 17H_2O$$

This last hydrolysis step is empirically inhibited by magnesium ions. The hydroxyapatite and phosphoric acid formed from the last step of hydrolysis could combine to produce anhydrous dicalcium phosphate. Because magnesium ions inhibit the hydrolysis, this inhibition is assumed to indirectly prevent the formation of anhydrous dicalcium phosphate. Thus, even the apparently simple removal of water molecules can be quite complex. For toothpaste production, such dehydration must be prevented.

Because we still do not know how the stabilizers work, you may wonder how they were discovered. Guy MacDonald discovered the trimagnesium phosphate stabilizer by trying many and various compounds as additives and finding that trimagnesium phosphate worked. Theoretically, this chemistry is still a great challenge. Considerable research is probably furrowing the brows of chemists in various laboratories as they strive to unravel the riddle of the missing water.

USE AS DENTIFRICE. Dicalcium phosphate dihydrate made its debut on toothbrushes in the early 1930s, replacing some of the then commonly used precipitated chalk in toothpastes. A dicalcium phosphate dihydrate dentifrice not only cleans teeth but also is less abrasive than chalk and puts a glossier shine on tooth enamel. This advantage is endlessly touted on television and in magazines and newspapers, usually with a picture of a beautiful smiling girl with pretty teeth. Despite its virtues as a dentifrice, however, DPD could not be used in toothpastes containing fluoride ions. Although dicalcium phosphate is relatively insoluble in water, sufficient solubility remains to produce enough calcium ions to remove all of the fluoride ions as a precipitate of the extremely insoluble calcium fluoride (CaF_2). Unfortunately, insoluble fluoride is unavailable to the teeth. Research by various investigators has suggested that soluble fluoride ions have a stabilizing effect against the slow but steady dissolution of human dental enamel. The mineral component of dental enamel is mostly impure calcium hydroxyapatite. The fluoride ion converts the enamel surface to calcium fluoroapatite, which is much more resistant to attack by food-derived acids. Researchers also found that fluoride ions provide a greater degree of remineralization.

As more and more people preferred fluoride-containing toothpastes to the non-fluoride-containing toothpastes, the demand for those polishing agents, such as dicalcium phosphate, that reacted with fluoride ions decreased. The fluoride-containing toothpaste formulations used polishing agents such as calcium pyrophosphate, insoluble sodium metaphosphate, or silica, all of which are more compatible (i.e., did not react) with fluoride. However, in 1967, a report stated that dicalcium phosphate was compatible, that is, did not react, with sodium monofluorophosphate (Na_2PO_3F), known more commonly as MFP (*3*). MFP was also found to be an effective caries inhibitor. On a fluoride-content basis, MFP is equivalent to the soluble fluoride ion containing compounds. However, of the dicalcium phosphate dihydrates available commercially, only some batches were compatible with MFP, and some batches were not as compatible. A great deal of research was therefore needed to identify what was unique about the compatible batches. This research showed that certain modifications of the dicalcium phosphate dihydrate manufacturing process resulted in a product that was consistently compatible with MFP. These process modifications included the careful scheduling of the various process steps, including control of the order of addition of reagents, the pH, the temperature, and the additives. Many of these findings were discovered by trial and error (*4, 5*). No theory allows the prediction of which process modifications result in a more compatible polishing agent, but the major phosphate-containing toothpaste in the United States today is made from MFP-compatible dicalcium phosphate dihydrate. This toothpaste has been approved as a "decay-preventive dentifrice" by the Council on Dental Therapeutics of the American Dental Association (ADA).

A typical formulation containing MFP is as follows (component, % by weight) (*6*): polishing agent—dicalcium phosphate dihydrate, 48.76; humectant—glycerin (95%), 22.00; organic surfactant—sodium lauryl sulfate, 1.50; binder—sodium (carboxymethyl)cellulose, 1.00; sweetener—sodium saccharin, 0.20; stabilizer—tetrasodium pyrophosphate, 0.25; flavoring—peppermint, 1.00; anticaries agent—sodium monofluorophosphate, 0.76; and distilled water, 24.53.

Note the high percentage of dicalcium phosphate dihydrate used as the polishing agent in this MFP formulation. Considering how much toothpaste we use every year, we can easily understand why so many millions of pounds of polishing agents are produced.

Anhydrous Dicalcium Phosphate ($CaHPO_4$). Anhydrous dicalcium phosphate can be manufactured in the same equipment used to make the dihydrate; the only difference is that the reaction is carried out above 70 °C to eliminate formation of the dihydrate.

Anhydrous dicalcium phosphate competes with the hydrated version as a polishing agent in toothpaste. However, this compound is so much more abrasive than dicalcium phosphate dihydrate that if used alone as the polishing agent, it would wear the teeth away to nothing. Thus, small amounts of anhydrous dicalcium phosphate are used in combination with the dihydrate to remove stains, for example, from the teeth in specialty items such as "smoker's toothpaste".

Anhydrous dicalcium phosphate is also combined with less abrasive polishing agents such as tricalcium phosphate. Although still relatively small, the market for anhydrous dicalcium phosphate could be larger if a process for controlling its abrasiveness were discovered. Research in this area is receiving special attention in many dentifrice laboratories.

Calcium Pyrophosphate. PREPARATION. Of historical interest is the fact that the first important fluoride toothpaste was made with calcium pyrophosphate, the most insoluble and inert of all calcium phosphates. This compound is prepared by the thermal dehydration of dicalcium phosphate dihydrate:

$$2(CaHPO_4 \cdot 2H_2O) \longrightarrow Ca_2P_2O_7 + 5H_2O$$

When this dehydration reaction is carried out under carefully controlled conditions in a humid atmosphere, water is lost rapidly at 120–160 °C; this value corresponds almost exactly to the theoretical value for complete removal of the two molecules of water of hydration. Upon further heating, at above 400 °C, anhydrous dicalcium phosphate begins to lose water and condenses into calcium pyrophosphate. (Several crystalline forms of calcium pyrophosphate exist. The γ form is made at around 530 °C; the β and α forms are made at successively higher temperatures.)

USE AS POLISHING AGENT. Because of its chemical inertness to fluoride ions, extremely low solubility in water, and proper abrasiveness, calcium pyrophosphate was the first polishing agent found to be compatible for use in fluoride toothpastes. As discussed earlier, dental researchers have known for years that a minute amount of fluoride ion strengthens teeth and helps prevent cavities, especially when teeth are being formed in children. Accordingly, addition of fluoride to toothpaste is an effective means of cavity prevention.

The first fluoride-containing toothpaste with calcium pyrophosphate as the polishing agent was Procter & Gamble's Crest toothpaste formulation (6). Research leading to the formulation of Crest indicated that stannous fluoride, SnF_2, in combination with stannous pyrophosphate, $Sn_2P_2O_7$, is effective in preventing cavities. Calcium pyrophosphate was chosen as the polishing agent. In the book *Accepted Dental Remedies* (7), issued by the American Dental Association in 1960, Crest toothpaste was listed as a dentifrice containing 0.4% stannous fluoride, 39% calcium pyrophosphate, 10% glycerin, 20% sorbitol (70% solution), 1% stannous pyrophosphate, 24.9% water, and 4.63% miscellaneous formulating agents.

The calcium pyrophosphate used in toothpaste is actually a mixture of its γ and β forms. From the standpoint of chemical inertness, the pure β and α forms prepared at the higher temperatures are ideal, but these forms are too abrasive for teeth; the selected mixture is a good compromise that is sufficiently inert and is abrasive enough but not too abrasive.

Insoluble Sodium Metaphosphate

Insoluble sodium metaphosphate (IMP) was also used as a polishing agent in toothpaste. The compound is especially suited to fluoride-containing anticavity toothpastes (*8*). Even though the IMP is slightly soluble in water, the positive cation liberated is the sodium ion. Thus, when IMP reacts with the fluoride, it only forms sodium fluoride, which is still soluble and reactive with the teeth. IMP-containing fluoride toothpaste was also accepted by the American Dental Association. This toothpaste offers, in addition to the fluoride ions, good polishing and cleansing action. However, the use of IMP in toothpastes has now declined substantially for economic reasons.

One of the earliest methods used to measure the abrasiveness of dental polishing agents was developed by the ADA. A specially constructed machine holds standardized toothbrushes against antimony metal strips. The polishing agent to be tested is formulated into a standard toothpaste. After the antimony strip is brushed mechanically with this paste for 10,000 back-and-forth strokes, the antimony strip is washed and weighed. The abrasiveness of the polishing agent is calculated by measuring how much antimony disappears from the strip—in other words, its weight loss. The higher number, the more abrasive the material. This test assumes that what is too abrasive for antimony would also be too abrasive for teeth. Typical abrasive values for various polishing agents obtained by this method are as follows (polishing agent, ADA method values): dicalcium phosphate dihydrate, 2–3; anhydrous dicalcium phosphate, 30–50; hydroxyapatite, 2; calcium pyrophosphate, 10–12; and insoluble sodium metaphosphate, 2–3.

Another commonly accepted method for determining abrasiveness is the radioactive dentin abrasion (RDA) procedure. Extracted human teeth are cut to remove the enamel-covered portions. The remaining root portions, known as dentin, are irradiated in a nuclear reactor. These irradiated dentin samples, after aging for 7–10 days, are polished with a slurry of the polishing agent under standardized conditions. In this process, some portion of the dentin is removed, and the amount removed is measured by the radioactivity of the slurry. The more abrasive the polishing agent, the more dentin is removed. All tests are run against a control, which is a standard sample of calcium pyrophosphate having an arbitrary RDA value of 100. Less abrasive polishes have RDA values less than 100; more abrasive polishes have values greater than 100. The desired value is around 100.

Dentifrice Market Changes

Although the less abrasive calcium phosphates had replaced much of the precipitated chalk as the polishing agent in dentifrice formulations, a new dentifrice polishing agent, hydrated silica, appeared in 1969. This agent was introduced by Lever Brothers in a translucent gel formulation, Close-Up. This gel-type dentifrice received immediate consumer acceptance. Therefore, this formulation was followed by the introduction of other dentifrices containing hydrated silica, and in 1981, Procter & Gamble introduced a new gel formulation of Crest, which used hydrated silica as a polishing agent and replaced the expensive stannous fluoride with sodium fluoride as the therapeutic agent for the inhibition of caries. The switch of dentifrice formulations from pastes containing insoluble phosphates to gels containing hydrated silica has led to a major shift in the market for polishing agents. Thus, in the United States, hydrated silica has replaced most of the dentifrice market for calcium pyrophosphate, and hydrated silica has intruded on the dentifrice market for the dicalcium phosphates. This market shift now appears to have stabilized; however, this shift is an example of the effect of changing consumer attitudes.

Phosphorus Compounds as Anticaries Agents or Plaque Inhibitors

The beneficial effects of fluoride in reducing the incidence of caries, or tooth decay, are well established. Unfortunately, fluoride is not a complete caries preventative. So the search continues for more effective methods to prevent this disease. An approach that has attracted perhaps the most attention by chemists and dental professionals is the use of phosphorus-containing compounds that have demonstrated activity as cariostatic agents. More than 100 studies have shown that a significant number of phosphorus-containing compounds have such cariostatic activity when fed to animals. The phosphorus-containing compounds showing such activity include monosodium phosphate, disodium phosphate, monocalcium phosphate, dicalcium phosphate dihydrate, sodium trimetaphosphate, calcium metaphosphate, glycerol phosphate, calcium sucrose phosphate, phytic acid derivatives, and several organic phosphonic acids (*9–12*).

Despite the evidence of cariostatic activity in tests on animals, problems in confirming this activity arise when clinical tests of these compounds are conducted on humans. One of the problems is that clinical tests with humans are much more difficult to control than tests with animals. A second problem is the flavor the phosphorus-containing additive contributes to the food being eaten. In many cases, people not only object to the taste of these phosphorus compounds, especially the water-soluble inorganic phosphates, but also are thereby able to recognize the test samples in double-blind tests. Regardless of why, the results of these clinical trials on humans lend support to the cariostatic effect of the phosphorus-containing additive. So far, the variation in the extent of benefit observed is too broad to allow the conclusion that these phosphorus-containing compounds are truly effective.

In the 1970s, product development activities were conducted at two well-known companies engaged in personal-care products. These activities indicated that certain phosphates provide sufficient anticaries activity to consider using them for this purpose in commercial products. At one of these companies, tests with dicalcium phosphate dihydrate showed efficacy, and dicalcium phosphate stabilized with tetrasodium pyrophosphate was found to be a more effective cariostat than that stabilized with trimagnesium phosphate (*13*). At about the same time, a second company reported they were considering manufacturing a toothpaste formulation using sodium trimetaphosphate as a cariostat (*14*). Also, during the same period, a candy manufacturer considered using calcium trimetaphosphate as a cariostat in candy (*15*), and a food manufacturer was developing the use of sodium phosphates on sugar-coated cereal (*16–21*). However, none of these activities reached the point where commercially useful products were identified.

In 1985, another product, an "antitartar" toothpaste, underwent market tests in the United States by a major dentifrice manufacturer. This product includes sodium acid pyrophosphate and tetrapotassium pyrophosphate as the antitartar agent (*22–24*). Tartar is a calcified form of plaque. Plaque is an almost invisible coating that is formed on teeth through the action of mouth bacteria on carbohydrates. Plaque consists of a mixture of glycoproteins, bacteria, and exocellular polysaccharides, for example, glucans and dextrans produced by the bacteria. The demineralization of tooth enamel, which initiates caries formation, is generally believed to result from the action of the bacteria within the plaque on soluble carbohydrates such as sugar; this reaction converts the soluble carbohydrates to organic acids. This reaction lowers the pH (i.e., increases the acidity) of the plaque, and this released acidity effects the dissolution or demineralization of the hydroxyapatite in the dental enamel. The removal of tartar and plaque could do much to reduce the incidence of caries.

How do the phosphates function as cariostats in animals? We do not know. All evidence indicates the cariostats act locally at the surface of the tooth. They do not appear to act systemically through ingestion. Are the phosphates a truly useful anticaries or antiplaque agent in humans? Until the clinical data on the latest antitartar toothpaste are made public, the answer will remain uncertain. However, importantly, leads such as these must continue to be followed. Despite the frustration from many years of research that have so far led only to equivocal results, the promise of phosphorus compounds as anticaries agents continues to stimulate scientists to do further research. The potential of finding a more effective inhibitor for caries formation, a disease that is probably the most prevalent one affecting humans, is well worth the continued effort.

Pharmaceutical Tableting

Pharmaceutically active materials such as drugs and vitamins are often administered orally in solid dosage forms such as tablets or capsules. Additives of supposedly inert materials are frequently included in these formulations to improve stability, to enhance appearance, to act as diluents, to assist in the disintegration of the tablet after administration, or to aid in release of the pharmaceutically active substance.

Tablets can be prepared by wet granulation, dry granulation, and direct compression. The most widely used method today is wet granulation. In this process, the components of a tablet are formed into granules from a wet mixture, or preblend, of the tablet ingredients. These granules are dried, screened, and compressed into tablets.

Dry granulation is primarily used with ingredients that are moisture- or heat-sensitive. Dry granulation entails compacting a preblend of the tablet ingredients into large granules, which are then crushed to a preferred size and screened, and those particles meeting the size specifications are compressed into tablets. The oversized and undersized materials are usually recycled to the compactor.

Direct compression entails dry blending the tablet ingredients and compressing this dry blend directly into tablets. Direct compression eliminates the need for granulation and therefore reduces labor cost, energy requirements, and manufacturing equipment.

In 1964, unmilled dicalcium phosphate dihydrate and, more recently, a granulated grade of tricalcium phosphate were found to be useful diluents or excipients, particularly for preparing tablets by dry compression. Their advantages are low cost, insolubility in water, flowability, compressibility, inertness to most dry ingredients, stability in storage, food-grade quality, and, therefore, safety.

Certain physical properties are critical in direct compression tableting. These properties include particle size and particle-size distribution, flowability, and compressibility. The correct particle size and particle-size distribution are essential for good flowability and compressibility. Commercial high-speed tableting machines produce tablets at rates as high as 15,000 tablets per minute. Thus, the physical properties must allow rapid and consistent flow to the tablet compressing machine dies.

The preparation of dicalcium phosphate dihydrate is described under Dicalcium Phosphate Dihydrate ($CaHPO_4{\cdot}2H_2O$). The unmilled material is dried to obtain the proper particle size for tableting (after preparation and isolation). The isolation of dicalcium phosphate dihydrate having the required particle-size properties is a result of careful control of both the crystallization and drying conditions. These controls include the order and rate of addition of reagents, the use of additives to control particle size, the control of the temperature and pH during the preparation, and the use of drying conditions that retain the desired properties.

An example of a typical particle-size analysis is shown in Figure 5.1. The particles range in size from 325 mesh (44 μm) to 20 mesh (720 μm) with a weight-average particle diameter of about 100 mesh (144 μm).

Although tricalcium phosphate, when prepared by adding phosphoric acid to a slurry of lime, is obtained directly without milling as fine particles with an average diameter of under 2 μm, granular grades of particle size and particle-size distribution similar to those of granular dicalcium phosphate dihydrate have recently become available.

The flowability of the dry-blended tablet ingredients is important, because this property affects the filling of the dies of the tablet compressing machine. The excipient, when blended with the other tablet ingredients, must flow freely and at a uniform rate. These factors are important for controlling the amount of material entering the die and, therefore, for controlling the tablet weight.

Compressibility relates to the force needed to compress the ingredients into a tablet. The goal is to prepare a tablet sufficiently hard to maintain its integrity but not so hard as to adversely affect the disintegration time required for the tablet, once it is eaten.

The excipient should be inert to the other ingredients in the tablet formulation. This property is particularly true for pharmaceutically active materials such as drugs or vitamins. Should the excipient react with these pharmaceutically active ingredients, the tablet would, of course, not provide the user with all the active ingredient promised.

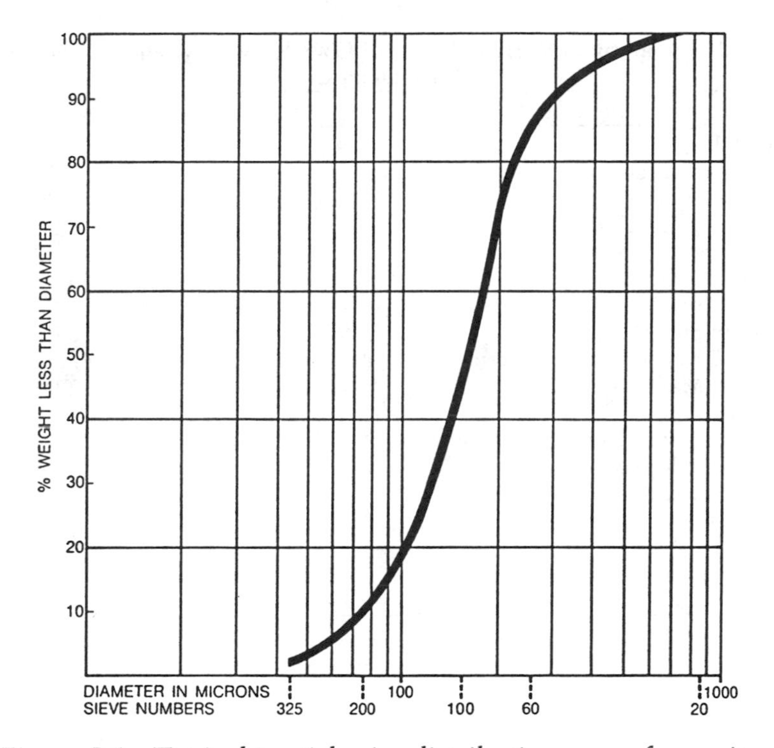

Figure 5.1. Typical particle-size distribution curve for excipient-grade dicalcium phosphate dihydrate. Twenty percent of the weight of the material has a diameter of less than 100 μm, and 50% has a diameter of less than 150 μm. (*Reproduced with permission from reference 25. Copyright 1981 Stauffer Chemical.*)

In addition to dicalcium phosphate dihydrate and tricalcium phosphate, other excipients or diluents used include lactose, microcrystalline cellulose, sugar, mannitol, and calcium sulfate.

The excipient must remain inert to the environment in which the tablet is stored. The excipient must not react with air to oxidize or adsorb moisture so as to cause disintegration of the tablet or deterioration of the components of the tablet. Thus, the excipient must be stable throughout storage, in the untableted formulation and in the tablet.

Because tablets are ingested, the components, including the diluent, must be safe to eat.

So that useful tablets are produced, a tablet formulation may contain other useful components in addition to the pharmaceutically active ingredients and the diluent or excipient: (1) Lubricants, such as stearic acid, hydrogenated vegetable oils, or magnesium stearate,

Table 5.I. Composition of High-Potency B Complex with Vitamin E and 750 mg of Vitamin C

Ingredient	*mg of Ingredient/Tablet*
Folic acid	0.50
Pyridoxine hydrochloride	5.38
Riboflavin, type S	11.00
Vitamin B_{12}, as 0.1% SD	7.20
Thiamin mononitrate	16.50
Calcium pantothenate	25.00
Starch 1500	15.00
Dicalcium phosphate dihydrate, unmilled	30.00
Vitamin E as dry vitamin E acetate 50%—SD	66.00
Cab-O-Sil M-7	2.00
Vitamin C	
C-90 (ascorbic acid, 90% granulation)	525.00
Niacinamide ascorbate	420.00
Avicel PH-102	81.42
Stearic acid	70.00
Magnesium stearate	8.00
Magnesium oxide	5.00
Total weight	1,288.00

SOURCE: Reproduced with permission from reference 26. Copyright Hoffman-LaRoche.

are added to improve the rate of flow of the dry formulation into the tableting dies and to prevent adhesion of the tablet to the surface of the tableting dies and punches. (2) Disintegrators, such as gums, starch, or clays, are used to aid in the breakup or disintegration of the tablet after ingestion. (3) Binders, which are often the same type of materials used as disintegrators, for example, gums, gelatin, starch, sugars, or (carboxymethyl)cellulose, are added to improve the cohesive properties of the dry powders and granules. (4) Colors and flavors are used to make the tablets esthetically more appealing. Any excipient must also be inert to these components of the formulation.

Table 5.II. Typical Tranquilizer Tablet Formulation

Ingredient	*% by Weight*	*mg of Ingredient/Tablet*
Chlorpromazine hydrochloride, USP	28.0	100.00
Avicel PH-102 cellulose	35.0	125.00
Dicalcium phosphate dihydrate, unmilled	35.0	125.00
Cab-O-Sil silica	0.5	1.74
Magnesium stearate, USP	1.5	5.25
Total		356.99

SOURCE: Reproduced with permission from reference 27. Copyright FMC.

Otherwise, some surprisingly strange-appearing or unpleasant-tasting tablets could be formed.

The compositions of a typical high-potency vitamin supplement and a typical tranquilizer tablet formulation, both using a phosphate excipient, are shown in Tables 5.I and 5.II.

Literature Cited

1. Brown, W. E.; Smith, J. P.; Lehr, J. R.; Frazer, A. W. *Nature (London)* **1962,** *196*, 1050.
2. Brown, W. E. *Nature (London)* **1962,** *196*, 1048.
3. Saunders, E. U.S. Patent 3 308 029, March 7, 1967 (to Monsanto).
4. Sherif, F. S.; Majewski, H.; Via, F. A. U.S. Patent 4 487 789, Dec. 11, 1984 (to Stauffer Chemical).
5. Michel, C. G. U.S. Patent 4 472 365, Sept. 18, 1984 (to Stauffer Chemical).
6. Pierce, R. C.; Mitchell, R. L. U.S. Patent 4 348 382, Sept. 7, 1982 (to Colgate-Palmolive).
7. Nebergall, W. H. U.S. Patent 2 876 166, March 3, 1959 (to Indiana University Foundation).
8. Manahan, R. D.; Richter, F. J. U.S. Patents 3 227 617 and 3 227 618, Jan. 4, 1966 (to Colgate-Palmolive).
9. Harris, R. S. *Proc. Inst. Med. Chicago* **1964,** *25*, 30.
10. Ostrom, C. A.; Van Reen, R. *J. Dent. Res.* **1963,** *42*, 732.
11. Nizel, A. E.; Harris, R. S. *J. Dent. Res.* **1964,** *43*, 1123.
12. Stookey, G. *Cereal Foods World* **1981,** *26* (1), 10.
13. Baron, H. Proceedings Workshop on Carcinogenicity of Foods, Beverages, Confections, and Chewing Gum, 1977, p 75.
14. *Business Week* **1972,** *Dec. 16*, 25.
15. Shaw, J. H. *J. Dent. Res.* **1980,** *59*, 644.
16. McClure, F. G.; Muller, A. *J. Dent. Res.* **1959,** *38*, 776.
17. Reussner, G. H.; Coccodrilli, G., Jr.; Thiessea, R., Jr. *J. Dent. Res.* **1975,** *54* (2), 365.
18. Stookey, G. K.; Muhler, J. C. *J. Dent. Res.* **1966,** *45* (3), 856.
19. Carroll, R. A.; Stookey, G. K.; Muhler, J. C. *J. Am. Dent. Assoc.* **1968,** *76* (3), 564.
20. Brewer, H. E.; Stookey, G. K.; Muhler, J. C. *J. Am. Dent. Assoc.* **1970,** *80* (1), 121.
21. Tepper, L. B. Statement before the Select Committee on Nutrition and Human Needs, U.S. Senate, April 16, 1973.
22. *Chem. Eng. News* **1985,** *March 25*, 31.
23. *Fortune* **1985,** *May 13*, 73.
24. Parran, J. J., Jr.; Sakkab, N. Y. U.S. Patent 4 575 772, May 7, 1985 (to Procter & Gamble).
25. Di-Tab Unmilled Dicalcium Phosphate Dihydrate, USP/FCC in *Compression Tableting*; Stauffer Chemical: Westport, CT, 1981.
26. Roche Technical Brochure TSR-90; Roche Chemical Division, Hoffman-La-Roche, Nutley, NJ.
27. FMC Bulletin PH-9; Avicel Department, FMC, Philadelphia, PA.

Triple superphosphate granules of different analyses and sizes are kept in separate piles inside this bulk storage building. Shuttle belt conveyors overhead distribute the product to the proper pile. A front-end loader reclaims for bagging or bulk shipment.

Chapter Six

Fertilizers

PHOSPHORIC ACID IS USUALLY SOLD AND SHIPPED as a clear, easy-to-handle liquid in 75%, 80%, and 85% concentrations. The remaining 25%, 20%, and 15% is water. A 100% phosphoric acid concentration tends to crystallize to a solid that melts at 42.35 °C. This solid is hygroscopic; that is, it grabs water from the air and becomes mushy. When dilute phosphoric acid is shipped, especially for large-scale operations, the 15–25% water represents a major expense. Obviously, a better way of handling higher concentrations of phosphoric acid is needed. This need has resulted in the development of superphosphoric acid.

Superphosphoric Acid

Various investigators, especially at the Tennessee Valley Authority where fertilizer research is an important project, studied the liquid–solid phase equilibrium for concentrated phosphoric acid to determine the solidification temperature and eutectic points of liquid phosphoric acids as the phosphorus pentoxide concentration increases. Figure 6.1 (*1*) shows what happens. The solidification temperature for a 75.0–76.0% P_4O_{10} acid is lower than that for a 72.6% P_4O_{10} concentration (equivalent to 100.0% H_3PO_4) and about the same as that for dilute 60.0% P_4O_{10} (equivalent to 82.0% H_3PO_4). The 75.0–76.0% P_4O_{10} containing phosphoric acid is at a eutectic point concentration in the freezing point curve. Because the solidification of this eutectic mixture is below 65 °F (18.3 °C), the eutectic mixture can be shipped and handled as a liquid above this temperature. Phosphoric acid at this concentration is called superphosphoric acid. This contains orthophosphoric acid, about 42% pyrophosphoric acid, and some tripolyphosphoric acid.

When the pyrophosphoric acid and tripolyphosphoric acid in superphosphoric acid are diluted with water and heated, they revert to

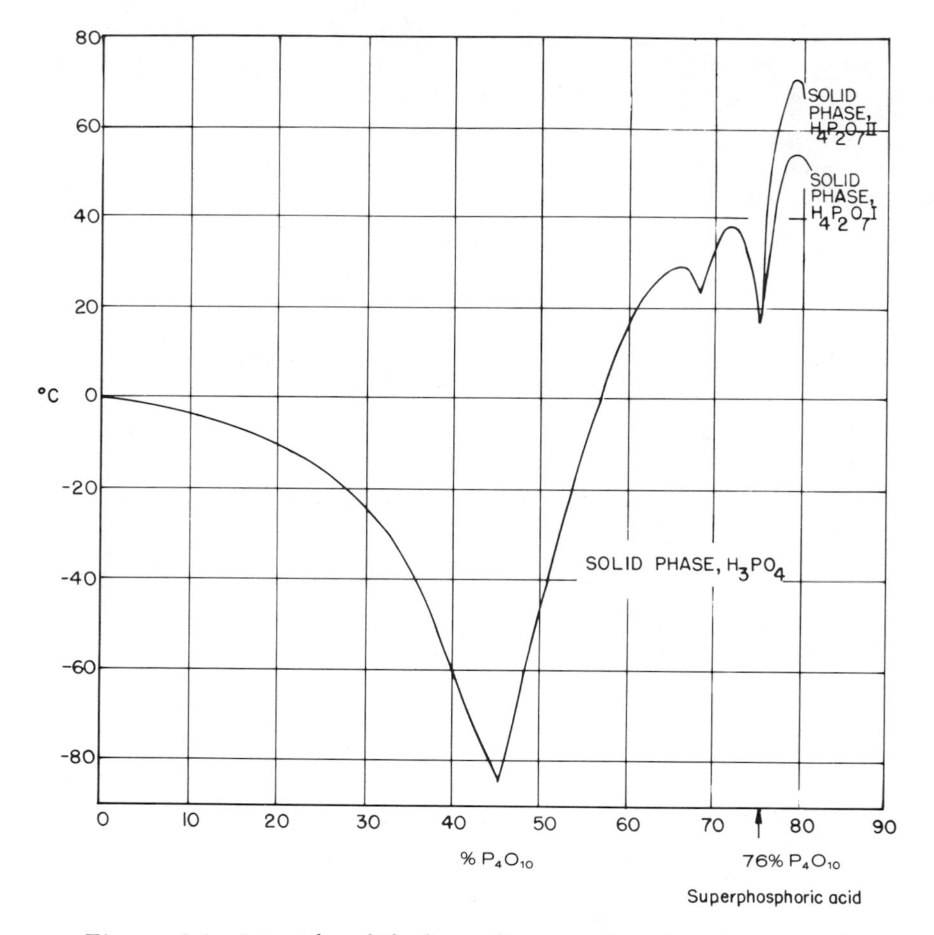

Figure 6.1. Liquid–solid phase diagram for phosphoric acid systems.

orthophosphoric acid. One hundred pounds of superphosphoric acid diluted with water gives about 105 lb of 72.6% P_4O_{10} phosphoric acid. Because the 72.6% P_4O_{10} phosphoric acid is 100.0% H_3PO_4, generally, superphosphoric acid is said to be equivalent to 105.0% H_3PO_4. Superphosphoric acid is used as a source of orthophosphoric acid for many applications—for example, for augmenting the phosphoric acid content of a bright dip bath for chemically polishing aluminum without adding water. Superphosphoric acid prepared from thermal phosphoric acid is also used to prepare liquid fertilizers, although not as extensively as superphosphoric acid prepared from the cheaper wet-process phosphoric acid.

Wet-process superphosphoric acid contains considerable metallic impurities such as iron and aluminum, and its P_4O_{10} content is around 77.0%, or 106.5% phosphoric acid. For liquid fertilizer, the pyrophosphoric acid content is most important.

This compound holds the metal impurities in solution through a complexing action known as sequestering. If pyrophosphate were absent, the metallic impurities would precipitate and foul the equipment. The high concentration of P_4O_{10} in superphosphoric acid also permits the manufacture of fertilizers with very high phosphorus content. Wet-process superphosphoric acid can be used directly as fertilizer. In one method, this compound is added directly to irrigation water.

Calcium Phosphates

Phosphorus is essential for plant growth. This element is available from phosphorus-containing plant foods—that is, from fertilizers, or from finely ground phosphate rock. However, the low solubility of phosphate rock (especially of the fluoroapatite) restricts its availability to plants.

Superphosphate. The idea of converting insoluble tricalcium phosphate or phosphate rocks into soluble monocalcium phosphate dates back to 1830. At that time, the German chemist Justus Liebig reported that acidulated bones (monocalcium phosphate) made a good fertilizer. By the 1870s, acidulated phosphate rocks were produced industrially to be used as fertilizer. Today, superphosphate is made by the reaction of accurately proportioned, finely ground phosphate rock and sulfuric acid. Tricalcium phosphate in phosphate rock is converted to hydrated monocalcium phosphate. Calcium sulfate, the other product, is mixed in, and the hydrogen fluoride leaves. The simplified reaction is

$$2Ca_5(PO_4)_3F + 7H_2SO_4 + 3H_2O \longrightarrow 7CaSO_4 + \underset{\text{superphosphate}}{3Ca(H_2PO_4)_2 \cdot H_2O} + 2HF$$

This reaction is exothermic, and the liberated heat drives off a large amount of water as steam. The amount of sulfuric acid added is less than that theoretically called for to ensure that the product is free of uncombined phosphoric acid or sulfuric acid (either of which would make the product hygroscopic and cause caking). As a result, a small quantity of dicalcium phosphate dihydrate appears in the product along with some undecomposed phosphate rock.

The reaction mixture is cured by storage, sometimes for up to 3 weeks, to complete the reaction. After the mixture is cured and dried, the fertilizer is ground and bagged.

In the original reaction and during the curing, some fluoride in phosphate rock evolves as hydrogen fluoride gas. Some fluoride comes off as silicon tetrafluoride (SiF_4) from the reaction of hydrogen fluoride with silica. However, about 50–75% of the fluoride remains in the fertilizer product, which also contains 18–20% available P_4O_{10}.

In the 1940s, superphosphate was used in about 90% of the domestic phosphate fertilizers. This amount decreased to about 30% in the 1960s, and presently about 8% of the domestic phosphate fertilizers contain superphosphate, as the usage of ammonium phosphate in solid, liquid, and suspension form continues to increase.

Triple Superphosphate. In superphosphate fertilizer, the presence of calcium sulfate byproduct acts only as an inert diluent; this byproduct has almost no nutrient value for plants. A more desirable product is monocalcium phosphate monohydrate, which contains very little useless calcium sulfate. The desire for this product led to the development of "triple superphosphate" fertilizer. Triple superphosphate is prepared by the decomposition of phosphate rock with phosphoric rather than sulfuric acid:

$$Ca_5(PO_4)_3F + 7H_3PO_4 + 5H_2O \longrightarrow 5Ca(H_2PO_4)_2 \cdot H_2O + HF$$

Thermal phosphoric acid (74–78% concentration) can be used, but it is generally too expensive for this purpose. Instead, wet-process acid is mixed with finely ground phosphate rock in correct proportions to produce monocalcium phosphate. The thick slurry discharges into the upper end of a kiln that is lined with firebrick and heated with gas or oil. The hot mixture, still in the form of a slurry, is then dumped into a concrete bin (called a den) where the mixture hardens into a porous mass. After the mass is aged to complete the reaction, the product is dried of residual moisture and is then ready for packaging and shipping. Triple superphosphate contains about 44–46% P_4O_{10}. This trend is in part a result of the industries' attempts to use lower grades of phosphate rock in the process.

The market for triple superphosphate, like that for superphosphate, is declining. Whereas in the 1960s triple superphosphate represented more than 30% of the fertilizer phosphate market, by the 1980s its market had fallen to less than 18%, and this market share continues to decline.

Ammonium Phosphates

Ammonium phosphates, which contain both nitrogen and phosphorus needed by plants, are often used in combination with potassium phosphates. These combination fertilizers, which promote root and leaf development, lead to better flower development and better fruit production.

Ammonium phosphates for fertilizer applications are prepared as solids, liquids, and suspensions.

The solid forms of monoammonium phosphate (MAP) are made by mixing ammonia with phosphoric acid in an NH_3-to-H_3PO_4 mole ratio of 1.4–1.5:1. The resulting slurry is then sprayed into a granulator, where additional phosphoric acid is added until the mole ratio is 1:1. The granulated material is then dried in a rotary dryer. A typical monoammonium phosphate prepared in this manner is a 10–50–0-grade fertilizer. (On commercial fertilizer packages, these three numbers represent the N–P_4O_{10}–K_2O content, that is, 10% N, 50% P_4O_{10}, and 0% K_2O.)

The preparation of diammonium phosphate (DAP) is similar, except, because diammonium phosphate, when heated, releases ammonia to form monoammonium phosphate, care must be taken to reduce the extent of this deammoniation and to recycle any released ammonia back to the product. Various devices have been developed to accomplish this procedure, such as feeding the ammonia beneath the surface of the reaction mixture in the reactor–granulator and scrubbing the off-gases to recover released ammonia.

Liquid solutions of ammonium phosphate are prepared from the direct reaction of wet-process phosphoric acid or superphosphoric acid with ammonia. In a typical example, ammonia is added to a tank of superphosphoric acid (containing 40% or more of the P_4O_{10} as pyro- or polyphosphoric acid) until the pH is 6. This reaction is very exothermic, and therefore, cooling is provided by use of an outside heat exchanger. The temperature is held below 90 °C to prevent hydrolysis of the pyro- or polyphosphates. Water is added as needed to maintain the desired product analysis. A typical fertilizer solution is 10–34–0; about 40% of the 34% P_4O_{10} (or about 13% P_4O_{10}) is present as ammonium polyphosphates.

The ammonium phosphate fertilizer suspensions are saturated solutions in which finely divided solid particles of the ammonium phosphates are suspended by incorporating a suspending agent such as a gelling clay. The main advantages of these suspensions are (1) a higher total plant nutrient content (in excess of the solubility limits), (2) more grade flexibility, and (3) the use of lower cost phosphate ingredients, such as off-grade phosphoric acid. Also, when KCl is used as the source of potash, the K_2O content cannot exceed 10% because of the solubility limits of the potassium salts. However, a suspension allows the formulation to exceed this 10% K_2O limit.

The ammonium salts of superphosphoric acid have a higher concentration of ammonia and phosphorus than diammonium phosphate and are favorites as fertilizers.

In 1985, of the estimated 5.2 million short tons of P_4O_{10} used in fertilizer, approximately half was derived from ammonium phosphates. An additional amount of ammonium phosphate fertilizer was prepared on the farm, in situ, from wet-process phosphoric acid and ammonia.

Potassium Phosphates

Potassium phosphates are valuable compounds for special applications: for example, in the catalyst system for synthetic rubber manufacture, for radiator coolants, as a soluble fertilizer, and for liquid detergents. Like their sodium counterparts, potassium orthophosphates are prepared by replacing the hydrogen atoms in phosphoric acid with potassium atoms. Typical reactions are

$$\underset{\text{phosphoric acid}}{H_3PO_4} + \underset{\text{potassium hydroxide}}{KOH} \longrightarrow \underset{\text{monopotassium phosphate}}{KH_2PO_4} + H_2O$$

$$H_3PO_4 + 2KOH \longrightarrow \underset{\text{dipotassium phosphate}}{K_2HPO_4} + 2H_2O$$

$$H_3PO_4 + 3KOH \longrightarrow \underset{\text{tripotassium phosphate}}{K_3PO_4} + 3H_2O$$

The preferred source for sodium in the preparation of sodium phosphates is sodium carbonate (soda ash). The more expensive sodium hydroxide is used only when sodium carbonate cannot be used for chemical reasons (*see* Chapter 7 on sodium phosphates). These equations show that the opposite is true for potassium. Potassium hydroxide is preferred over potassium carbonate as the source of potassium because the former compound happens to be cheaper than the latter compound.

Some of the industrially available potassium phosphates are monopotassium phosphate (KH_2PO_4), dipotassium phosphate (K_2HPO_4), tripotassium phosphate (K_3PO_4), tetrapotassium pyrophosphate ($K_4P_2O_7$), potassium tripolyphosphate ($K_5P_3O_{10}$), and potassium polymetaphosphate [$(KPO_3)_n$].

Because of their higher cost, the production of potassium phosphates is limited. For example, the annual production of sodium tripolyphosphate in 1985 is more than 1.4 billion pounds, but yearly production of potassium tripolyphosphate is practically nil. Nevertheless, potassium phosphates have unique properties that are needed in special applications.

Monopotassium phosphate is prepared industrially by adding 1 mol of a water solution of potassium hydroxide to 1 mol of phosphoric acid. The solution is concentrated by boiling off water. After the mixture is cooled, crystals of monopotassium phosphate separate and are removed by centrifuging. The mother liquor is recycled for the next batch. At 30 °C, the solubility of monopotassium phosphate is a little over 20 g/100 g of solution.

Manufacturers of everything from gin and vodka to antibiotic wonder drugs use monopotassium phosphate as a mineral nutrient for the microbes in their fermentation tanks. For the rapid growth of microbes (molds, yeasts, and bacteria), both potassium and phosphorus are usually essential. The organisms use the phosphate in energy-transfer reactions. For example, in the fermentation of glucose (a simple sugar), glucose 1-phosphate is the first compound formed. Depending on the nature of the yeast used, this reaction can lead finally to the formation of alcohol or other fermentation products. Experiments have shown that the rate of the fermentation can be gauged by the rate of orthophosphate disappearance. Phosphate plays such an important role in all biological processes that if it is absent from the culture medium, the microbes simply will not grow.

Because monopotassium phosphate goes readily into solution, it is ideal for liquid fertilizer. A special commercial fertilizer, $KH_2PO_4 \cdot (NH_4)_2HPO_4$, consists of equal parts of monopotassium phosphate and diammonium phosphate. This formulation contains about 10.5% N, 52.0% P_4O_{10}, and 17.5% K_2O (potassium oxide) and thus is rated as 10–52–17. (Dried cow manure is 2–1–2.) This potassium–ammonium phosphate mixture is very soluble in water and extremely effective in minimizing shock when seedlings such as tomatoes, tobacco, peppers, and cabbages are transplanted. The shock from transplanting usually causes the plants to wilt and sometimes die. Several days are required for the transplants to resume their normal growth.

Literature Cited

1. Shen, C. Y.; Callis, C. F. *Preparative Inorganic Reactions*; Jolly, W. L., Ed.; Interscience: New York, 1965; Vol. 2, pp 139–167.

Today's shopper has an amazing array of cleaners and detergents from which to choose. Sodium phosphates are used widely in these compounds.

Chapter Seven

Phosphates for Cleaners, Detergents, Dispersants, and Chelants

THE MOST WIDELY USED PHOSPHATES for cleaning and detergent compositions are the sodium phosphates. These phosphates are also very important in metal cleaning, water softening, various food processing, and toothpastes. As a class, however, sodium phosphates represent many compounds. Sodium orthophosphates are sodium salts of phosphoric acids. Hydrogen atoms in phosphoric acids are replaced by sodium ions. Sodium orthophosphates are prepared by successively replacing the hydrogen atoms in orthophosphoric acid by one, two, or three sodium ions. For example, the formation of monosodium phosphate is

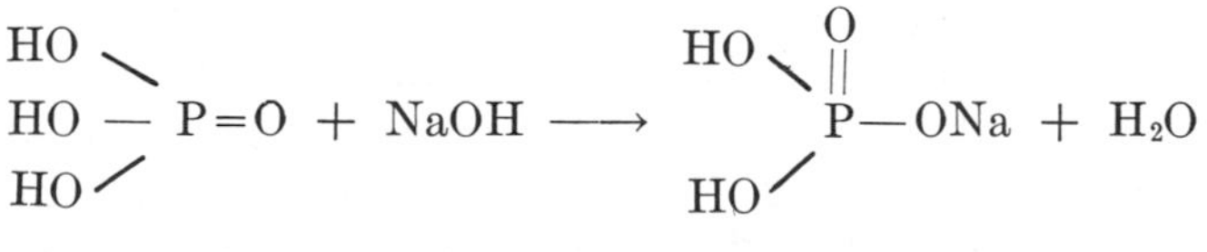

$$(HO)_3P{=}O + NaOH \longrightarrow (HO)_2P({=}O){-}ONa + H_2O$$

Sodium phosphates containing more than one phosphorus atom in a single molecule—as in sodium pyrophosphate (or sodium diphosphate) and sodium tripolyphosphate—can also be made by replacing the hydrogen atoms in the corresponding pyrophosphoric acid or tripolyphosphoric acid with sodium ions. Commercially, this class of sodium polyphosphates is not manufactured by this route. Their preparative procedures are described under their respective sections.

Many sodium phosphates form hydrates. Some sodium phosphates are available in both anhydrous and various hydrated forms. Because so many commercially important sodium phosphates are available, the box on page 86 will help differentiate them.

1002–0/87/0085$06.00/1

Twelve different sodium phosphates are listed in the box. Even though they are all from the same family, each is unique, and sometimes they behave as though they are not at all related. Chemists take advantage of these differences and use each compound for quite different duties depending on the individual talents of each.

	Sodium Phosphates
NaH_2PO_4	Monosodium phosphate
Na_2HPO_4	Disodium phosphate
Na_3PO_4	Trisodium phosphate
$(Na_3PO_4 \cdot 12H_2O)_5 \cdot NaOH$	Trisodium phosphate dodecahydrate–sodium hydroxide complex
$(Na_3PO_4 \cdot 11H_2O)_4 \cdot NaOCl$	Trisodium phosphate–sodium hypochlorite complex
$Na_2H_2P_2O_7$	Sodium acid pyrophosphate
$Na_4P_2O_7$	Tetrasodium pyrophosphate
$Na_5P_3O_{10}$	Sodium tripolyphosphate
$-(NaPO_3)_{15-20}-$	Glassy sodium polymetaphosphate
$-(NaPO_3)_n-$	Insoluble sodium metaphosphate
$(NaPO_3)_3$	Cyclic trimetaphosphate
$(NaPO_3)_4$	Cyclic tetrametaphosphate

Trisodium Phosphate for Cleaning Painted Surfaces

Many forms of trisodium phosphate are commercially available. One form is the anhydrous form, Na_3PO_4, in which all three of the hydrogens in phosphoric acid have been replaced by sodium ions. Another is trisodium phosphate dodecahydrate. This compound also contains about 1 mol of sodium hydroxide for every 5 mol of trisodium phosphate [$(Na_3PO_4 \cdot 12H_2O)_5 \cdot NaOH$]. Trisodium phosphate, or TSP, is the compound purchased for home use.

In the preparation of trisodium phosphate, sodium atoms from sodium carbonate replace hydrogen atoms from phosphoric acid. However, only about two of the hydrogen atoms in phosphoric acid can be replaced this way. Sodium carbonate is not basic enough to replace the third atom. In commercial manufacturing practice, the third hydrogen atom in phosphoric acid is replaced by sodium from sodium hydroxide—a very strong base; thus, about two-thirds of the sodium comes from sodium carbonate and one-third from the more expensive sodium hydroxide.

Trisodium phosphates are strongly alkaline, especially the dodecahydrate, which contains the extra sodium hydroxide. Many trisodium phosphate applications depend on this high alkalinity. For example, in cleaning painted walls with trisodium phosphate, the alkalinity of the solution causes a thin layer of paint that has been oxidized by air to be removed from the wall surface. The clean underlayer is thus exposed. If a hot solution containing about 10% trisodium phosphate is used, all of the painted coating is removed from wooden or metal surfaces. The high alkalinity of trisodium phosphate saponifies (splits apart) the compounds (fatty acid esters) that hold the oil-based paint together.

In many ways, paint removal is closely related to soap making because fats are broken up in both processes. Chemically, fat is a glyceride, that is, a compound of glycerine and high molecular weight organic acids; these acids are often called fatty acids. Their nature varies depending on the type of fat from which they are derived. For example, fatty acids of beef tallow are mostly stearic acid with 18 carbon atoms, palmitic acid with 16 carbon atoms, and oleic acid with 18 carbon atoms and a double bond. The following reaction shows the saponification of a fatty acid ester (glyceride) into glycerin and the sodium salt of palmitic acid (soap) with trisodium phosphate:

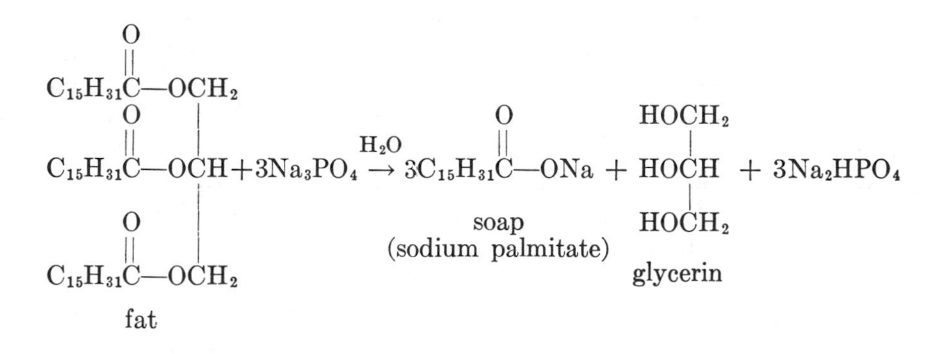

The ability of trisodium phosphate to break fats and greases into water-soluble glycerin and soap makes it extremely useful in scouring powders. When combined with abrasives and chlorine-generating bleaches, trisodium phosphate removes cooking grease and stains.

Trisodium Phosphate–Sodium Hypochlorite Complex for Cleaning, Bleaching, and Sanitizing

In many industries, such as in food processing and dairy plants where cleanliness is important, the high alkalinity of trisodium phosphate makes it the cleansing agent of choice. After being cleaned, however, the equipment must still be sterilized. One sanitizing agent used is sodium hypochlorite ($NaOCl$); this potent oxidizer and chlorinating compound is prepared by the action of chlorine (Cl_2) on a water solution of sodium hydroxide ($NaOH$):

$$2NaOH + Cl_2 \longrightarrow \underset{\text{sodium hypochlorite}}{NaOCl} + NaCl + H_2O$$

Sodium hypochlorite is familiar to us as liquid laundry bleach (Clorox, Purex, etc.). This compound is stable only in a solution containing excess sodium hydroxide; as a solid, it quickly decomposes. For cleaning and sanitization of food-handling equipment, a one-step process is highly desirable. If trisodium phosphate and sodium hypochlorite could be combined into one compound, cleansing and sanitization could be done in one operation.

In the early 1920s, L. D. Mathias, a chemist with Victor Chemical Works, working on this problem, theorized that because sodium hypochlorite is prepared by the action of chlorine on sodium hydroxide and trisodium phosphate dodecahydrate is actually a complex of trisodium phosphate with sodium hydroxide, possibly a reaction between the chlorine and the sodium hydroxide in the trisodium phosphate complex could be obtained. A new trisodium phosphate–sodium hypochlorite complex would result. The compound Mathias envisioned would have the formula $(Na_3PO_4 \cdot 12H_2O)_5 \cdot NaOCl$, and the reaction would be

$$(Na_3PO_4 \cdot 12H_2O)_5 \cdot NaOH + NaOH + Cl_2 \longrightarrow (Na_3PO_4 \cdot 12H_2O)_5 \cdot NaOCl + NaCl + H_2O$$

Mathias' hunch was 99% right. After much laboratory work, a new compound with the structure $(Na_3PO_4 \cdot 11H_2O)_4 \cdot NaOCl$ was discovered—not exactly the compound Mathias had planned, but very close. The compound was given the trade name Cl-TSP for chlorinated trisodium phosphate. One process developed to manufacture chlorinated trisodium phosphate involves the addition of a sodium hydroxide solution of sodium hypochlorite to a concentrated solution of disodium phosphate. Disodium phosphate reacts with sodium hydroxide to form trisodium phosphate in solution. Trisodium phosphate, in turn, reacts with sodium hypochlorite to form the trisodium phosphate–sodium hypochlorite complex, and then the complex (Cl-TSP) precipitates as a stable solid.

Subsequent investigators discovered that trisodium phosphate forms a complex not only with sodium hydroxide and sodium hypochlorite but also with many other sodium salts. One interesting complex is that formed with sodium permanganate $(Na_3PO_4 \cdot 11H_2O)_7 \cdot NaMnO_4$. Because this complex is deep purple, it was instrumental in eliminating a problem of mistaken identity. Cooks in the kitchens of dining cars on trains had been using Cl-TSP for cleaning and sanitizing equipment. Because Cl-TSP is a white powder like sugar and salt, careless waiters sometimes filled sugar bowls or salt shakers with it by accident. Cl-TSP tastes like soap and when shaken on meat or scrambled eggs produced enraged roars from Pullman passengers.

So that such accidents could be prevented, research came to the rescue. One obvious solution—adding a dye to Cl-TSP—posed problems because sodium hypochlorite in Cl-TSP bleaches most organic dyes. The difficulty was finally solved by adding a small amount of sodium permanganate to the reaction mixture. The entire product is colored by the trisodium phosphate-sodium permanganate complex to an attractive purple-pink, yet the cleaning and sanitizing action of the main component, Cl-TSP, is unaffected.

Even though the theoretical available chlorine content of Cl-TSP is 4.66%, Cl-TSP contains 3.4–3.6% available chlorine as produced commercially. (Available chlorine is the chlorine content calculated as available for bleaching and sanitizing.) Above this level of available chlorine, the material is not stable, and Cl-TSP decomposes back to the 3.4–3.6% range. However, we have found that if about 2% of sodium silicate is added during the manufacturing process, a Cl-TSP containing 4.0–4.2% available chlorine is obtained (*1*).

When mixed with an abrasive, such as a silica, and a detergent, such as sodium alkylbenzenesulfonate, Cl-TSP is a very effective cleanser for hard surfaces such as kitchen sinks and bathtubs. The chlorine liberated upon addition of water bleaches the stains almost immediately. The compound is also effective in automatic dishwashing detergents. Cl-TSP is formulated with 40–55% sodium tripolyphosphate and a surfactant, along with sodium metasilicate. Cl-TSP provides high alkalinity, which disperses grease. Because it is water-soluble, Cl-TSP is easily rinsed from glass surfaces. This compound provides a sheeting action on glass surfaces that helps remove soil and cleaning formulation components during the washing and rinsing cycles. This action cleans glasses, leaving them streak free and spotless. The chlorine liberated from the Cl-TSP also bleaches stains from dishes.

Tetrasodium Pyrophosphate (TSPP) in Detergents

Tetrasodium pyrophosphate is made by removing 1 mol of water from 2 mol of disodium phosphate:

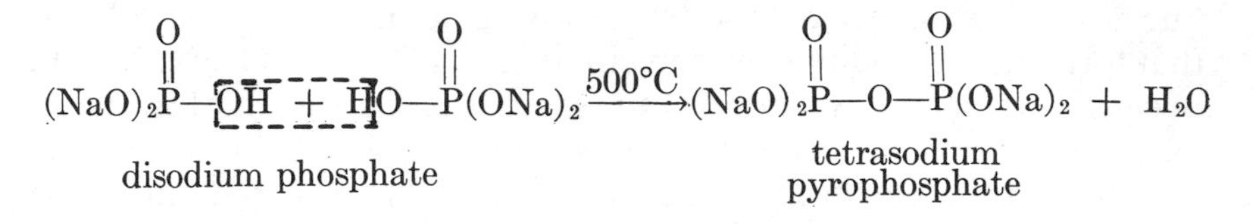

This reaction is carried out at about 500 °C. The product is a white powder that is soluble in water to the extent of about 10%. This relatively low solubility limits its use in some applications, such as in liquid detergents. However, TSPP is still very useful in soaps where the 10% solubility in water is more than enough.

TSPP found its first major use as a builder for soap during the 1930s. Chemically, the term builder means a compound, such as TSPP, that when added to soap produces a more efficient cleaning agent. The action of each ingredient enhances that of the other ingredient. Other builders still used to some extent include soda ash (sodium carbonate, Na_2CO_3), sodium silicates, and trisodium phosphate. As builders, however, these compounds are not quite as effective as tetrasodium pyrophosphate. Recently, new builders such as sodium nitrilotriacetate (SNTA) and certain zeolites have been found to be quite effective replacements. SNTA will be described in more detail under Chelation. Zeolites and the current status of SNTA and zeolites as alternates to phosphates in detergents are discussed in Chapter 8.

From the standpoint of chemistry and physics, the cleaning process is extremely complicated. One reason for the effectiveness of tetrasodium pyrophosphate as a builder is its effect on the "critical micelle" concentration of the detergent. The theory behind the cleaning action of soap (or organic surfactant compounds that do the same job) is that a soap molecule has a long organic chain at one end. The other end, called the polar end, consists of an ionic grouping, $(-CO_2^-)(Na^+)$. At a certain critical concentration in water, the soap molecules agglomerate into a micelle—that is, about 60–80 molecules of soap aggregate into a small jelled lump with the organic or nonpolar end on the inside and the inorganic, polar end on the outside. The organic groups on the inside dissolve the organic oily and greasy portion of the dirt.

The presence of a builder lowers the concentration of the soap or organic surfactant in water necessary for micelle formation; that is, now less than 60–80 soap molecules are required to aggregate into a micelle. Thus, less soap or surfactant is needed for the same cleaning power because many small micelles clean more efficiently than one large micelle.

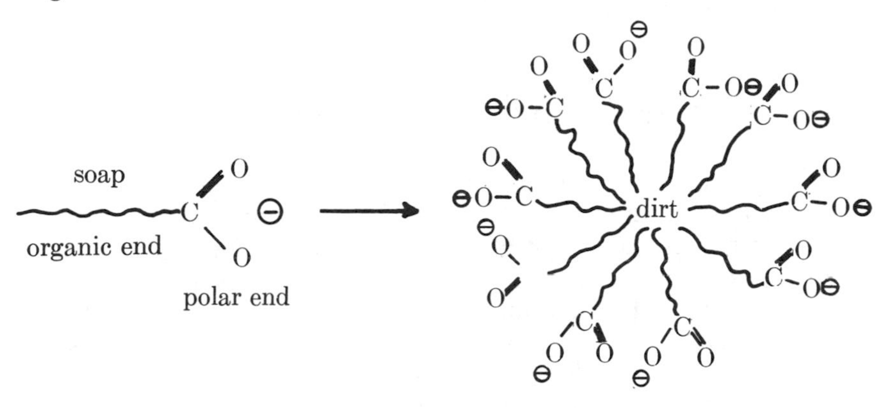

Most dirt particles are attached to fabric by an electrostatic force or by an oily film. In cleaning, the detergent loosens this link so that the particles become suspended in the water. TSPP is a good dispersant, so it helps soap hold the dirt particles in suspension.

In hard water, as described in Chapter 11 on water softening, calcium and magnesium ions react with the soluble sodium soap to form insoluble calcium and magnesium soaps. These insoluble soaps form a gray scum that can redeposit on the clothing. Insoluble calcium and magnesium soaps also decrease the concentration of the soluble sodium soap in the cleaning solution. TSPP is an effective builder for soap also because at the right concentration it reacts with calcium and magnesium ions to hold them tightly as a complex water-soluble compound; this reaction is called sequestration. The calcium and magnesium ions thus sequestered are no longer available to react with soap to form insoluble calcium and magnesium soaps.

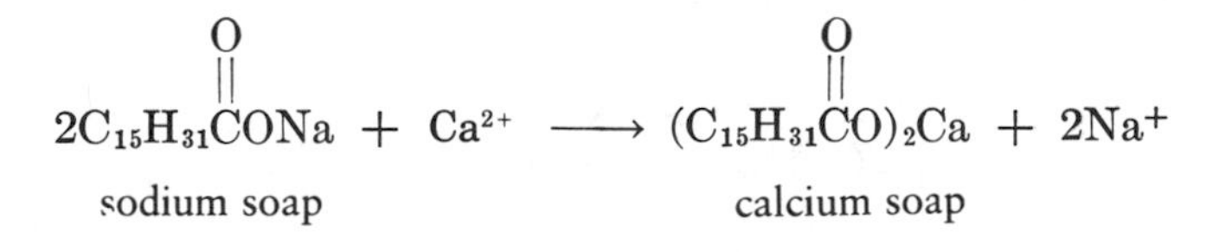

$$2C_{15}H_{31}\overset{\overset{O}{\|}}{C}ONa + Ca^{2+} \longrightarrow (C_{15}H_{31}\overset{\overset{O}{\|}}{C}O)_2Ca + 2Na^+$$

When TSPP was introduced commercially as a soap builder, sodium tripolyphosphate was not yet available. Since the introduction of synthetic surfactants with sodium tripolyphosphate as the builder, TSPP-built soap has become much less important. One major reason TSPP is no longer used extensively as a commercial sequestering agent is that even though it has a strong complexing action, if present in large excess, TSPP forms extremely insoluble pyrophosphates of Ca^{2+}, Mg^{2+}, and Fe^{2+} if these ions are present in excess. This precipitate formation tends to counteract the complexing ability of the pyrophosphate ion.

Sodium Tripolyphosphate as Detergent Builder

At this writing, sodium tripolyphosphate (STPP) continues to be a much-maligned compound. STPP's use in detergents has been blamed as the chief cause of pollution that leads to the eutrophication of lakes and streams. As a component in detergents, it has had a profound influence on that industry. On the positive side, STPP has provided users with a new order of magnitude in the cleanliness of their laundry. "Tattletale gray" shirts and the dull look on colored dresses are now a thing of the past. At present, about 1.4 billion pounds of sodium tripolyphosphate is produced each year in the United States. This figure is a significant drop from the 2.2 billion pounds per year sold at the height of STPP's usage during the early 1970s. Actually, this 1.4 billion pounds per year volume is still large. This volume is understandable when we realize that in over half of the U.S. households some sodium tripolyphosphate is still used in washing machines. Today, many packages of synthetic detergents on the market contain up to 34% STPP. The primary purpose for its use is as a builder for the synthetic surfactant (detergent) sodium alkylbenzenesulfonate.

Sodium alkylbenzenesulfonate is similar to soap in that it is also a molecule with an organic end and a polar end. The organic end is usually a benzene ring to which is attached an organic alkyl group with an average of 12 carbon atoms. The polar end is the sodium salt of a sulfonic acid group ($-SO_3Na$). Sodium alkylbenzenesulfonate can be represented by the formula

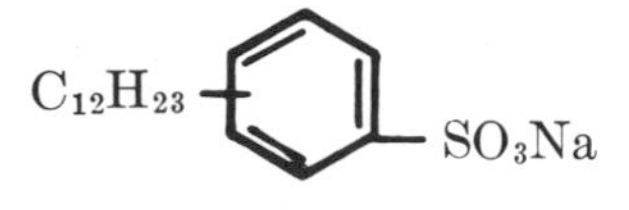

or pictorially as

$CH_3CH_2CH_2CH_2CH_2CH$ $CH_2CH_2CH_2CH_2CH_2CH_3$

organic group

SO_3Na

polar group

The alkyl chain attached to the benzene ring can be a straight chain, as illustrated, or a branched chain. The straight-chain compounds are now preferred because they can be biologically "chewed up", that is, destroyed in sewage plants by bacterial action. In other words, the straight-chain compounds are biodegradable, and this property eliminates foaming problems in rivers and streams.

Sodium tripolyphosphate is manufactured by heating a mixture of 2 mol of disodium phosphate and 1 mol of monosodium phosphate. Two moles of water is eliminated:

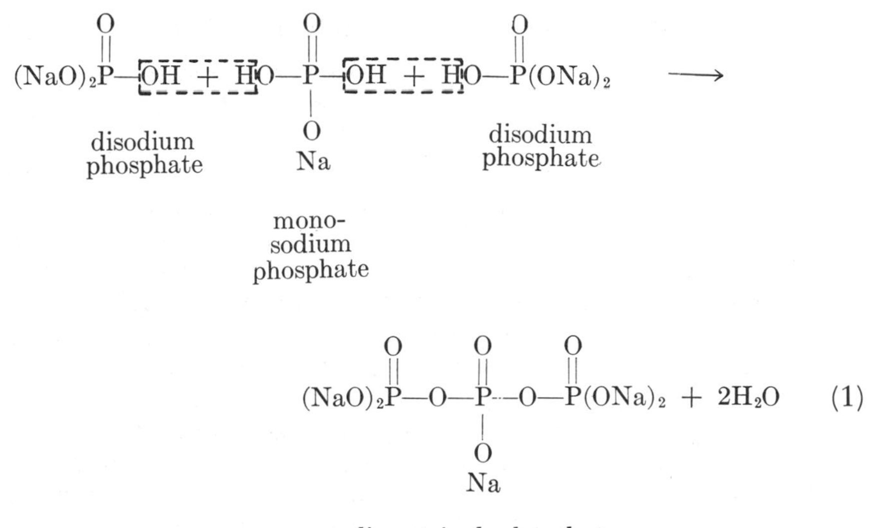

The two anhydrous forms of sodium tripolyphosphate are called type I and type II. These forms are chemically the same, but because their crystal structures are different, these forms exhibit somewhat different behavior. Type I is formed by carrying out reaction 1 at 540–580 °C. A pure type II can be obtained by heating a 1:2 mixture of the mono- and disodium phosphates to a molten state at about 620 °C, keeping it at 550 °C for some time, and then rapidly cooling it to room temperature.

As the mixture cools to 150–100 °C, the glassy product disintegrates into a fine type II powder. Type I is the stable crystalline phase of sodium tripolyphosphate, and type II is regarded as the metastable phase. Type II can be converted to type I by heating but not vice versa.

In industry, a mixture of types I and II is produced. For those detergent formulations needing a higher percentage of type I, a temperature range of 415–490 °C is used. When mostly type II is desired, a temperature range of 350–400 °C is used. At room temperature, commercial-grade type I has a water solubility of about 15–16 g/100 g of solution, and type II has an immediate solubility of about 31 g/100 g of the solution. The higher solubility of type II is short-lived. Within minutes, type II reacts with water to form the hexahydrate ($Na_5P_3O_{10}{\cdot}6H_2O$), which has a solubility of only about 14 g/100 g of solution. This conversion to the hexahydrate can also occur with type I:

$$Na_5P_3O_{10} + 6H_2O \longrightarrow \underset{\text{sodium tripolyphosphate hexahydrate}}{Na_5P_3O_{10}{\cdot}6H_2O} \qquad (2)$$

Most synthetic detergent producers mix a water slurry of solid sodium tripolyphosphate with sodium alkylbenzenesulfonate and other ingredients. Fortunately, with the mixture of types I and II, sodium tripolyphosphate does not form the low-solubility hexahydrate even after a few hours. Thus, a slurry can be obtained with higher solids content, which is more economical (less water to remove) when the slurry is subsequently spray-dried to the finished detergent composition.

In the spray-drying process, used by all major detergent manufacturers, sodium tripolyphosphate is eventually converted entirely to the hexahydrate. Some of the hexahydrate is degraded to tetrasodium pyrophosphate and a small amount of byproduct sodium orthophosphate. This byproduct has no practical value as a detergent builder.

For specialty formulations, large, expensive spray-dry equipment is uneconomical for drying detergent mixtures. Thus, a water solution of sodium alkylbenzenesulfonate is sprayed into a mixer containing a porous sodium tripolyphosphate and other ingredients. This porous compound absorbs aqueous sodium alkylbenzenesulfonate solution, and the absorbed water is used to form dry sodium tripolyphosphate hexahydrate. In this way, a solid mixture product containing water of crystallization is produced ready for packaging. Hence, the compound must be made with the correct balance of type I and type II and with the correct porosity.

Another method for making STPP-built detergents without the necessity of the expensive spray-drying equipment is the batch fluff process of Monsanto. This process is based on the reaction of a solution of cyclic sodium trimetaphosphate (STMP) (Chapter 4) in water with sodium hydroxide to form STPP (2).

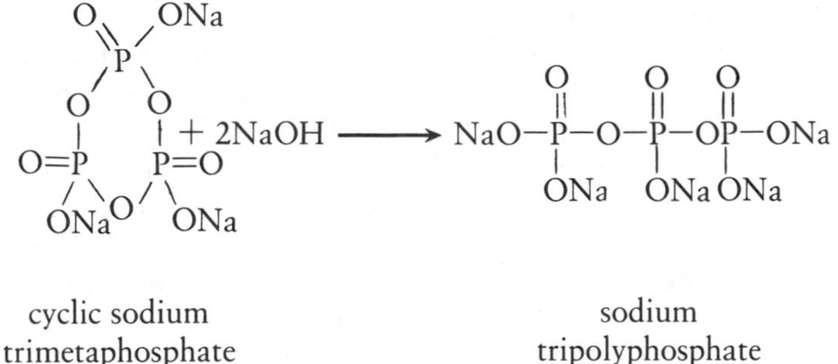

cyclic sodium trimetaphosphate (STMP) — sodium tripolyphosphate (STPP)

The background for the development of this process is quite interesting. The practicality of this process for commercialization is another case of keen observation by the research scientist on a chemical effect. More important is their translation of the observed phenomenon into an industrial process.

In the manufacture of STPP on an industrial scale, the purity of the product is usually in the order of 94–95% or better. One byproduct is STMP. However, STMP is generally not found in the STPP-built detergent product. According to C. Y. Shen, the inventor of this batch fluff process, the STMP must be converted to STPP during the detergent manufacturing process. During the late 1950s, the detergent manufacturers wanted STPP having an assay close to the upper 90%. Shen believed that if his reasoning was correct, then the STMP byproduct should be considered as part of the STPP, because the byproduct is converted to STPP during the detergent manufacturing process. To check his hypothesis, Shen prepared an appropriate detergent slurry with all the regular detergent ingredients. Instead of STPP, he used STMP and sodium hydroxide. Shen was able to prepare this detergent slurry without difficulty. However, when the conversion of STMP to STPP hexahydrate began, the detergent slurry temperature increased rapidly and soon reached the boiling point. The escaping steam expanded the solidifying STPP hexahydrate–detergent mixture into a light, fluffy solid. Afer evaporative cooling of the fluff product, an easily handleable solid was obtained. Shen recognized that this combination of events could be used to develop a process for the preparation of light-density detergents. However, this process did not catch on immediately, and Shen had to invent a simple machine to make the process acceptable to detergent manufacturers. This process is now used in many developing countries for the manufacture of STPP-built light-density detergents, because the process does not require a large investment for spray-drying equipment.

A main function of STPP as a builder is its chelating action—that is, its ability to sequester ions such as Ca^{2+} and Mg^{2+} (which are responsible for water hardness) into a stable water-soluble complex. One mole of STPP binds 1 mol of Ca^{2+}. In this way, hard water is softened. One theory for the stability of the STPP–metal complex is that the metal ions not only bond to the ionic oxygen atoms in STPP by ionic force but also are held tightly to the double-bond oxygens by electrostatic force.

In its other functions as a builder, STPP (1) increases the efficiency of the surfactant, probably by lowering the critical micelle concentration [discussed under Tetrasodium Pyrophosphates (TSPP) in Detergents]; (2) furnishes the proper alkalinity for cleaning and yet is not too alkaline, so it does not burn eyes and sensitive skin; (3) provides resistance to change in alkalinity during washing; and, (4) because of its high negative charge and its ability to absorb on dirt particles, it builds up negative charges on dirt particles and the repulsion between like charges keeps the dirt in suspension.

Because of its water-softening action and cleansing power, STPP is also used in automatic dishwashing compounds. A typical product may contain as much as 60% sodium tripolyphosphate. The rest consists of sodium metasilicate (as a corrosion inhibitor) and a trisodium phosphate—sodium hypochlorite complex or other chlorine-generating compounds such as potassium dichlorocyanurate (bleaching agent for stain removal). This general phosphate-containing formulation is so effective that thus far very few substitutes or suggestions have been advanced to replace it.

Tetrapotassium Pyrophosphate in Liquid Detergents

Tetrapotassium pyrophosphate is the most important potassium phosphate. An extremely water-soluble compound (187 g/100 g of water at 25 °C), this compound is also very resistant to hydrolysis. These properties, along with the fact that it is a pyrophosphate, make tetrapotassium pyrophosphate ideal as a builder for liquid detergents despite its relatively high cost. In contrast, tetrasodium pyrophosphate is only soluble to the extent of 10%, which is insufficient for a concentrated liquid detergent.

In the 1950s, manufacturers introduced a liquid detergent for heavy-duty laundering. This detergent was originally designed for automatic washing machines because a liquid is easier to measure and pour than powdered forms. Its convenience and popularity make liquid detergent an important item in the detergent industry today. A typical liquid detergent is about 11–20% organic surfactants, about 20%

builders, and 7–10% of a "coupling agent", along with other minor ingredients. (A coupling agent such as sodium or potassium toluenesulfonate causes the organic surfactants and the aqueous phase, which normally do not mix, to form a homogeneous solution.) The remainder of the formulation (about 50%) is water.

Tetrapotassium pyrophosphate, the builder, is made by heating a solution of dipotassium phosphate to dryness; heating is continued to about 400 °C until a white powder is obtained. For use in liquid detergents, this product can be sold as a dry powder or as a 60% water solution.

$$(KO)_2P(=O)OH + HOP(=O)(OK)_2 \longrightarrow (KO)_2P(=O)\text{—}O\text{—}P(=O)(OK)_2 + H_2O \quad (3)$$

dipotassium phosphate → tetrapotassium pyrophosphate

Besides its use as a builder for liquid detergents, tetrapotassium pyrophosphate is also used as a component in catalyst systems to polymerize butadiene and styrene to a superior grade of synthetic rubber, sometimes called "cold rubber" (because the polymerization is generally done at around − 10 °C). Tetrapotassium pyrophosphate, unlike tripotassium phosphate, depends for its function on the pyrophosphate anion, which complexes the ferrous iron used in the polymerization. The ferrous–pyrophosphate complex, in its interaction with cumene hydroperoxide, acts as the catalyst that permits the polymerization to be carried out at low temperature. Because the active ion is the pyrophosphate ion, tetrasodium pyrophosphate may also be used. However, many of the recipes call specifically for tetrapotassium pyrophosphate.

Chelation

A chelating agent is a compound containing donor atoms, such as trivalent nitrogen and oxygen, that can combine with a metal atom by coordinate bonding to form a cyclic structure, called a chelation complex or a chelate. An example is the complex of glycine, an amino acid containing both oxygen and nitrogen on the same molecule, and cupric ion, Cu^{2+}. In solution together, the following reaction occurs:

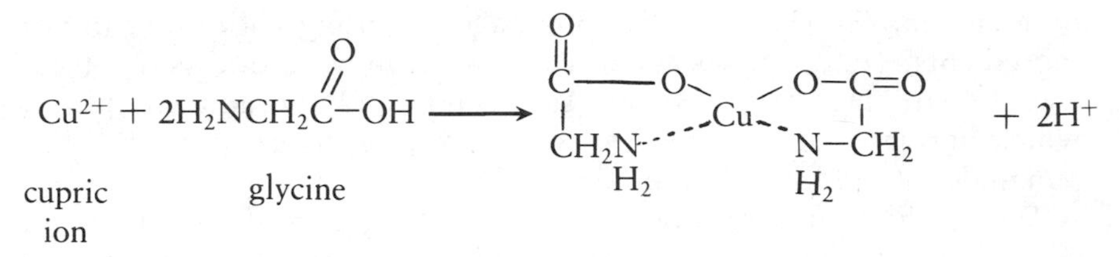

In this example, the cupric ion reacts with two moles of glycine to form a complex containing two five-membered rings. The coordinate bonds are indicated by the dashed lines. Ligands, such as glycine, which can coordinate through two polar groups on the same molecule, are called bidentate ligands (these ligands can essentially "bite" twice at the metal ion). In this case, two hydrogen ions are released. Therefore, the extent of the reaction depends on the hydrogen ion concentration of the solution, usually measured by the pH (*see* p 24).

Chelation agents capable of forming more than two coordinate bonds with a single metal ion also exist. These agents are commonly described as tridentate, tetradentate, and polydentate ligands. The tendency to form ring structures, particularly five- and six-membered rings, is quite common. When the structure is drawn as shown, the early discoverers of this effect considered the bidentate ring closure about a metal ion as analogous to the closing of the two claws of a crab—thus, the name chelate, derived from the Greek word for claw, "chela".

The importance of chelating agents arises from their ability to control or modify metal ions by forming such complexes. The complexes can have properties markedly different from those of either the chelating ligand or the metal ion. This effect can result in reducing the undesirable effects of metal ions, by sequestering them, or by generating desirable effects such as making the metal ions soluble.

This feature, the ability of chelating agents to control the concentration of the free forms of metal ions in solution, is of great importance relative to their commercial applications. An example is dissolving boiler scale such as calcium sulfate or calcium carbonate where the chelating agent solubilizes the otherwise insoluble calcium compound. Other examples include cleaning films of calcium salts from dairy equipment, preventing the precipitation of magnesium or calcium soap scum during laundering, and cleaning metal oxide films from metal surfaces.

Chelates can also form pigments such as in the case of the intensely colored phthalocyanine pigments. These compounds also can react with certain metal ions to form catalysts. In biochemical processes, certain enzymes, coenzymes, and vitamins possess chelate structures, structures that are often required to effect their biochemical function. Another useful property of chelates is displacment of metal ions, as in the case of one of the treatments for lead poisoning. Here the chelate formed from calcium ion and ethylenediaminetetraacetic acid (EDTA) is used to remove lead from body fluids. When the calcium form of EDTA encounters the lead ions, the calcium ion is displaced by the lead, to form the more stable lead chelate and release the calcium ion. Because no calcium ion is removed from the body fluids by this process, an excess of the EDTA chelate of calcium can be used to effect a complete removal of the lead. The lead chelate is then discharged from the body through the urine.

Chelating agents find extensive use in laundering, metal cleaning, bottle washing, and alkaline cleaners. These agents also find extensive use in textile processing (to prevent iron contamination and discoloration during mercerizing), dyeing, aluminum etching, photography, agriculture (for micronutrients), and some food products.

An important class of commercially useful organic chelating agents is aminocarboxylic acids. Examples are EDTA [$(HO_2CCH_2)_2NCH_2CH_2N(CH_2CO_2H)_2$] and nitrilotriacetic acid [NTA; $N(CH_2CO_2H)_3$].

Hydroxycarboxylic acids represent another commercially useful class. Examples are citric acid, gluconic acid, and tartaric acid.

Why is chelation included in a book on phosphorus compounds? For one reason, some phosphorus compounds such as sodium tripolyphosphate (STPP) and sodium hexametaphosphate are themselves considered chelating agents. Their ability to control the concentration of metal ions is believed to be derived from their ability to form chelate rings.

A second reason relates to the recent public effort to remove phosphates from laundry formulations. Because a major function of detergent phosphates, such as STPP, is to control the concentration of the free forms of calcium and magnesium ions in laundering solutions, if phosphates must be removed, perhaps the organic chelating agents could be useful replacements. To some extent, replacement will be successful. Although the organic chelating agents may not provide all of the functions of STPP to the laundering solutions, some of these chelating agents function well enough to be considered reasonable replacements. On the basis of cost and performance, both sodium citrate and the sodium salt of NTA have been found useful replacements. NTA as the sodium salt has been used for many years in

Canada, and NTA is gaining in use in the United States and Europe (more about NTA in Chapter 8 on eutrophication). Because these organic chelating agents are major competitors to the detergent phosphates, knowledge about them is required to fully understand the future prospects for the detergent phosphates.

The third reason, and perhaps the most important reason, is that a number of organophosphorus compounds have been discovered that are effective chelating agents.

Historically, when the chelation process became known, researchers sought chelating agents that were even more effective or more selective. In the course of this search, organophosphorus analogues of the amino acids and the hydroxycarboxylic acids were studied. Several very effective organophosphorus chelating compounds were thus identified.

The first reported preparation of an (α-aminomethyl)phosphonic acid, $H_2NCH_2P(=O)(OH)_2$, analogous to glycine, $H_2NCH_2C(=O)OH$, was by Pikl and co-workers of E. I. du Pont de Nemours & Co. (*3, 4*). This discovery was made in 1941 by the hydrolysis of the reaction product of *N*-methylolstearamide and phosphorus trichloride. In 1948, Chavanne (*5*) prepared the same compound from bromophthalimide and sodium diethylphosphite. Although Pikl claimed his products had metal-deactivating properties, his patents failed to present data showing the effectiveness or selectivity. Rather, confirmation that the $>NCH_2PO_3H_2$ group provided efficient chelating action had to wait for the work of Schwarzenbach et al. in 1949 (*6*). These researchers showed that two phosphorus derivatives of NTA in which some of the $-COOH$ groups in NTA were replaced with $-PO_3H_2$ groups were effective chelating agents for Cu^{2+}, Mg^{2+}, Ba^{2+}, and Sr^{2+}, with compound **I** being more effective than compound **II**. In 1956, Martell and Westerback (*7, 8*) showed that the phosphorus analogue of EDTA (compound **III**) having only $>NCH_2PO_3H_2$ groups, with no $-C(=O)OH$ acid groups at all, was a very effective chelating agent for Ca^{2+}, Mn^{2+}, Cu^{2+}, and Fe^{3+}.

$$\text{I:}\quad (HOC(=O)CH_2)_2NCH_2PO_3H_2$$

$$\text{II:}\quad (HOC(=O)CH_2)_2NCH_2CH_2PO_3H_2$$

$$\text{III:}\quad (H_2O_3PCH_2)_2NCH_2CH_2N(CH_2PO_3H_2)_2$$

Many additional aminophosphonic acids were subsequently prepared. However, despite their effectiveness, except for some very special uses, the processes available made these chelating agents commercially noncompetitive with such compounds as EDTA. A breakthrough occurred in 1966 when two chemists from Monsanto Chemical Co., Irani and Moedritzer (*9, 10*), discovered and described a simple, economic process for preparing (aminomethyl)phosphonic acids. This process entailed the reaction of phosphorous acid with ammonium chloride and formaldehyde in the presence of aqueous hydrochloric acid:

$$\underset{}{NH_4Cl} + \underset{\text{formaldehyde}}{3CH_2O} + \underset{\text{phosphorous acid}}{3H_3PO_3} \xrightarrow{HCl}$$

$$\underset{\text{nitrilotris(methylenephosphonic acid) (NMPA)}}{N(CH_2PO_3H_2) + 3H_2O + HCl}$$

This reaction is quite versatile. The reaction also works with the hydrochlorides of primary and secondary amines to yield the corresponding (aminomethyl)phosphonic acid derivatives. This route is commercially important to a large family of chelating agents based on (aminomethyl)phosphonic acids.

Besides the (aminomethyl)phosphonic acids, 1-hydroxyethylidene-1,1-diphosphonic acid (HEDPA) is also a very effective chelating agent. This compound was prepared in 1897 by Bayer and Hofman (*11*) from the reaction of phosphorus trichloride and acetic acid. However, this work remained unnoticed until the 1960s when the chelating properties of this compound became apparent (*12–16*).

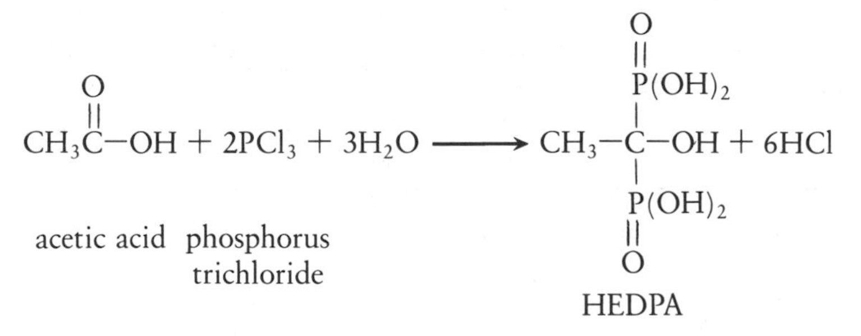

Many other organophosphorus chelating agents have been prepared, but these two, NMPA and HEDPA, have become the most prominent in industrial applications.

In the field of cleaning agents, these organophosphorus chelating agents have gained acceptance in applications such as caustic bottle-washing formulations, liquid detergents, toilet bowl cleaners, industrial and institutional hard-surface cleaners, and stabilizers in bleaching solutions. These agents are also used in allied technologies such as textile processing and paper making. In these applications, these agents are used to control heavy metal ion concentrations, reduce water hardness, prevent or remove iron stains, and prevent deposition or redeposition of unwanted soils.

These compounds function in many ways like sodium tripolyphosphate, in that they provide control of metal ions. These compounds also have the capability of dispersing soils. An important application is their use for the "threshold" effect, whereby a very small quantity can inhibit nucleation of crystals from saturated solutions. This inhibition prevents precipitation of unwanted materials during processing. In addition to those listed previously, organophosphorus chelating agents have one other very important property: they are hydrolytically stable. This property allows them to be used in many applications where conditions are such that sodium tripolyphosphate would be hydrolyzed and become ineffective. Several examples of such applications are metal finishing, boiler-scale removal, and oil-well drilling muds. These applications will be described in Chapters 12, 13, and 15. Some compounds prepared as organophosphorus chelating agents were found to have not only the expected ability to solubilize micronutrient metal ions but also unusual and unexpected biological activity. These properties are described in Chapter 21.

So far, we have dealt only with organophosphorus chelating agents that function in aqueous solutions. Many agents exist that are effective in nonaqueous systems such as lubricating oils and in organic solvents such as metal extractants. Another interesting type of organophosphorus chelating agent is the ethylenebisphosphines as exemplified by $(C_6H_5)_2PCH_2CH_2P(C_6H_5)_2$. Compounds such as this one coordinate with transition metal ions to form effective catalysts for such processes as the hydrogenation and carbonylation of olefins. Examples of these compounds are discussed in Chapter 15.

Currently, the market for organophosphorus chelants for aqueous systems is dominated by NMPA and HEDPA. In 1984, the total usage was estimated at 25–27 million pounds per year. These products are sold as aqueous solutions of 40–60% either in the form of the free acid or as sodium salts. Current prices are as follows ($/lb): for NMPA as a 40% solution of the acid, 0.57; for NMPA as a 50% solution of the pentasodium salt, 0.67; and for HEDPA as a 60% solution of the acid, 1.07.

Literature Cited

1. Toy, A. D. F.; Bell, R. N. U.S. Patent 3 656 890, April 18, 1972 (to Stauffer Chemical).
2. Shen, C. Y. *J. Am. Oil Chem. Soc.* **1968**, *45*, 510.
3. Pikl, J. U.S. Patent 2 328 358, Jan. 23, 1941 (to E. I. du Pont de Nemours).

4. Engelmann, M.; Pikl, J. U.S. Patent 2 304 156, Dec. 8, 1942 (to E. I. du Pont de Nemours).
5. Chavanne, V. *Bull. Soc. Chim. Fr.* **1948,** *27*, 774.
6. Schwarzenbach, G.; Ackermann, H.; Ruckstuhl, P. *Helv. Chim. Acta* **1949,** *32*, 1175.
7. Westerback, S; Martell, A. E. *Nature (London)* **1956,** *178*, 321.
8. Westerback, S.; Rajan, K. S.; Martell, A. E. *J. Am. Chem. Soc.* **1965,** *87*, 2567.
9. Moedritzer, K.; Irani, R. R. *J. Org. Chem.* **1966,** *31*, 1603.
10. Irani, R. R.; Moedritzer, K. U.S. Patent 3 288 846, Nov. 29, 1966 (to Monsanto).
11. Baeyer, H. V.; Hofmann, K. A. *Ber. Dtsch. Chem. Ges.* **1897,** *30*, 1974.
12. Blaser, B.; Worms, K.-H. U.S. Patent 3 122 417, Feb. 25, 1964.
13. Blaser, B.; Worms, K.-H. U.S. Patent 3 214 454, Oct. 26, 1965 (to Henkel & Cie G.m.b.H.).
14. Rogovin, L.; Brown, D. P.; Kalberg, J. N. U.S. Patent 3 400 147, Sept. 3, 1968 (to Procter & Gamble).
15. Kabachnik, M. I.; Lastovskii, R. P.; Medved, T. Y.; Medyntsev, V. V.; Kolpakova, I. D.; Daytlova, N. M. *Dokl. Akad. Nauk SSSR* **1967,** *177*, 582.
16. Kabachnik, M. I.; Medved, T. Y.; Daytlova, N. M.; Rudomino, M. V. *Russ. Chem. Rev. (Engl. Transl.)* **1974,** *43*, 733.

This ugly scene along a bank of Lake Minnetonka, north of Minneapolis, shows dead and dying algae that prohibit any life in these waters. Algae feed off nutrients poured excessively into lakes from cesspools, sewage treatment plants, and runoff from agricultural lands. Just as the alga receives excessive nutrients, so too does it grow excessively, devouring all oxygen within its reach, eventually causing its own death. (Courtesy of Minneapolis Star.*)*

Chapter Eight

Phosphates in Eutrophication

PHOSPHATE HAS BECOME A VERY POPULAR or unpopular subject, depending on your point of view. Some item on it appears almost every day in the news. Ecologists condemn it, and public officials ban it. Some newspaper editors even regard phosphorus as a deadly element.

The main focus of this attention, of course, is the use of phosphate in detergents. In this role, phosphate has been targeted as the pollutant that has caused rapid eutrophication of our lakes, streams, and rivers. All over the United States and in many European countries, the call to ban phosphate in detergents is sounded every day. The subject has become so emotional that well-meaning and intelligent people find discussion of this topic in a rational and scientific manner difficult. This chapter presents some scientific facts about eutrophication and the role phosphorus and other elements play in it.

Many people use the terms pollution and eutrophication interchangeably. Here, pollution will mean the act of introducing a contaminant into a normally clean water system, and eutrophication will mean the process of overfertilizing such bodies of water as streams, ponds, and lakes with nutrients so as to cause a rapid and excessive growth of aquatic plants and algae.

Eutrophication, as a natural geological process, involves a body of water such as a lake, where organic life develops and multiplies over the years; fish, insects, shellfish, bacteria, algae, and various aquatic plants appear and flourish. With time, the lake bottom accumulates the remnants of organic life and other sediment and builds up. As the lake becomes more shallow, the marine life changes character and so do the aquatic plants. Eventually, the lake becomes so shallow that it is a marshland or swamp; finally, it may become dry land. This eutrophication process normally takes thousands of years for a large body of water. Our present concern is that with the inflow of excess rich nutrients, eutrophication is rapidly accelerated. For

1002–0/87/0105$06.00/1

example, many people believe that the shallow western basin of Lake Erie has eutrophied an equivalent of 15,000 years in the past 50 years.

One of the best indicators of advanced eutrophication is the presence of a great deal of algae. Because most environmentalists and lawmakers regard the rapid and excessive growth or "blooming" of algae as the indicator of eutrophication, we shall also use algae, and the nutrients that supply their growth, in the same context.

Algae are small plants that live suspended and free floating in our waters. These plants are probably the most obnoxious organisms in lakes and ponds. The most common species are called blue-green algae; some of them are truly blue-green in color, others are pale yellowish green, and some are actually red. About 2500 species of blue-green algae are known. Under the right conditions (proper nutrients, right temperature, and correct pH), the rapid growth of these microscopic phytoplankton can easily result in water so opaque that an underwater swimmer cannot see more than a few inches in front of him. Depending on the prevailing algal species, a lake may look like green pea soup or be almost blood red. The Green Bay and the Red Sea get their color primarily from the color of the algae present. During periods of heavy growth, some algal species rise to the surface, where wind blows them onto the shore. Bacterial decomposition of the accumulated dead algae creates an unpleasant stench. Much of the dead algal cells settle slowly, always decaying, to the lake bottom where decay continues. This bacterial decay consumes a large quantity of dissolved oxygen, sometimes to such an extent that normal diffusion of oxygen from air to water is not sufficient to replenish the supply. The resulting oxygen depletion kills fish and promotes the growth of weeds and more algae.

Both the scientific and nonscientific communities agree that if the present rate of pollution continues unabated, eutrophication will indeed become very rapid. These groups do disagree, however, as to the source of the nutrients that promote the algal growth and the most expeditious methods for eliminating or reducing them.

The enrichment of lakes with nutrients is a very complicated process. Many lawmakers believe that algal bloom is a modern phenomenon, coinciding with the large-scale use of phosphates in detergents that began in the late 1940s; they think that a simple way to stop eutrophication is to remove phosphates from detergents. The lawmakers tend to forget that when human habitation and land cultivation increase, the inflow of decaying organic matter and animal and human wastes into our waters also increases. As a result, algal bloom is much more frequent. Perhaps much of our present dilemma can be traced to the invention of the flush toilet in the 1840s, and this problem has accelerated in time with increasing population growth.

To show that periodic algal blooming is not an exclusively modern phenomenon, Willy Lange of the University of Cincinnati has cited recorded occurrences in biblical times. One interesting example he cited occurred in 1825 in Lake Murtansee in Switzerland. After a heavy rainfall that washed organic matter and nutrients into the lake, the lake suddenly turned red, like blood. The Swiss could not explain what they saw, but they recalled that on that day two centuries earlier, they had won a bloody battle against the Burgundians. They thought that the lake's turning red was nature's reminder of the slaughter and superstitiously called the phenomenon "Burgunder Blut", or Burgundian blood. Now the red coloration was determined to be the result of excess algae growth. When Professor Lange obtained a sample of the algal specimen, *Oscillatoria rubescens*, from Switzerland and cultured it in his laboratory, he found that it does color the water red. Today's hysteria by some, giving the impression that the noxious algal bloom in our lakes could be eliminated by removing or reducing phosphates in detergents, is not the result of superstition but an oversimplification of a very complex problem.

First, let us look at what makes algae grow. Phosphorus is an element essential to all human, animal, and plant life. All plant and animal cells contain 3–5% phosphorus in both organic and inorganic form. However, though essential, phosphorus is not the only critical element needed for growth. When Dr. H. Clyde Eyster was at Monsanto Research Corp., he and his co-workers showed that besides hydrogen and oxygen, blue-green algae require about 16–17 other essential elements for growth. The ratios of these elements are shown in Figure 8.1. The sizes of the circles representing the elements show the relative amounts of the various elements needed. Note that the

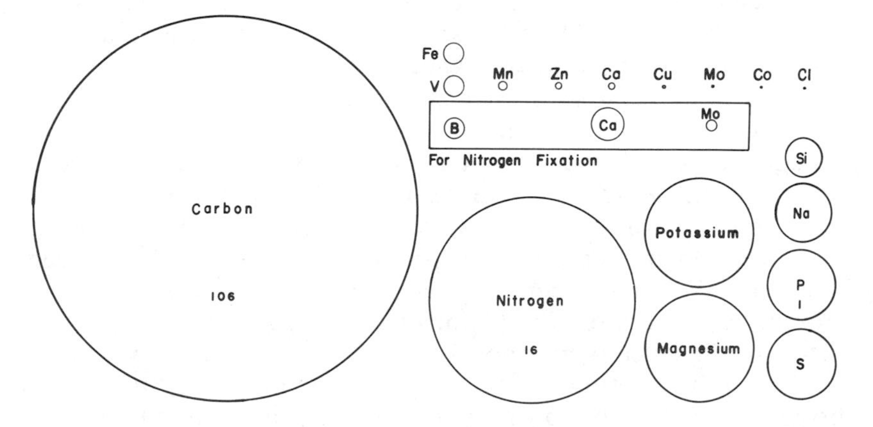

Figure 8.1. Ratios of nutrient elements required for algal growth. (Reproduced with permission from H. Clyde Eyster.)

ratio of the number of atoms of carbon to nitrogen to phosphorus needed is approximately 106:16:1. Most of our waters contain all of these elements to a greater or lesser extent. If a lake contains a minimal amount of phosphorus as phosphate, when all of the phosphate is consumed, the algal population will cease to increase, regardless of the amount of other nutrients present. Thus, phosphorus is the limiting element for algal growth in that lake. The same finding is true for the other essential elements. In other words, the element that is present in an amount that is depleted first by the growing algae is the limiting element in that particular lake. Some lakes could be nitrogen limiting, and some could be iron or carbon limiting. If carbon is the limiting element, when it is all consumed, no increase in algae will occur, no matter how much more phosphate is added.

Where do the elements that are required for algal growth come from, and how can we control them? Most of the metallic elements, such as potassium, magnesium, sodium, and calcium, and the trace elements (those elements needed for growth in very small amounts) are normally present in the soil and are washed into the lakes. These elements are quite difficult to control. Nitrogen can come from the nitrogenous waste in human and animal excrement as nitrites, nitrates, and ammonia. It can also come from the nitrogen in the atmosphere, which is converted into a usable form by the so-called "nitrogen fixation" organisms. Figure 8.1 also shows that these nitrogen fixation organisms need boron, calcium, and molybdenum. Of course, runoff from fertilized fields can also contribute to the nitrogen content of lakes or streams. Carbon, the element required in the largest amount, can originate from the diffusion of carbon dioxide from the atmosphere into the water; it can come from decaying vegetative matter, including the dead algae that grew in the lake; it can also arise from rotting organic matter carried into the lake by land drainage. Other important sources are household sewage and agricultural and industrial wastes.

One of the largest sources of phosphates is sewage effluent; the other source is land drainage and agricultural runoff. Most authorities agree that each source contributes about half of the phosphates in our waters. Because phosphates from land drainage and agricultural runoff are difficult to control, let us consider the phosphates in sewage effluent. If most of the phosphate in sewage effluent came from detergents, its elimination from that source would solve the problem simply and easily. Unfortunately, in the 1970s, depending on the locality, detergent phosphates contributed only about 28.5–70.0% to sewage effluent. When the production of sodium tripolyphosphate was 2.2 billion pounds per year, the generally accepted figure is 66%, or about two-thirds of the total. The other third comes mostly from human waste. Every day we consume phosphates in our food and excrete a corresponding amount. This consumption and excretion also apply to cats, dogs, and other animals. Some phosphates also enter sewage systems through the waste disposal units in kitchen sinks.

The range of phosphate concentrations in sewage systems contributed by detergents, as reported in the literature, is based on various surveys and actual measurements in selected communities in this country. In 1974, I checked these figures with some rough calculations to see whether my values were significantly different—that is, by an order of magnitude—from those I read. Physiology texts say that with an average American diet, each of us consumes and excretes approximately 1.4 lb of phosphorus per year. With a U.S. population of approximately 200 million, this value amounts to about 280 million pounds of phosphorus per year. The total sodium tripolyphosphate manufactured per year in this country is approximately 2200 million pounds. Some is for export, and some goes into uses other than detergent. Let us assume that about 90%, or 2000 million pounds of sodium tripolyphosphate, goes into detergent. Because each mole of sodium tripolyphosphate contains 25.3% phosphorus, 2000 million pounds of it is equivalent to 505 million pounds of phosphorus. The sum total of phosphorus contributed by human excrement and detergents therefore is 280 million plus 505 million pounds—that is, 785 million pounds of phosphorus. Dividing 505 million pounds into the total of 785 million pounds shows that the contribution of phosphorus from detergents is 64%. On the basis of these rough calculations, the figures reported in the literature were within reason. In 1984, when the production of sodium tripolyphosphate was about one-third less, that is, 1.4 billion pounds, a similar calculation shows the contribution from detergent phosphates was about 47%.

How much phosphorus is needed for algal growth? Dr. Lange found that when other required nutrients are present in excess, he could grow blue-green algae in as little as 7 parts of phosphorus per 1 billion parts of water (7 ppb). This figure agrees well with the findings of others. For example, Clair N. Sawyer, one of the most often-quoted authorities, reported that undesirable algal growth will occur at approximately 10 ppb of phosphorus. With these figures in mind, the amount of phosphorus that is available to our streams and lakes can be calculated. A task group on Nutrient Associated Problems in Water Quality and Treatment of the American Water Works Association, headed by P. L. McCarty of Stanford University, reported that the amount of phosphorus in the average effluent from a sewage treatment plant (which removed about 30% of the phosphorus) ranges from 3.5 to 9.0 mg/L. This amount is equivalent to 3500–9000 ppb of phosphorus. Assuming that 66% of this phosphorus is from detergents and is removed, about 1200–3100 ppb is left. Assume that this sewage effluent, on its way through streams to lakes, is diluted with water of lower phosphorus concentration. Assume also that some phosphorus is lost by reaction with minerals in the stream so that only 1% of the original phosphorus arrives at the lake. With these assumptions, still 12–31 ppb of phosphorus would enter our lakes from sewage effluent, which is of nondetergent origin. These calculations support those sanitary engineers, researchers in environmental-quality laboratories, civil engineers, zoologists, and biologists who state that even if all the phosphate were removed from detergents, more than several times the amount needed to support healthy algal growth would be present from domestic sewage alone. More specifically, Dr. William J. Oswald, Professor of Public Health and Sanitary Engineering, University of California, Berkeley, said:

> Independent of any detergent source, the average domestic sewage contains sufficient phosphorus from uncontrollable origin to support the growth of 1000 milligrams per liter of blue-green algae. Such an algal concentration is 50 times that ever found in Clear Lake, California, and 100 times that found in Lake Erie, and 1,000 times that found in the oceans.

The problem of reducing algal growth by limiting the phosphorus concentration in our waters is further complicated by the fact that other sources of phosphorus exist in addition to domestic sewage—for example, land drainage. Streams in three forested areas in Washington state had an average soluble phosphorus concentration of 0.007 mg/L, or 7 ppb.

Phosphorus from agricultural runoff varies with locality. The minimum comes from normal land drainage, medium amounts occur where fertilizers are heavily used, and maximum concentrations are found in runoff from animal feedlots. Modern chicken farms are also great contributors of nutrients for algal growth. The Environmental Pollution Panel of the President's Science Advisory Committee (1965) showed that a cow generates as much waste as 164 humans, one hog as much as 1.9 people, and seven chickens present as great a disposal problem as one person. Altogether, U.S. farm animals produce 10 times as much waste matter as people.

To make the situation even worse, algae are extremely efficient users of phosphates. When they die, 30–70% is readily decayed by associated bacteria; phosphates and other nutrients to support new growth are released. The remaining debris settles to the bottom and decays gradually over a period of years. A lake with a long history of algal growth can actually continue to support new growth even if no fresh nutrients, phosphate included, are introduced.

Some of the laws proposed or passed require a total ban on phosphates in detergents, and others limit phosphorus content in detergent to 8.7%. Because many nonchemists cannot differentiate phosphorus from phosphate, let us recall that the content of phosphorus in sodium tripolyphosphate (STPP), the phosphate normally used in detergents, is 25.3%. An 8.7% phosphorus content, therefore, is equivalent to 34% STPP. The average concentration of STPP in detergent is around 45%. The new laws thus propose a cut of STPP in detergent from around 45 to 34% or a reduction of the phosphate contribution from detergents to our sewage systems from 66 to 50% or a reduction of the phosphorus in our sewage to 84% of the original total. To expect that this reduction will help solve the algal growth problem is scientifically unsound. However, many municipalities have accepted detergents having the reduced (34%) level of STPP and others have totally banned its use. As a result, in 1985, the usage of STPP had been reduced to 1.4 billion pounds. Fortunately the ban on phosphates does not include dishwashing compositions. Modern dishwashing machines just will not work efficiently without phosphates.

This discussion assumes that phosphorus is the limiting element for algal growth. This situation is not always true for many bodies of water in the United States or, in fact, in the world. Dr. Lange used Lake Erie water to study more than 1260 separate cultures of algae. He simulated the influx of each of the 16 essential elements by enriching each culture in the Lake Erie water with one of the elements. His results indicated that influx of nitrogen was growth-enhancing twice as often as that of phosphate and that iron (as a soluble complex) and cobalt stimulated algal growth just as often as did phosphate.

Algal growth depends on photosynthesis, and photosynthesis does not occur without carbon dioxide. Thus, carbon could be the limiting element. If luxuriant algal growth depended on atmospheric CO_2 alone, this source of carbon would be the limiting factor. During a short period of vigorous growth, algae in the upper 12 in. of a lake can increase from 5 to 55 mg/L. The amount of CO_2 needed has been calculated to be insufficient from the diffusion of atmospheric CO_2 into the water; diffusion would be too slow. The present theory is that a large portion of CO_2 is supplied by the bacterial decay of the dead algae and other organic matter. The symbiotic relationship between bacteria and algae is interesting. Algal growth by photosynthesis produces oxygen needed by the bacteria. Bacterial decay of organic matter produces CO_2 needed by algae.

Another source of carbon is the dissolved bicarbonates and carbonates in water. Many researchers on algal growth, including Pat C. Kerr and her co-workers, of the Federal Water Quality Administration of the U.S. Department of the Interior, have shown that an increase in the availability of inorganic carbon such as CO_2 and bicarbonate ions (HCO_3^-) rather than phosphate is responsible for algal growth in the waters they studied. For their studies, therefore, carbon is the limiting element.

Many entrepreneurs have introduced nonphosphate detergents. Most of these products use soda ash (sodium carbonate, Na_2CO_3) and sodium silicates to replace phosphate. More recently, products have been introduced in which zeolite (*see* page 113) was used to replace part of the phosphate (*1*). Many lawmakers even propose a return to the use of sodium carbonate–soap combinations. When this extra supply of carbon from soap and the carbonates enters bodies of water where carbon is the limiting element, it will, of course, enhance algae growth.

Sodium tripolyphosphate, STPP, is an important builder in synthetic detergents. Among other functions, STPP softens water by tying up hardness-causing metal cations into water-soluble complexes that do not interfere with cleaning and can be easily rinsed off. STPP also gives a washing solution with a pH of around 9.0–9.8, that is, alkaline enough for efficient dirt removal but not caustic enough to injure the skin. The result is extraclean clothing. The preference of Americans for this new detergent over soap has practically eliminated laundry soap from the market. Soap is still very effective, however, in soft or softened water.

With the growing publicity on phosphate in detergents as a possible cause of eutrophication, detergent manufacturers first used the sodium salt of nitrilotriacetic acid (NTA) as a substitute builder. This compound is also a water softener. However, in 1970, government reports suggested that when NTA is undecomposed in our soil or water, it can form complexes with mercury or cadmium. If these poisonous complexes entered our food, they might pose a health hazard. Many scientists agreed that the government report was based on a study that was not statistically significant. However, because doubt was cast on the safety of NTA, the Surgeon General of the Public Health Service of the United States suggested that the detergent industry withhold the use of NTA until more data are available. Therefore, as a result of these government reports, the detergent industry in the United States voluntarily suspensed the use of NTA in household detergents. Although the evidence of these government reports was later refuted, during the mid-1970s added concern regarding possible carcinogenicity arose. In 1977, an International Joint Commission in the Great Lakes Region concluded no danger to health or to the environment existed from NTA at low use levels, and in May 1980, the Environmental Protection Agency announced that on the basis of available research, no reason to regulate NTA in detergents existed. The test data indicated that the risk of NTA in detergents was minimal. Therefore, in 1980, at least one of the major U.S. detergent manufacturers introduced household detergents containing the sodium salt of NTA. As of this writing, such a detergent is being sold statewide in Indiana. However, in the United States, the status of NTA is still unclear. In November 1982, after an NTA-containing detergent was test-marketed in upstate New York, the New York Department of Environmental Conservation proposed to ban the use of NTA in laundry detergents. Although the final decision was to have been made in June 1984, the decision has not yet been made. Should the New York State ban be imposed, this ban will be considered to be a major commercial setback for the use of NTA (*2–4*).

Meanwhile, similar studies have been conducted in Europe. In June 1981, the West German Ministry of the Interior established a commission to assess the impact of the use of NTA in West German laundry detergents. In September 1983, this commission (known as the Bernhardt Commission) gave a tentative, restricted, and highly controlled approval of NTA. As a result, no more than 25,000 t/year could be used in West Germany. This value corresponds to an average concentration in the detergent of 3.4%. The Ministry of the Interior requested soapers to link the introduction of NTA with a further reduction in the use of STPP, indicating that if no adverse effects are seen from the use of NTA, then limited increases beyond the 25,000 tons of NTA are possible. Small amounts are also allowed in Holland (6000 t/year) and Italy (2000 t/year).

Meanwhile, since 1972, Canada has allowed full use of NTA in laundry detergents, with no evidence of an adverse effect. Their consumption has risen to 25,000–30,000 t/year.

This story on NTA illustrates the fact that, like STPP, the future success of NTA in the United States may have little to do with the material itself. A great deal has to do with politics and consumer emotion. Current usage in the United States is under 5000 t/year, and should the New York ban be imposed, NTA usage is predicted to remain in this range. However, should no ban be imposed, the U.S. market is projected to rise to the 100,000 t/year range (*5*).

With the requirement for reduction of the phosphates, many other alternates to STPP were explored by the detergent manufacturers. One novel approach was the use of zeolites. Zeolites could be used as a partial replacement for STPP. Zeolites are crystalline aluminosilicates. They have a large, regular pore structure that lends itself to adsorption. It also has a crystalline structure and chemical composition to allow it to act as an ion exchanger to remove the calcium ions, which are responsible for some of the hardness in water.

However, the zeolites had three defects. These defects are (1) the exchange reaction is not instantaneous, (2) the zeolites are insoluble and therefore could leave deposits on the garments, and (3) the zeolites are effective in removing calcium ions but not magnesium ions from solution. Because most hard water contains both calcium and magnesium ions and both ions interfere with the laundering process, zeolites alone are not sufficiently effective.

So that these problems could be corrected, the zeolite was made as a fine powder, sufficiently fine that even if deposited on the garment during the laundry process, it was not visible to the eye. This situation also helped solve the rate of ion exchange problem, and these fine particles of zeolite were found to effectively remove the calcium ion during a 15-min laundering cycle. Finally, to solve the magnesium ion problem, STPP was added, but at a reduced level, 24% versus 34–50%. Today, such zeolite-containing detergents have had widespread usage throughout those areas of the United States where reduced phosphate concentration is allowed.

Most of the other nonphosphate detergents contain soda ash (sodium carbonate, Na_2CO_3) and sodium silicates (sodium disilicate, $Na_2Si_2O_5$; and sodium metasilicate, Na_2SiO_3) as builders. These compounds soften water by removing the "hardness"-causing cations as precipitates, for example, calcium carbonate ($CaCO_3$), magnesium silicate ($MgSiO_3$), and calcium metasilicate ($CaSiO_3$). Not all of these precipitates are rinsed off; some are trapped in the fabric. These trapped salts interfere with the flame-proofing property of the phosphorus-containing flame retardants. Many of the available flame retardants developed for children's pajamas and work clothes contain phosphorus compounds. The theory is that, upon pyrolysis, the phosphorus compounds decompose into phosphoric acids, which in turn catalyze the decomposition of the cellulose fabric into very slow burning chars. The trapped salts on the cloth from nonphosphate detergent convert the phosphoric acids into phosphate salts, which are not effective as catalysts for converting burning cellulose to char. Thus, the cloth burns.

How well do the nonphosphate detergents clean? A limited test was done by a consumer organization using about 300 families. The families were given unlabeled phosphate and nonphosphate detergents. After about a month, most of the families using the nonphosphate detergents were about as satisfied as those using the phosphate detergents.

In the laboratory, however, cleanliness can be measured more precisely under carefully controlled conditions. Artificially soiled cloths are washed with various phosphate and nonphosphate detergents, and cleanliness or whiteness is measured by a reflectometer. Such data are reproducible. Results from many laboratories show that phosphate-built detergents in hard water indeed clean better. Friends of mine who travel extensively in the Far East where phosphate detergents are not prevalent did not notice much change in their shirts laundered abroad until they returned and found that the shirts left at home were much whiter. However, this finding is not scientific proof that phosphate detergents give whiter shirts.

From the health-hazard point of view, commercially available phosphate detergents are less caustic or less basic than those built with soda ash and sodium silicates. The pH of laundry water solutions that contain phosphate detergents is around 9.0–9.8, and those with soda ash and sodium silicate are around 10.5–11.5. Because the pH scale is logarithmic, an increase of 1 unit, for example, from 10 to 11, means an increase of 10 times in basicity. Highly caustic or basic material is injurious to the sensitive skin of our eyes, nose, and throat.

In September 1971, Dr. Jesse L. Steinfeld, then the Surgeon General of the Public Health Service, told Congress that the federal government might eventually have to ban detergents built with soda ash because of their health hazard. He also advised consumers to use phosphate detergents because they are now the safest. Dr. Steinfeld's statement was supported by Russell E. Train, at that time Chairman of the President's Council on Environmental Quality; Dr. Charles C. Edwards, the head of the Food and Drug Administration; and William D. Ruckelshaus, then and again in 1984 the administrator of the Environmental Protection Agency. These men strongly urged state and local authorities to reconsider laws and policies that unduly restrict phosphates in detergents.

Eutrophication of streams and lakes cannot be solved by simply banning phosphates. Temporarily expedient solutions may cause more harm than good. As the then Secretary of Commerce Maurice H. Stans said succinctly in his address to the National Petroleum Council on July 15, 1971:

> . . . Laws to ban phosphate detergents may give the public the notion that the problem is solved, while nutrients, including phosphates, continue to flow into the lakes and rivers from other sources—agricultural and natural as well as manmade. And some of these cannot be controlled. So if people assume that just a legal ban on phosphate detergents will do the job, they may only lull themselves into neglecting far more significant scientific efforts to help purify our waters through phosphate removal techniques in municipal waste treatment plants.

Do we now possess the technology to remove phosphates and the other major nutrients for algal growth in sewage treatment plants? The answer is yes. Are we using this technology? Very little. Sewage treatment practices, though they vary with locality, consist in general of three separate steps. The primary step is the removal of large solid particles from the raw sewage by settling or by filtration through coarse screens. The effuent from this step, which still contains the fine solids, organic waste products, and dissolved minerals, is passed into the secondary treatment step. Here the organic matter is decomposed by air and bacterial action. The secondary treatment step employs one of two general methods. The first, a trickling filter process, brings air and the liquid waste together and passes the mixture through a pebble and sand filter bed loaded with attached microorganisms. Much of the organic waste is thus decomposed by bacterial action and air oxidation. The second method, the activated sludge process, destroys organic waste by bubbling air through the sewage liquor containing a suspension of microorganisms. The sludge thus formed settles and is discarded as waste.

When carried out properly, the secondary treatment step removes most of the soluble organic matter as volatile carbon dioxide gas; thus, if the treated liquor is discharged into streams and lakes, it is not a source of organic carbon to support algal growth. The liquor also does not decay further, using up dissolved oxygen in the water. In other words, the liquor would have a low biological oxygen demand (BOD). The secondary treatment step also removes as much as 20–30% of the phosphates as precipitates (mostly as calcium and magnesium phosphates).

The treated liquor from the secondary treatment step still contains much phosphate and other dissolved minerals. The liquor can now go to another plant for the tertiary treatment step if such a plant is available. Phosphate can be removed here as the insoluble tricalcium phosphate or hydroxyapatite by the addition of lime or as the insoluble iron phosphates by the addition of water-soluble ferrous or ferric salts or as the insoluble aluminum phosphates by the addition of a soluble aluminum salt. Such treatments, coupled with filtration through a bed of carbon in which the carbon has been specially treated or activated for absorbing impurities, can remove up to 98% of the phosphates. Such treated water, after chlorination to kill the residual bacteria, is suitable for drinking. A sewage treatment system that uses modern technology with good results is that of Lake Tahoe, on the California–Nevada border.

Unfortunately, many municipalities have no sewage treatment plants, and their waste is discharged directly into the nearest stream, river, or lake. In a large city like New York, an insufficient number of sewage treatment plants exist: In 1985, this city still released 200 million gallons per day of raw sewage into its surrounding waters (6).

In some cities, only primary and secondary treatments are available, and some cities have only primary treatment. These sewage treatment facilities cannot keep pace with the population growth and are overloaded. Consequently, much of their effluent is only partially treated.

In many areas, domestic sewage goes into individual septic tanks, where organic waste is decomposed by bacteria, and the effluent seeps into the ground. Here, most of the phosphate reacts with the soil and precipitates as insoluble salts. However, in some cases bacterial decomposition of the organic waste is incomplete. For example, in Suffolk County, Long Island, New York, where more than 1 million people live, most of the household sewage goes into individual septic tanks. Sometimes, because of incomplete decomposition, some of the sewage water carrying residual organic matter percolates through the sandy soil and enters the water supply. If some of the organic waste is from the surfactants in the detergents, water drawn from a faucet will be foamy. Thus, an early 1970 law forbids the use of synthetic detergents altogether in Suffolk County. Unfortunately, this law removes only the indicator showing that contamination from sewage has occurred but not the contamination itself.

Some localities are reluctant to build sewage treatment plants because they are expensive. In 1973, depending on the percentage of phosphate to be removed from sewage effluents, the cost of the many systems proposed varies from 2.5¢/1000 gal. for 90% phosphate removal to 3.7¢/1000 gal. for 98% phosphate removal. Ruckelshaus stated that based on a survey of 12,500 sewage plants, the cost would probably be from $1.70 to $3.35 per year for each person living near the sewage plant to operate the phosphate treatment system. A report prepared in November 1980 for the Soap and Detergent Association and written by James M. Folsum and Lloyd E. Oliver of Glassman-Oliver Economic Consultants, Inc., estimated the June 1980 cost for reducing phosphates in municipal sewage as ranging from $1.51 to $4.31 per million gallons or $0.81 to $2.10 per household. They compare this cost with a recurring annual cost generated by a phosphate ban of $11.10 per household. This cost estimate includes only the increased costs for the cleaning, bleaching, and fabric softener steps. The estimate does not include costs related to such factors as reduced garment service life or the use of water softeners. More recently, Mary Purchase, laundry chemist for Cornell Coopertive Extension, estimated the phosphate ban cost for each household as $15/year (7). Estimates for building sufficient sewage treatment plants in this country run as high as tens of billions of dollars. In fact, the estimate by EPA in 1973 put the total at $60 billion by 1990.

Many laboratories are now working to develop more efficient and economical methods for treating sewage waters. Phosphate removal is only part of their program because phosphate is only one of the many nutrients in sewage waste that supports algal growth. Finding better and cheaper methods for cleaning our sewage certainly is a challenge to chemists, biologists, and engineers.

The development of an efficient, safe, and economical builder to replace phosphate in detergent is also a challenge. With such a builder available, provided that it does not present a removal problem, the size of the equipment and the amount of chemicals needed for phosphate removal in a sewage treatment plant would be reduced.

Chemical precipitation–tertiary treatment of sewage has a long history as an extremely effective means of removing phosphorus from water streams. However, recent innovative approaches have placed emphasis on both phosphorus removal and recycling the sludge for other uses. As examples, Swedish scientists are exploring the use of algae to purify the waste water at Rigsjon, Sweden. The algae absorb the nutrients, including the phosphorus. The resultant biomass is reported useful for energy, horticulture, and aquaculture. The Royal Dutch Institute for the Purification of Effluent is reported to have developed techniques for recovering the phosphate for reuse in detergents. As another example, in the United States, Air Products and Chemicals, Inc., is operating a 9 million gallon per day plant in the city of Largo, Florida, to remove phosphate from waste water and to recover the phosphate value. In this process, water undergoes a 0.5-h anaerobic bacterial treatment. During this stress treatment, the bacteria release the phosphorus content to the liquor and oxidize the organic materials. The whole mixture is then subjected to aerobic digestion. During this stage, the bacteria readsorb the phosphorus. The mixture is then passed into a settling cone, where the bacteria precipitate as sludge. The processed waste water, essentially free of organic material and phosphates, is released. Ninety-seven percent of the sludge is recycled to the anaerobic section for treatment of incoming waste water. The 3% sludge is filtered, and the 20% solids filtercake is dried in a kiln. The dried sludge contains 98% solids, up to 10% phosphates, primarily in the form of linear polyphosphates, 7–8% nitrogen, and 2–3% potash. The dried sludge is an effective fertilizer, because 90% of the phosphorus and 75% of the nitrogen are available. The costs per unit of fertilizer value are substantially less than those of commercial fertilizers. The city of Largo plans to enlarge the plant to 15 million gallons per day, and at least 10 similar plants are being designed for other cities.

Can agricultural runoff be controlled so that we can return to the environmental standards of our forefathers? I think such controls are possible. For example, large basins might be built to collect runoff from areas of intensive cultivation for treatment before allowing it to flow into the neighboring streams and lakes. Similar collection basins might also be built to collect runoff from animal feedlots.

The waste effluent from animal feedlots is a large order of magnitude greater than that from agriculture runoff. Because the feedlots are in specific locations, the waste control is probably more readily accomplished. Even so, the control of animal feedlot runoff is not a minor task. On a rainy day, a 50-acre feedlot is estimated to produce runoff equivalent to the raw sewage produced by a city of 60,000 people, and annually more than 1 billion tons of manure is produced by the animal feedlots. Antipollution laws are gradually forcing livestock men to devise ways to collect and use animal waste and prevent it from contaminating the environment. To comply with those laws, some feedlot operators have resorted to housing the cattle in huge metal barns with pens bedded with a few inches of raw sawdust or ground bark. Periodically, the manure and bedding mixture are collected and, through bacteria breakdown, converted into pasteurized and deodorized garden fertilizer. Other methods under development include collection of the cattle manure and leaching out of its protein content. The residue is then fermented into roughage. The combined protein and roughage obtained are claimed to be comparable in flavor and food value with corn silage. Poultry manure has been similarly processed into cattle feed. All these indoor manure treatment plants represent billions of dollars nationwide. Eventually, we, the consumers of agricultural commodities, will have to pay for it by paying much more for our food.

Meanwhile, after more than 15 years since the initial phosphate ban proposals, many states and cities have imposed restrictions or total bans on the use of phosphates in laundry detergents. However, the results have not necessarily been those expected. Eutrophication continues in many of the places where the ban is imposed. In at least two locations, the ban was rescinded. In June 1982, Wisconsin, one of the first states to ban phosphates, rescinded the ban. However, the ban was restored in January 1984. The Wisconsin Center for Public Policy, in commenting on this action, stated "At best, the ban produced only limited results in decreasing the amount of phosphorus entering Wisconsin's waterways. At worst, the increased household laundry costs to the consumer resulting from the ban are irritating and discriminatory. Neither effect seems significant enough in relation to the overall question of water quality to warrant the attention it has received and the acrimony it has generated" (*8*).

In January 1985, Dale County, Florida, rescinded its 12-year-old ban. The county now falls under the state law, which limits phosphorous content in household detergents to 8.7% (*9*). In addition, more consideration is being given by state and county governments to the cost of the phosphate ban versus its benefits, perceived or actual. Meanwhile, pressure, particularly consumer pressure, continues in many areas of the county to impose bans or to further restrict use. The final outcome remains to be seen.

Eutrophication is a complex process. We do have the technology to build efficient sewage treatment plants to remove phosphates and other nutrients that support algal growth. We should continue our research to develop improved technology for an improved environment.

Literature Cited

1. Layman, P. L. *Chem. Eng. News* **1982,** *Sept. 27,* 10.
2. Werner, M. *Household Pers. Prod. Ind.* **1984,** *Oct.,* 129.
3. *Soap Detergent Assoc. Newsl.* **1984,** *Oct.,* 2.
4. *Chem. Bus.* **1983,** *May 2,* 31–34.
5. *Phosphorus Potassium* **1984,** *129,* 23–27 (this review summarizes much of the history of STPP and NTA in detergents).
6. *New York Times* **1985,** *Dec. 31,* B1.
7. *Soap Detergent Assoc. Newsl.* **1984,** *Oct.,* 5.
8. *Soap Detergent Assoc. Newsl.* **1984,** *Sept.,* 3.
9. *Soap Detergent Assoc. Newsl.* **1985,** *Feb.,* 3.

Aerial view of the $16 million biological wastewater treatment facilities of the Rahway Valley Sewerage Authority—one of the largest secondary wastewater treatment complexes built in New Jersey. At left, foreground, are two new aeration tanks, larger than a football field. At right, foreground, are four new 120-ft-diameter final settling tanks. At center is the new Pump and Blower Building, housing the principal control center for the complex.

Chapter Nine

Metal Treating and Cleaning

Phosphatizing Metals

Modern cars, refrigerators, washing machines, and other electrical appliances with painted or enameled surfaces all wear phosphatized undercoatings between the metal surface and the paint to prevent the paint from blistering and peeling. The preparation of such a coating is called phosphatizing. Phosphatizing entails treating the metal surface to provide it with a compact, adherent coating of insoluble metal salts of phosphoric acid. Many metal salts of phosphoric acid are insoluble in water but soluble in a mineral acid such as phosphoric acid. This property is a basis for the formation of this adherent metal phosphate coating. Commercial phosphating solutions are carefully balanced solutions of a metal phosphate in a dilute solution of phosphoric acid. As long as the acid concentration remains above a critical point, the metal phosphate remains in solution. When a reactive metal, such as steel, is immersed in the phosphatizing solution some dissolution, or "pickling", of the metal occurs and the acid concentration is reduced at the liquid–metal surface interface. In the case of steel, iron is dissolved to release ferrous (Fe^{2+}) ions into the interface. The acidity at the interface is reduced, and metal phosphate salts, both those salts initially in the bath plus secondary ferrous phosphate ($FeHPO_4$), precipitate on the surface of the metal. The term "conversion coatings" is used to describe integrally bonded coatings such as phosphate coatings that are formed in place at the metal surface. When this process is done properly, the orginial steel surface, which is electrically conductive and susceptible to corrosion, is converted to a surface that is nonconductive and corrosion resistant. A major purpose of this treatment is to make the metal surface resistant to corrosion. Other purposes include improving the adherence of paints to the metal surface and creating a surface that can be more effectively lubricated.

1002–0/87/0121$06.00/1

Many variations of the phosphatizing process exist; each variation causes a different effect. Examples of effects sought are control of coating thickness, coating crystal size, coating composition, and, of course, coating performance. Methods for varying the phosphatizing process include the use of additives, such as surfactants, accelerators, nucleating agents, or nucleation inhibitors; the nature of the metal ions used in the phosphatizing solution; the concentration of the components of the solution; and the method of application, that is, either dip or spray.

Modern phosphatizing is a result of technologies developed during the last century. As early as 1869, a British patent was issued to prevent corset stays (thin metal strips) from rusting because of damp air or perspiration. The stays were plunged red hot into a phosphoric acid solution, but the resulting coating was thin and not very durable. In 1907, T. W. Coslett patented the first improvement on this process. He dissolved iron filings in dilute phosphoric acid. This coating consisted of a mixture of corrosion-resistant iron phosphates. Other modifications were later introduced. One significant improvement was the replacement of iron filings with zinc or manganese salts. A zinc salt produced a smooth coating of fine zinc–iron phosphate crystals. A manganese salt produced a coating chiefly of relatively coarse (absorbent) crystals of manganese phosphate along with iron phosphate.

Phosphate conversion coating processes have been classified into three general types on the basis of the metal ions in the phosphate salt that precipitates to form the coating. These types are manganese, zinc, and iron phosphatizing. The order given is the order of decreasing coating weight (milligram per square feet) formed on the metal surface.

Manganese Phosphate Coatings

A phosphoric acid solution containing manganese salts is used to coat objects that will not be painted. The bath, containing a solution of 6–10% manganese phosphate, is operated at 140–200 °F. The coating is thick and absorbent and provides excellent corrosion resistance. So that protection is increased, a thin film of oil or wax is usually applied to the absorbent surface. This black, slightly oily surface is seen on nuts, bolts, screws, tools, and machine parts.

Manganese phosphate coatings are often used for engine parts such as pistons, piston rings, tappets, cams, and gears. The film, wetted with machine oil, acts as a wear-resistant coating on surfaces subject to constant friction. The wear then produces a very smooth burnished surface, and scoring is reduced to a minimum. In an engine, the phosphatized coating may be completely worn off during the break-in period, but the coating is replaced by phosphorus-containing materials, usually zinc dialkyl phosphorodithioates, added in motor oils (Chapter 15). Today, with improvements in metal alloys and lubricants, the use of manganese phosphate conversion coatings to provide a lubricant interlayer is, in many close-tolerance situations, no longer needed.

Manganese phosphate coatings, as well as heavy zinc phosphate coatings, are also used to coat sheet metal that is subsequently drawn, stamped, or cold-worked (e.g., bottle caps). Here, the phosphated coatings act as lubricants during the press operations and keep the metal from cracking.

The chief features of manganese phosphate conversion coatings are as follows: (1) these coatings are almost invariably applied by immersion because spray applications tend to deposit too light a coating and (2) the manganese phosphate coating structure is coarser and more porous than other phosphate coatings. Coating weights are 1000–4000 mg/ft^2. The coatings have a high retention for lubricating or rust-proofing oils.

Zinc Phosphate Coatings

When steel is immersed in a dilute solution of zinc phosphate in phosphoric acid, the primary zinc acid phosphate is converted to the less soluble tertiary zinc phosphate [$Zn_3(PO_4)_2$]. As indicated earlier, secondary ferrous phosphate also coprecipitates as the steel dissolves.

Three basic types of zinc phosphate coating baths exist: for heavy zinc phosphate coatings (1000–6000 mg/ft^2), for conventional coatings (100–1000 mg/ft^2), and for calcium–zinc phosphate baths or microcrystalline zinc phosphate baths. The last type provides coating weights similar to those of the conventional zinc phosphate coating; however, this type provides a microcrystalline coating that is associated with improved corrosion inhibition. Conventional zinc phosphate processes are usually operated at 100–140 °F, and the heavy zinc phosphate process is operated at 160–180 °F. Also, the heavy zinc phosphate baths have a higher concentration of primary zinc phosphate and phosphoric acid.

Conventional zinc phosphate coatings represent by far the largest share of the phosphate conversion coating market. The main features are a relatively heavy, crystalline coating structure that provides excellent paint adhesion and good corrosion resistance to the painted metal. Heavy zinc phosphate coatings are thick, crystalline coatings that, like manganese phosphate coatings, act as binders for lubricants and rust-inhibiting compounds. Heavy zinc phosphate coatings are applied only by immersion techniques.

The main use for conventional and microcrystalline zinc phosphate coatings is for painted metal surfaces. The absorptive, crystalline structure of the conventional coating permits the application of heavy paint finishes. As indicated earlier, microcrystalline structure coating is used where maximum corrosion protection is required.

Iron Phosphate Coatings

So that a metal surface with less rigid final requirements can be phosphatized, a formulation containing monosodium phosphate is used. Monosodium phosphate is prepared by the reaction between sodium carbonate and a concentrated solution of phosphoric acid in water:

$$2H_3PO_4 + Na_2CO_3 \longrightarrow 2NaH_2PO_4 + H_2O + CO_2$$

Because only one hydrogen atom in the phosphoric acid is replaced with a sodium atom, the remaining product with two hydrogens is still an acid. The major applications for this compound are based on the fact that it is a solid acid that dissolves readily in water. For iron surfaces, monosodium phosphate is generally used in combination with a surfactant. The surfactant is absorbed into solid monosodium phosphate particles. So that the surfactant can be absorbed, the monosodium phosphate must be produced in a porous form by spray-drying or drum-drying its water solution. Monosodium phosphate obtained by crystallization from a solution would not be porous and therefore would not be suitable for this application.

For many indoor applications—for example, office furniture—where weathering is not a major problem, a high-quality protective undercoating such as that from zinc-phosphoric acid phosphatization is not necessary. Only a treatment that cleans the metal surface and makes it chemically acceptable to paint is needed. Thus, the metal surface is sprayed with or dipped in an aqueous solution containing about 2 oz/gal. of a solid mixture of monosodium phosphate, sodium acid pyrophosphate, and some detergent. This solution is acidic (pH 3.8–4.1) and reacts well with the iron surface. The iron phosphate coating formed is extremely thin (about 50 mg/ft^2), but this thickness is enough to make the surface cling to the paint. The surfactant in the formulation modifies the coating characteristics as well as cleans any dirt spots.

The Phosphatizing Process

The phosphatizing process is part of a multistep process. As conducted on a commercial scale, the process consists of the following steps:

1. pretreatment of the metal surface
2. application of the phosphate-coating process
3. posttreatment and drying

The pretreatment step, which has a major effect on the crystal size of the phosphate coating, entails cleaning the metal and rinsing to remove the cleaning agents, followed by special treatments. The cleaning steps can include a vapor degreasing with a chlorinated hydrocarbon solvent to remove oil and grease, an alkaline cleaning bath to remove residual organic soils, and an acidic pickling step using mineral acids such as 20% phosphoric acid to remove rust. After the acid pickling and rinsing, an activating step is often used. The special treatment entails applying a dilute solution of disodium phosphate containing about 0.01% of a titanium phosphate complex. Although the mechanism of the operation of this titanium phosphate–disodium phosphate is incompletely understood, this complex is thought to activate the metal surface by providing sites to nucleate the metal phosphate crystal formation.

The phosphatizing step can be conducted by immersing the metal in a hot phosphatizing bath or, except in the case of a heavy zinc or manganese phosphatizing, by spraying the phosphate coating solution onto the metal surface. Temperatures vary from 100 to 200 °F; the actual temperature depends on the type of coating. After the coating is formed, the coated metal is again rinsed several times with warm water.

A very common posttreatment step is immersion in a dilute (0.25%) chromic acid bath, either in dilute phosphoric acid or in alkali. This step is passivating and exerts great influence on the corrosion resistance and paint adhesion of the treated metal. Another posttreatment can be a simple acid rinse with a dilute acid solution to remove the traces of unreacted chemicals from the surface. The metal is then rinsed and dried.

Other types of phosphatizing systems have been developed. Two examples include low-temperature phosphatizing, which can operate at 60–120 °F, and solvent phosphatizing. Low-temperature phosphatizing was developed to reduce the amount of energy consumed in the process. Formulations are proprietary, but most likely they entail high doses of accelerators and perhaps higher bath concentrations. Solvent phosphatizing uses an organic solvent in place of water.

One such process consists of a vapor degreasing step in boiling trichloroethylene, a phosphatizing step using a solution of an organic acid phosphate in trichloroethylene, and a coating step with a trichloroethylene solution of an organic finish.

Phosphate conversion coatings are also applied to metals such as zinc, aluminum, cadmium, and tin. Cadmium and zinc are phosphated largely for improvement of paint adhesion. Paint does not adhere well to these metals, particularly zinc, without such pretreatment. Phosphate coatings on aluminum are generally applied from phosphatizing solutions containing either fluoride (F^-) or chromate (CrO_4^{2-}) ions. The phosphate–chromate coatings are amorphous and range in color from nearly colorless to dark green. These coatings are electrically nonconductive and attractive. They are used extensively in an unpainted form for decorative purposes. Applications include aircraft components, siding, roofing, fencing, and screens. Tin is phosphatized by using an alkaline phosphate bath. The phosphating of tin is done to prevent formation of a black sulfide discoloration on the inside of cans as well as to retard rusting of the can exterior.

A major drawback of the phosphatizing processes for iron and steel is that these processes take too long—usually more than 2 h at 180–200 °F (82–93 °C). When a solution is freshly prepared, complete coating only takes about 10–15 min.

As the bath is used, coating of each successive metal piece takes longer. The slowdown results from the accumulation of ferrous ions that have migrated from the metal–solution interface into the bulk solution of the bath. As more iron pieces are phosphatized, more ferrous ions accumulate. The net effect is a gradual change in the composition of the acid solution from its original zinc or manganese phosphate content to one with zinc–ferrous phosphate or manganese–ferrous phosphate.

Thus, ferrous phosphate in phosphoric acid retards coating formation. A pure zinc phosphate–phosphoric acid solution coats fastest; next fastest is the manganese phosphate–phosphoric acid solution. The obvious answer is to remove ferrous ions from the phosphatizing solution, but phosphatizing an iron surface inevitably means that some iron will dissolve to give ferrous phosphate.

Fortunately, *ferric* (Fe^{3+}) phosphate is insoluble in the phosphatizing bath. The method developed to remove ferrous ions, therefore, involves converting them to insoluble ferric phosphate by adding a small amount of an oxidizing agent (sodium nitrate). Thus, the coating time is reduced to less than 5 min; the actual time depends on the operating temperature and other solution characteristics.

A large number of substances have been found to act as accelerators for the phosphatizing process. These include (1) oxidizing agents (nitrites, chlorates, and peroxides), (2) reducing agents (sulfites and hydroxylamines), (3) organic compounds (quinoline and nitrophenol), and (4) certain metal salts (copper, nickel, and chromium). At the commercial level, only the oxidizing agents have achieved significant industrial importance. These agents are believed to act by converting the ferrous ion passing into the bulk solution to ferric ion and by depolarizing the hydrogen formed on the metal as a result of the dissolution of the metal. If oxidants are not present, ferrous ion concentration rapidly increases to such a level that it adversely affects the quality of the phosphate coating as well as the cycle time.

Various other additives have been included in phosphatizing baths, for example, polyelectrolytes to control nucleation of the phosphate crystals, surfactants to effect efficient wetting of the metal surface, and chelating agents to control metal ion concentrations. Because of the severe operating conditions of the phosphatizing process, the organophosphorus chelating agents, such as NMPA and HEDPA, are often preferred (*see* Chapter 7). Other additives control crystal size and crystallinity. These additives are often referred to as "mummy dust" additives, and they are sometimes treated as closely guarded "secrets" by the purveyors of phosphatizing formulations.

Metal Cleaning

Prior to electroplating or painting, metal surfaces must be thoroughly cleaned. This cleaning can be done in a number of ways, including abrasive blast cleaning, molten salt bath descaling, vapor degreasing using a volatile solvent, solvent cleaning by immersion, alkaline cleaning, and acid cleaning. This cleaning process can be multistep, using several or all of the above techniques. In addition, rinsing and drying steps are usually included.

Phosphorus compounds find their main use in metal cleaning in commercial acid and alkaline cleaners. These cleaners are cleaning solutions that can be applied by spray or immersion. The immersion baths can be soaking tanks, or they can have added mechanical work performed, through agitation of the bath, through ultrasonic agitation at the metal surface, or through electrolytic action.

Acid Cleaning with Phosphoric Acid. In acid cleaning, the metal surface is cleaned with a solution of an acid, often in conjunction with a surfactant and perhaps an organic solvent, such as an alcohol. The acidic material is present to remove inorganic smut including oxides or rust. The surfactant is present to remove residual organic soils, such as oils or grease. The user of such cleaners tries to use the least expensive cleaner that is effective in providing acceptable cleaning performance, and therefore, mineral acids less expensive than phosphoric acid are used whenever possible. However, several types of cleaners exist in which phosphoric acid provides significant advantages.

One of the most popular acid cleaners based on phosphoric acid uses mixture of organic solvents such as alcohol or butyl-Cellosolve ($C_4H_9OCH_2CH_2OH$, an ether alcohol) with phosphoric acid plus a wetting agent. The treatment removes greasy contaminants and destroys light rust and annealing scale on a metal surface. This cleaner also gives a mild phosphatic coating to the surface that promotes paint adhesion. Thus, this cleaning process is actually a mild form of phosphatizing. This cleaning solution can be applied by immersion or by wiping onto the metal surface.

Other, less complex phosphoric acid cleaners for ferrous metals include dilute aqueous solutions of phosphoric acid and dilute solutions of sodium acid pyrophosphate and sodium bisulfate. Surfactants, either anionic or nonionic, are added. Because acids can attack the surface of the metal, inhibitors are often added so the cleaning action is limited to removing the dirt and metal oxides without attacking the base metal. These inhibitors are often long-chained organic nitrogen or organosulfur compounds.

Such acid cleaners are used at temperatures ranging from room temperature to 160 °F, with hot solutions being the more common. These cleaners can be used in immersion baths, as sprays, or as wiping cleaners. The performance of the immersion baths can be augmented by ultrasonic action, which, through cavitation, provides vigorous mechanical action on the metal surface. Phosphoric acid cleaning solutions are seldom if ever used for electrolytic cleaning because such solutions are associated with an excessive evolution of gas.

A little known but ingenious use of dilute phosphoric acid is to remove rust safely from chrome strips on older cars. The most readily available source of this solution is the cola drinks. As described in Chapter 4, a cola drink contains 0.057–0.084% of 75% phosphoric acid. To see a person on a hot summer day take a drink of cola and then pour some of the same drink on a rag and use it to polish the rusty trim parts of their car is quite a sight.

Cleaners Containing Alkaline Phosphates. Alkaline salts of phosphoric acids are found in many alkaline solutions used for cleaning metals. The types of alkaline phosphate cleaning agents used in such systems and their function as cleaning agents were described earlier (Chapter 7). Such phosphates include trisodium phosphate, tetrasodium pyrophosphate, tetrapotassium pyrophosphate, sodium tripolyphosphate, and, more recently, the hydrolytically stable organophosphorus chelants and dispersants such as the sodium salts of NMPA and HEDPA (*see* Chapter 7). These phosphorus-derived materials provide alkalinity and buffer capacity, sequester metal ions such as calcium and magnesium, disperse soils, and act as inhibitors for the nucleation of precipitates.

The alkaline cleaners are often blends of the phosphate with sodium hydroxide, sodium carbonate, sodium silicate, or sodium bicarbonate. The first three compounds are used when a high degree of alkalinity is desired such as on ferrous metals. Sodium bicarbonate is used for cleaning metals like aluminum and zinc that are attacked by strong alkali. Either ionic or nonionic surfactants are added. Operating temperatures of such solutions are used usually in the 160–220 °F range. The solutions can be used as immersion, spray, or wipe cleaners. As immersion cleaners, these solutions can be agitated mechanically or by ultrasonic energy. When the cleaner is used as a spray, the surfactant level is often lowered to reduce foam formation. Alkaline cleaners can also be used in electrolytic cleaning. In this process, the metal is soaked in the hot alkaline cleaning solution while an electric current is applied. Oxygen and hydrogen evolve at the anode and cathode, and this action results in strong agitation at the metal surface, which assists mechanically in removing the dirt. Sometimes the metal is electrically cleaned first as the cathode and then as the anode. Electrolytic cleaning has some attendant hazards such as excess foam formation and noisy explosions when the oxygen and hydrogen react.

Time cycles for the soak cleaners are in the range of 3-5 min. However, when mechanical or electrical assistance is used, such as in the case of spray, ultrasonic, and electrolytic cleaning, the time cycle can be reduced to 0.5–1 min.

Cleaning Boilers with Phosphoric Acid. An example of the use of both phosphorus-based acid and alkaline cleaners is a recently found new use in cleaning power-plant boilers. In new boilers, phosphoric acid removes rust and mill scales; in old ones, it removes mineral water deposits such as calcium and magnesium carbonates. Formerly, inhibited hydrochloric acid (HCl) was used. The inhibited acid is formed by the addition of an organic amine, which retards acid attack on the metal surface.

Hydrochloric acid presents disadvantages. If the temperature is not kept below 150–160 °F (66–71 °C), the amine inhibitor does not work. Because HCl is quite volatile, its vapor is not inhibited, and it can attack the metal. This volatility also produces a corrosive acid atmosphere in the power plant that is unpleasant for the workers and endangers delicate and expensive control instruments. Finally, a surface cleaned by hydrochloric acid tends to rust rapidly when exposed to a humid atmosphere because of minor residues of chloride ion.

Phosphoric acid is an effective boiler cleaner and does not have these disadvantages. A 5% solution can be boiled in a power-plant boiler at atmospheric pressure to remove deposits. The cleaned surface, which now wears a thin coating of iron phosphate, then resists rusting. Statistics on new boiler cleaning are impressive. A boiler commonly used by electric power plants has a normal steaming capacity of about 600,000 lb/h and a steam pressure of 900 psi at 900 °F. For a boiler with such a capacity, 14,850 lb of inhibited 75% phosphoric acid is required. Before acid treatment, the boiler must first be heated with an alkaline cleaning solution containing 500 lb of trisodium phosphate dodecahydrate [$(Na_3PO_4 \cdot 12H_2O)_5 \cdot NaOH$] along with sodium sulfite. After acid cleaning, the boiler is rinsed with 0.1% phosphoric acid and then conditioned with a solution of tetrapotassium pyrophosphate ($K_4P_2O_7$) and potassium hydroxide (KOH) along with some potassium sulfite. The acid boil-out takes 8 h, and the drained solution contains 1 part/1000 parts of iron, equivalent to more than a ton of iron oxide (Fe_2O_3).

Mirrorlike Finishes on Metals

Chemical Polishing of Aluminum. Polishing gives metal surfaces a bright, shiny, mirrorlike finish. Because mechanical polishing requires considerable labor, it is expensive as well as impractical for irregularly shaped articles. Consider, however, chemical polishing, or "bright dip". This process is especially suited to objects made of aluminum or its alloys—for example, automobile trims. These trims have now almost completely replaced chrome-plated trims. Other bright-dip objects are some automobile grills, light reflectors, and handles for refrigerators, washers, and freezers.

In the bright-dip process, the article is immersed in a hot solution (91–99 °C) that is about 95 parts of 85% phosphoric acid and about 4–5 parts of 68% nitric acid (HNO_3), along with 0.01–0.04% copper nitrate. Supposedly, in some plants, the operators simply throw in a handful of pennies although the legality of this practice is questionable.

Actually, many aluminum alloys contain enough copper so that no extra copper is needed once the operation is on-stream. Sometimes a trace of nickel and a little wetting agent are also added. Because of the severe operating conditions, these wetting agents must be fluorocarbon-based surfactants. The exposure of the metal part to the chemical solution in the bright-dip process is less than 3 min.

Additives are important in polishing high-purity aluminum pieces. Without the additives, the process tends to emphasize attack at the grain boundaries of the metal, and the surface is not as shiny as it could be. The wetting agent brings the polishing solution and the metal surface into closer contact and increases the reaction rate; this action improves the metallic luster. Because the bright-dip bath is thick and sticky, each piece drags out some solution after it is dipped. Wetting agents in the solution lower the surface tension and thus reduce drag-out. Although nitric acid in the bath evolves toxic brown fumes of nitric oxide during operation, additives such as ammonium compounds eliminate most of the fumes. Silver ions aid the polishing effect of the bath more so than copper ions. However, unless the bath is chloride ion free, the silver ions will precipitate as silver chloride. Thus, this additive is not used commercially.

The composition of the bright-dip bath determines the quality of the finish. One important factor is aluminum content, and its optimum concentration is 25–30 g/L. Because chemical polishing of aluminum means that some of the metal dissolves, the aluminum content of the bath increases with usage. The higher the aluminum content, the higher must be the concentration of phosphoric acid for optimum brightness. However, an acid concentration that is too high will pit the surface. If the concentration is too low, the surface tends to be dull. The effect of dissolved aluminum content at various acid concentrations (as indicated by the density of the solution) on the brightness of the article is shown in Figure 9.1.

The operator of the bright-dip bath must measure the concentrations of aluminum, phosphoric acid, water, and nitric acid and adjust them to keep the bath operating at optimum conditions. Chemists are still looking for additives that will permit them to operate this type of bath at a wider range of aluminum or acid concentrations.

What actually happens during bright-dipping? Some metallurgical chemists believe the action is based on nitric acid oxidation of the aluminum surface to give a porous aluminum oxide film that is then dissolved by phosphoric acid. In this reaction, the microscopic mountains on the surface dissolve faster than the valleys. The process thus converts a relatively dull surface to an ultrasmooth, mirrorlike finish.

In commercial practice, the bright-dipped article, after rinsing, is generally anodized in an electrolytic bath to give the bright surface a protective coating of almost transparent aluminum oxide.

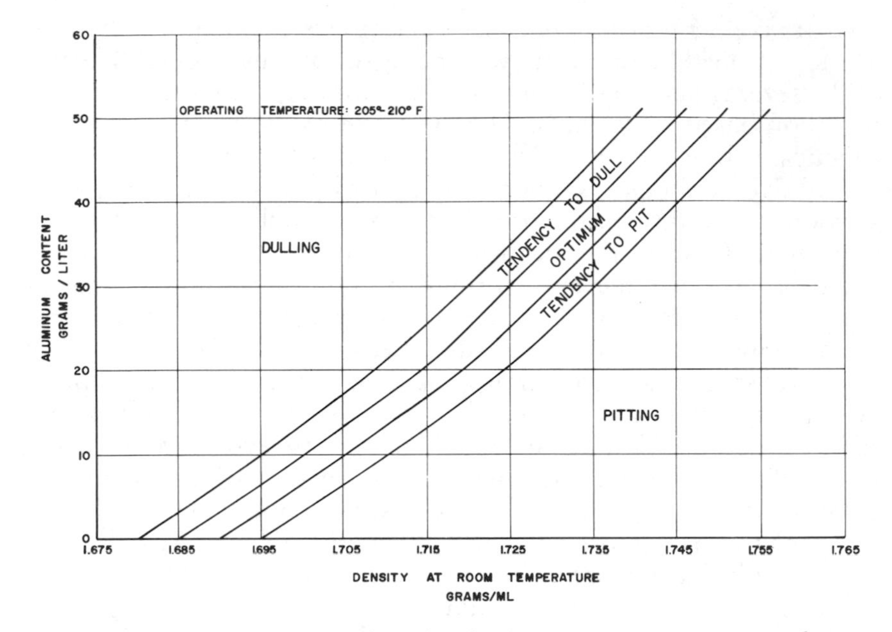

Figure 9.1. Effect of dissolved aluminum content on the brightness of an article at various concentrations (shown by the density of the solution) in a bright-dip bath.

Aluminum surface highly magnified, showing mountains and valleys.

Aluminum surface oxidized by nitric acid to form nodules of aluminum oxide (Al_2O_3). The mountains are attacked more rapidly than the valleys.

$$2Al + 2HNO_3 \longrightarrow Al_2O_3 + 2NO + H_2O$$

Aluminum surface after dissolution of aluminum oxide by phosphoric acid.

$$Al_2O_3 + 6H_3PO_4 \longrightarrow 2Al(H_2PO_4)_3 + 3H_2O$$

Chemical Polishing of Copper and Stainless Steel. Bright-dipping techniques are also used on copper. Although the most commonly used baths, which operate at room temperature, are based on blends of sulfuric acid, nitric acid, and hydrochloric acid or on sodium cyanide and hydrogen peroxide, a bath based on 55% phosphoric acid, 20% nitric acid, and 25% acetic acid is sometimes used. This bath operates at 130–175 °C. Sometimes small amounts (0.5%) of hydrochloric acid or chromic acid (up to 10%) are added. A key requirement for good performance is that the water content be kept below 10%.

A bright-dip system for stainless steel entails a blend of nitric acid, hydrochloric acid, phosphoric acid, and acetic acid. This bath operates at 160 °F. The stainless steel must be depassivated before bright-dipping. This process is done by immersing the stainless steel in hot 5% sulfuric acid.

Electropolishing of Metals. Like mechanical polishing and chemical polishing, electrolytic action can also polish metal surfaces. This third method—electropolishing—is especially suited to stainless steel articles. Shapes to be treated can be regular or irregular. Examples are forks, knives, spoons, surgical instruments, and refrigerator shelves. Electropolishing of ferrous metals is usually done in an electrolytic bath containing either a solution of 75% phosphoric acid or a mixture of phosphoric acid with a small amount of sulfuric acid. A mixture of phosphoric, sulfuric, and glycolic acids ($HOCH_2COOH$) is used in one commercial solution. The article is suspended from the anode of this bath, and an electric current is passed. The temperature is kept between 160 and 180 °F (71–82 °C), and the time varies from 1 to 10 min; the actual time depends on the roughness of the surface. The microscopic mountains on the surface are leveled by the combined action of the acid and electric current to give the desired finish. Nickel, nickel alloys, and carbon steel can be electropolished in mixed phosphoric acid–sulfuric acid baths. An unusual electropolishing bath for carbon steel consists of 400 g/L of pyrophosphoric acid in ethyl alcohol. This bath is anhydrous. It operates at 20 °F and 300 A/ft^2.

Electropolishing can also be used for aluminum. In fact, if done properly, the resulting electropolished aluminum surface is usually more specular or brighter than similar chemically polished pieces. The process entails cleaning the metal in a series of steps analogous to preparing the piece for chemical polishing and then immersing the metal into an electropolishing bath. Both acid and alkaline electropolishing baths are used. A well-established alkaline electropolishing system for aluminum is the Brytal process. In this process, the piece is first electropolished in hot, 165–190 °F, solution containing 5%

trisodium phosphate and 15% sodium carbonate at a current density of 50–60 A/ft^2. The piece is then anodized in a 25% solution of sodium bisulfate by using a current density of 6–8 A/ft^2, followed by a hot water step to seal the pores in the anodic coating.

An example of an acid electropolishing bath for aluminum is the Battelle process. As an illustration, aluminum is immersed in a bath containing 75% phosphoric acid, 4.7% sulfuric acid, and 6.5% chromic acid. The electropolishing is conducted at 175–180 °F by using a current density of 150 A/ft^2. The hexavalent chromium from the chromic acid, which is reduced to trivalent chromium, reduces the etching rate. The sulfuric acid increases the etching rate but also reduces the cell resistance, a beneficial effect. Other phosphoric acid–sulfuric acid electropolishing baths have been developed by using nitric acid in place of the chromic acid. However, the relative concentrations of the components in the bath are significantly different. Again, after electropolishing, the aluminum is anodized and sealed.

Although the electropolishing of aluminum can lead to a brighter finish, the cost of equipment is much higher than that for chemical polishing, and therefore the chemical polishing is used more often.

Stripping Metallic Coatings

Sometimes, in a typical electroplating system, a chrome–nickel–copper plate can be improperly deposited on either aluminum or steel. Rather than discard the plated part, the plating can be stripped from the aluminum or steel and the basic metal part recovered. A useful bath for such stripping is made from 3 parts by volume of phosphoric acid and 1 part of triethanolamine. When the plated part is immersed into this bath at 150–200 °F and the bath is operated with the part to be stripped as the anode, the undesired metal layers can be stripped effectively, beginning with the top layer, the chromium.

General References

1. Spring, S. In *Kirk-Othmer's Encyclopedia of Chemical Technology*, 3rd ed.; Wiley: New York, 1981; Vol. 13, pp 284–292.
2. Schoffman, L. F. In *Kirk-Othmer's Encyclopedia of Chemical Technology*, 3rd ed.; Wiley: New York, 1981; Vol. 13, pp 292–303.
3. Schneberger, G. L. In *Kirk-Othmer's Encyclopedia of Chemical Technology*, 3rd ed.; Wiley: New York, 1981; Vol. 13, pp 304–312.
4. *Metal Finishing* **1981**, *79*, 132–158, 173–183, 674–687.
5. *Metals Handbook*; American Society for Metals: 1964; Vol. 2, pp 307–370, 531–547, 616–617.

Chapter Ten

Flame Retardants

THE STORY WAS TOLD that in Papua New Guinea, nearly half of the people hospitalized for severe burns were victims of what doctors there called "grass skirt burns". The so-called grass skirts were actually made from dried coconut or banana leaves. The skirts are extremely flammable. So when the wearer got too near an open fire, which the natives use for cooking, the skirts could easily catch fire. In the United States, Halloween brings the familiar sight of children dressed as goblins, monsters, and ghosts in brightly colored costumes. Because the children sometimes carry candles and jack-o'-lanterns, why their paper costumes do not catch fire is a wonder to some people.

Paper dresses for everyday wear was also a fad. Because the women who wear them often smoke, one might also wonder why no accidental fires have started from these dresses. The answer is that Halloween costumes and paper dresses share a common characteristic—they are difficult to burn. The agents responsible for this property are the ammonium phosphates. Also used as fertilizers and in dyeing textiles, ammonium phosphates are today's most widely used flame retardants. If the grass skirts in New Guinea were also treated with ammonium phosphates, no more injuries from grass skirt fires would occur.

Ammonium Phosphates

Ammonium orthophosphates are prepared in ways similar to those for sodium and potassium orthophosphates (discussed earlier). The hydrogen ion in orthophosphoric acid is replaced with a monovalent cation—in this case, the ammonium ion. In many other respects, however, the preparation is quite different. Aqueous ammonia is a much weaker base than sodium or potassium hydroxide and forms an unstable salt with phosphoric acid. When ammonium salts are

1002–0/87/0135$08.50/1

heated, ammonia gas is released. Monoammonium and diammonium phosphate are the only ammonium orthophosphates stable enough toward decomposition to be commercially important.

Triammonium phosphate can be prepared at low temperatures, but at room temperature it reverts rapidly to ammonia gas and diammonium phosphate. Diammonium phosphate, when heated to about 70 °C, decomposes to ammonia and monoammonium phosphate. Although the monoammonium phosphate is the more stable of the two, it also decomposes to ammonia and phosphoric acid when heated. For example, at 125 °C a measurable pressure of ammonia is observed (0.05 mmHg). This thermal decomposition of ammonium phosphates is not always undesirable. As shown later, many applications for these compounds depend on it.

Commercial quantities of monoammonium phosphate are made by pumping dried ammonia gas into an 80% solution of phosphoric acid. Product composition is controlled by maintaining the pH at 3.8–4.5. (If the pH is too high because of excess ammonia, some diammonium phosphate forms.) When the mixture is cooled, crystals of monoammonium phosphate precipitate. They are separated by centrifugation and dried, and the mother liquor is recycled for the next charge.

Diammonium phosphate is prepared by bubbling 2 mol of ammonia gas into 1 mol of 80% phosphoric acid solution. The second mole of ammonia is added at temperatures below 50 °C to prevent thermal decomposition; pH is controlled to about 8. Crystals formed on cooling are centrifuged and then dried at below 50 °C.

Tetraammonium pyrophosphate can also be synthesized but not by direct thermal conversion of diammonium phosphate (as with tetrasodium and tetrapotassium pyrophosphates). When diammonium phosphate is heated to a high temperature, it dissociates into ammonia gas and monoammonium phosphate; further heating produces an amorphous mixture of polyphosphates from which most of the ammonia gas has been lost. Tetraammonium pyrophosphate can be made by neutralizing crystalline pyrophosphoric acid with ammonium hydroxide.

Ammonium polyphosphate, $(NH_4PO_3)_n$, is also difficult to prepare by the high-temperature conversion of monoammonium phosphate. However, crystalline ammonium polyphosphate with a chain length of greater than 50 can be prepared by the reaction of phosphoric acid with urea (*1*) at a temperature range of 170–220 °C. The urea acts as both a dehydrating and an ammoniating agent. During the reaction, an atmosphere of ammonia gas must be maintained over the reaction mixture. Several crystalline forms of the long-chain crystalline ammonium polyphosphates exist. Each form is prepared under different temperature conditions. Ammonium polyphosphate can also be prepared by neutralizing polyphosphoric acid with ammonium hydroxide.

Unlike ammonium orthophosphates or pyrophosphates, ammonium polyphosphate is practically insoluble in cold water. This water insolubility makes it useful as a flame retardant in paint coatings and in nonwoven fabrics where it can be bonded permanently by polymeric binders.

The use of ammonium phosphates as a flame retardant was first proposed by the French chemist Gay-Lussac in 1891 (*2*).

Because monoammonium phosphate and diammonium phosphate can be used interchangeably in most of their applications, these compounds are discussed together here. To understand how ammonium phosphates work as flame-retarding agents, we should consider some theories on reducing flammability. Most of these theories are based on what happens when cellulosic materials [cotton, paper, viscose rayon (regenerated cellulose), and wood] burn. First, most of the material is converted to liquid tarry depolymerization fragments called levoglucosan. Further heating converts these fragments into volatiles that then burn. Only a small amount of the original cellulose decomposes into solid material that, upon further heating, decomposes mostly to a slow-burning char (carbon).

In the presence of an acid, the process is somewhat different. Most of the cellulosic material is first converted into solid fragments. Further heating changes these fragments to a slow-burning char. An acid seems to catalyze the thermal dehydration of the cellulose molecule—that is, long chains of repeating glucose units ($C_6H_{12}O_6$)—into carbon and water. This char or carbonaceous residue insulates the remaining portion from further burning. Therefore, only a small portion of the cellulose is converted to flammable gases (*3*):

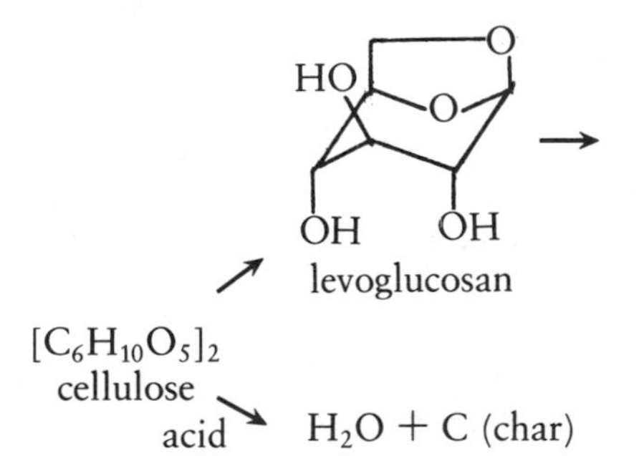

→ CO, alkanes, alkenes, alcohols, aldehydes, ketones, and other combustible volatiles

Acids are the best agents for catalyzing the thermal degradation of cellulose in the direction of slow-burning char and less flammable gases. However, acids disintegrate cloth or paper if applied directly.

So that this problem can be circumvented, the cloth or paper is treated with a nonacidic chemical that when heated decomposes into an acid. Mono- and diammonium phosphates decompose on heating to phosphoric acids. These acids then catalyze the decomposition of cellulose in the desired manner and extinguish the fire.

Another salt that generates nonvolatile acids is ammonium sulfamate ($H_2NSO_3NH_4$). When heated, it liberates ammonia and sulfamic acid (H_2NSO_3H).

When antimony oxide is combined with certain chlorine compounds, such as chlorinated wax, and heated, it forms the acidic antimony trichloride, an effective flame retardant. Antimony trichloride catalyzes burning cellulose to form as great a percentage of the slower burning char or carbon and less of flammable gas as the other salts mentioned previously, but it is not an entirely satisfactory flame retardant. The char continues to burn (afterglow), and just as nasty a burn can be received from a glowing char as from a flame. However, phosphoric acid generating flame retardants are also afterglow retardants; the char formed does not continue to burn.

One theory advanced to explain this phenomenon is that the phosphoric acid flame retardant catalyzes a burning process that favors the formation of carbon monoxide rather than carbon dioxide. Formation of carbon monoxide gives off only 24.4 kcal of heat per mol and the formation of carbon dioxide gives off 96 kcal of heat per mol. The greater heat liberated with CO_2 formation supports further combustion and, in this case, afterglow burning. (A calorie is the amount of heat needed to raise 1 g of water 1 °C. A kilocalorie is 1000 cal. Incidentally, a dietician's calorie is actually a chemist's kilocalorie. So a man of average size needs a daily intake of food that when burned gives off 3200 kcal of heat.)

When cloth and paper are flameproofed, ammonium phosphate can be applied by spraying as a water solution or by dipping. A 3–5% dry weight gain is generally needed for effective flame retardancy. However, because ammonium phosphate is water-soluble, clothing loses its flameproofing after laundering. This problem is not serious for draperies and theatrical scenery, which are not laundered often and which can be re-treated after each cleaning. Neither is it a problem for disposable paper dresses and Halloween costumes. Thus, ammonium phosphates are used extensively to flameproof these articles as well as building materials, such as wood and cellulose insulations, and the interior of wallboard. This usage is the reason old lumber from demolished buildings often makes poor fuel.

As indicated in Chapter 2, paper book matches and wooden matches are all treated with ammonium phosphates to prevent afterglow. Christmas trees and decorations are occasionally treated to reduce fire hazards.

Many of the chemical formulations used as fire extinguishers contain ammonium phosphates. These formulations usually also contain a thickener such as sodium (carboxymethyl)cellulose that forms a gel when water is added. In fighting forest fires, these flame-retardant gels can then be dropped from airplanes or helicopters in globs over the target area. They retard combustion by coating trees in the path of an advancing fire.

Urea Phosphate

Another industrially available flame retardant is urea phosphate. It is prepared by the reaction of 1 mol of urea [$H_2NC(=O)NH_2$] with 1 mol of phosphoric acid:

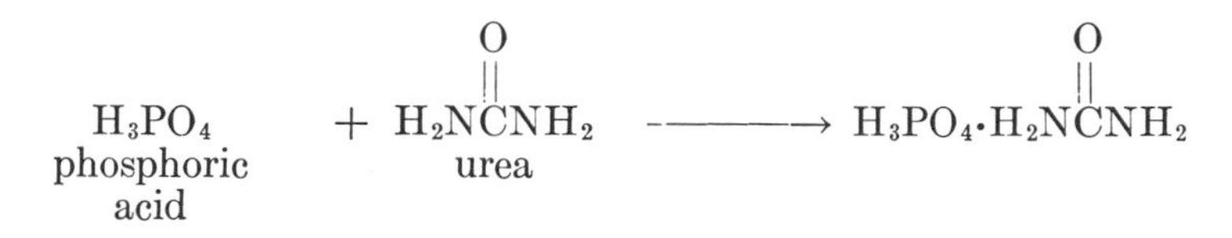

So that cotton fabrics can be flameproofed, the material is soaked in a solution of urea phosphate so that upon drying, the add-on, or gain in weight, is about 15% above the dry weight of the fabric. The dried, treated fabric is then cured at around 160 °C. Heat causes a chemical reaction between the phosphate and the cellulose molecules in the fabric. Because the phosphate is now chemically bonded to the fabric, it does not wash off. This treatment, however, partially degrades the fabric, causing it to lose about 20–50% of its tensile strength and even more tear strength.

Although the phosphate is bound more or less permanently to the fabric, after repeated washings the ammonium ion (from the decomposition of the urea) attached to the phosphorus atoms exchanges with the sodium ions from detergents and calcium ions from water. When enough ammonium ions have been replaced, sodium and calcium phosphates are left attached to the fabric, and flame resistance is lost because neither sodium nor calcium phosphate becomes acidic on heating.

Another interesting flame retardant is obtained by the high-temperature reaction of P_4O_{10} with gaseous ammonia. The resulting compound has a complicated and uncertain structure and hence no chemical name. In one form, it is known by the trade name Victamide. It contains phosphorus–nitrogen bonds and $PONH_4$ groups. It dissolves very slowly in water, first forming a gel. Because it is less soluble than ammonium phosphate in water, it does not wash off as quickly in water. It is used to some extent in place of ammonium phosphates to flameproof paper, fabrics, and backings for plastic sheets.

Phosphines

Phosphine. Phosphine, PH_3, is such a poisonous gas that 2.8 mg in a liter of air is lethal within a few minutes. When prepared by most common methods, phosphine contains an impurity ($H_2P–PH_2$) that makes it ignite spontaneously in air. A compound with these properties hardly seems a likely candidate for use in wearing apparel, yet a commercially available process for flameproofing cotton is based on a compound prepared from phosphine.

Phosphines, which are compounds of phosphorus and hydrogen, are also known as phosphorus hydrides or hydrogen phosphides. Of the many members of this family, the most well-known is the first, PH_3, usually called phosphine. The others are diphosphine, triphosphine, and tetraphosphine:

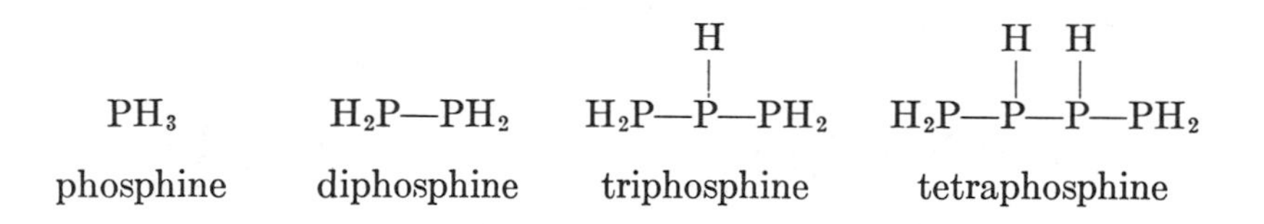

Simple phosphine is fairly easy to prepare. However, as the number of phosphorus atoms in the compounds increases, preparation and isolation become increasingly difficult.

Some people believe phosphine occurs in nature. They have speculated that "will-o'-the-wisp", the flickering light seen over marshland at night, may result from the spontaneous ignition of marsh gas (methane, CH_4) by the presence of trace amounts of phosphines. Both marsh gas and phosphines evolve from decaying vegetable matter.

Chemists are not yet able to prepare phosphine in a practical way by combining phosphorus and hydrogen directly. The most common preparative methods are (1) reaction of certain metal phosphides with water; (2) hydrolysis of phosphorus in a base such as sodium hydroxide or calcium hydroxide, $Ca(OH)_2$ (hydrated lime); and (3) reduction of red phosphorus in hot phosphoric acid solution (*4*).

Metal phosphides are compounds of a metal with phosphorus. Aluminum phosphide (AlP) is made by igniting an equimolar quantity of powdered aluminum and red phosphorus. Zinc phosphide (Zn_3P_2) is made similarly by combining zinc and phosphorus directly. Calcium phosphide (Ca_3P_2) is synthesized by reducing calcium phosphate with aluminum powder at high temperature. The products are calcium phosphide (Ca_3P_2) and aluminum oxide (Al_2O_3). These three metal phosphides will liberate phosphines when water is added. For example

$$2AlP + 3H_2O \longrightarrow 2PH_3 + Al_2O_3$$

Aluminum oxide is the byproduct; the volatile gas, which contains some diphosphine, burns spontaneously in air. This spontaneous-ignition property of impure phosphine gases is used in sea flares based on calcium phosphide.

Phosphine preparation by the hydrolysis of phosphorus in the presence of a base such as sodium hydroxide or calcium hydroxide is a complex reaction. Other products such as hydrogen (H_2), hypophosphite $\{[-OP(=O)H_2]^-\}$, and phosphite $\{[H-P(=O)O_2]^{2-}\}$ are also formed.

The industrial method for preparing hypophosphite is based on the same reaction (*see* Chapter 12). In other words, this reaction cannot produce phosphine without the hypophosphite and vice versa. The following equation for the reaction of phosphorus with sodium hydroxide shows only the major possible reactions that occur.

$$3NaOH + 4P + 3H_2O \longrightarrow \underset{\text{sodium hypophosphite}}{3NaH_2PO_2} + \underset{\text{phosphine}}{PH_3}$$

The preparation of phosphine from phosphorus, steam, and phosphoric acid entails first converting white phosphorus to a very reactive form of red phosphorus by passing phosphorus vapor at 350 °C though an electric arc and quenching the resulting vapors in warm water, where a slurry of reactive red phosphorus is formed. When this reactive red phosphorus is suspended in phosphoric acid and heated at 275–285 °C, phosphine is formed. A modification of this process entails treating white phosphorus at 275–285 °C with steam in the presence of phosphoric acid. The white phosphorus first converts in part to a high surface area red phosphorus, which then reacts to form phosphine.

Phosphine Flame Retardants for Cotton Fabrics. TETRAKIS-(HYDROXYMETHYL)PHOSPHONIUM CHLORIDE (THPC). The research that led to the use of phosphine as an intermediate for the preparation of a flame-retarding composition is an interesting story. The original research was carried out by Wilson Reeves and John Guthrie at the Southern Regional Research Laboratory of the U.S. Department of Agriculture in New Orleans. John told me how he and Wilson arrived at their process. During the Korean conflict in the early 1950s, a research program was initiated at their laboratory to develop a process for flameproofing cotton for use in military clothing. On the basis of his previous knowledge, Dr. Guthrie felt that new flameproofing compounds should not wash off. Further, they should not be ionic—that is, they should not pick up sodium and calcium ions from laundering solutions. In the literature, Guthrie found that trimethylene trisulfone, $(CH_2SO_3)_3$, reacts at room temperature with an alkaline water solution of formaldehyde to form a water-insoluble polymer. If this polymer were deposited on a fabric, it might impart permanent flameproofing.

When he carried out his experiment, however, the treated fabric had only low flame resistance. It also lacked glow resistance—that is, the char that formed on burning continued to burn by afterglow. Guthrie knew that this afterglow did not happen to fabrics flameproofed by phosphorus compounds. His next thought was to synthesize a phosphorus analogue of trimethylene trisulfone to see if it would form a phosphorus-containing polymer on reaction with formaldehyde. Because trimethylene trisulfone is made from trimethylene trisulfide, $(CH_2S)_3$, the first step was to make a phosphorus analogue of trimethylene trisulfide.

Trimethylene trisulfide is made by passing hydrogen sulfide (H_2S) gas into a water solution of formaldehyde (H_2CO) and hydrochloric acid (HCl). Trimethylene trisulfide is then oxidized to trimethylene trisulfone.

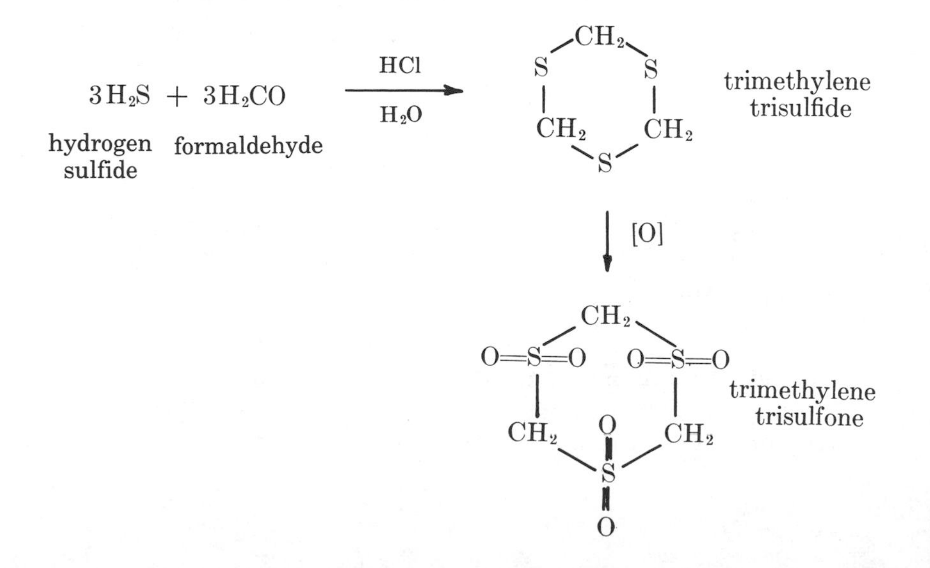

Because no phosphorus analogue of trimethylene trisulfide was reported in the literature, Guthrie assumed that a method analogous to that for trimethylene trisulfide preparation would work. Phosphine, PH_3 (or hydrogen phosphide, the phosphorus analogue of hydrogen sulfide), would be passed into a water solution of formaldehyde and hydrochloric acid. The reaction envisioned is

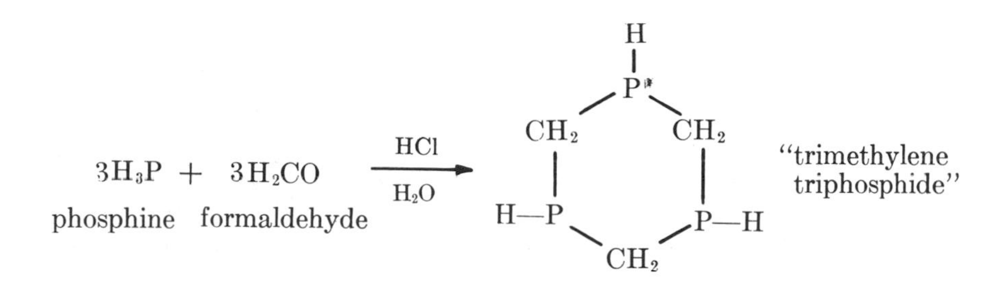

However, when this reaction was carried out, "trimethylene triphosphide" was not found. The reaction mixture was dried, and a white crystalline solid was obtained. This solid was identified as tetrakis(hydroxymethyl)phosphonium chloride, [$(HOCH_2)_4PCl$], an unwieldy name soon shortened to THPC. A literature check showed that THPC had been made in 1921 by Alfred Hoffman, an independent American research chemist:

$$H_3P + 4H_2CO + HCl \xrightarrow{H_2O} \underset{\text{THPC}}{(HOCH_2)_4PCl}$$

Because THPC was not the compound Guthrie and Reeves were looking for, less persistent and astute workers might have quit. However, Wilson Reeves, who did the actual experimental work, did not quit. He tested the reaction of THPC with various classes of chemicals.

At the same time, in connection with another project, Guthrie had made some partially aminoethylated cotton in which some of the hydroxy (–OH) groups in the cellulose molecule were replaced by an aminoethyl group ($-OCH_2CH_2NH_2$). Reeves found that THPC reacted with the amino group ($-NH_2$) of the aminoethylated cotton and was thus chemically bonded to the cotton. The first step of this reaction is shown at the top of p 144. Here, a cellulose molecule was connected chemically, though indirectly, to a phosphorus compound. Because the phosphorus compound was not an ammonium salt and would not exchange sodium and calcium ions upon laundering, it would not wash off. A fabric containing such treated cellulose mole-

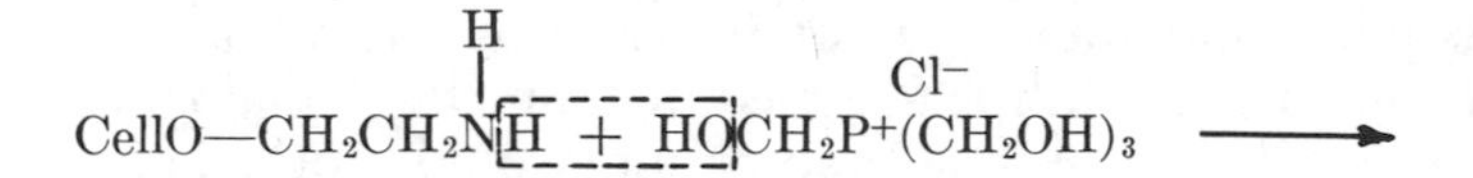

$$\mathrm{CellO{-}CH_2CH_2N(H)H + HOCH_2P^+(CH_2OH)_3\,Cl^- \longrightarrow}$$

aminoethylated cellulose THPC

$$\mathrm{CellO{-}CH_2CH_2N(H){-}CH_2P^+(CH_2OH)_3\,Cl^- + H_2O}$$

cules should also have some flame resistance. This situation was indeed the case. The project was a success, theoretically. Practically, it was not. The intermediate step, the joining of the aminoethyl group to the cellulose molecules in the cotton fabric, was simply too expensive. Chemically, THPC can react directly with cellulose to yield a phosphorus derivative of cellulose. However, the conditions necessary for this reaction are so drastic that the fabric is tenderized. Although clothing made from such fabric is flame-resistant, it tends to fall apart.

In recent years, another treatment for cotton fabric was developed that imparts a wash-and-wear quality (permanent press) to the finished garment. In the early days of this treatment, methylol urea, $HOCH_2NHC(=O)NHCH_2OH$, or methylol melamine was used:

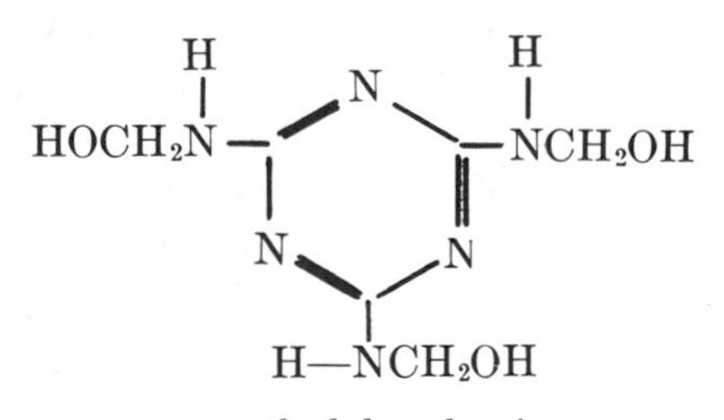

methylol melamine

These compounds are made from urea or melamine and formaldehyde. The cloth is dipped in a water solution of these chemicals, dried, and cured at about 330–340 °F. During treatment, the methylol groups ($-CH_2OH$) react with the hydroxyl groups ($-OH$) in the cellulose molecule. Because each molecule of methylol urea has two methylol groups ($-CH_2OH$), it can link two cellulose molecules together. Accordingly, each methylol melamine contains three methylol groups, so it can link three cellulose molecules together. When the cellulose molecules are thus cross-linked, the fibers become more rigid and harder to crease or, when creased, easier to smooth again. This property is wrinkle resistance.

This discussion shows that the methylol group (–CH_2OH) in methylol urea or in methylol melamine reacts with the hydroxyl group (–OH) of cellulose and is chemically bonded to it by the following reaction:

$$\text{CellOH} + \text{HOCH}_2\overset{\text{H}}{\overset{|}{\text{N}}}\text{—}\overset{\text{O}}{\overset{\|}{\text{C}}}\text{—}\overset{\text{H}}{\overset{|}{\text{N}}}\text{—CH}_2\text{OH} \longrightarrow \text{CellO—CH}_2\overset{\text{H}}{\overset{|}{\text{N}}}\text{—}\overset{\text{O}}{\overset{\|}{\text{C}}}\text{—}\overset{\text{H}}{\overset{|}{\text{N}}}\text{CH}_2\text{OH} + \text{H}_2\text{O}$$

Reeves found that THPC, besides being able to react with the amino group of (aminoethyl)cellulose, can also react with the –NH_2 group in other compounds: for example, the H_2N– groups in urea, H_2N–C(=O)–NH_2. He also found that THPC reacts with methylol groups ($HOCH_2$–) in methylol urea [$HOCH_2NHC(=O)NHCH_2OH$] and also the methylol groups in methylol melamine. The reaction of $(HOCH_2)_4PCl$ occurs through the three methylol groups. The fourth $HOCH_2$– group splits off with the Cl group as HCl and formaldehyde (H_2CO). Because three reactive methylol groups are present in THPC, two reactive H_2N– groups in urea, and three reactive methylol groups in trimethylol melamine, the whole mixture should be able to react together to form a polymer. It does.

If the ratio of reactants used is controlled so that some methylol groups are left from the methylol melamine, these should then react with the hydroxyl groups of the cellulose molecule in the cotton. The treated cotton should then withstand laundering. This hypothesis was experimentally confirmed. The first flame-retardant formulation developed consisted of a water solution of THPC, urea, methylol melamine, and triethanolamine. Because triethanolamine is a base, it ties up the HCl liberated from the THPC. After the THPC flame-retardant finish is chemically bonded to the cloth, the cloth is further treated with a hydrogen peroxide solution to oxidize the phosphine ($\geqslant$P) in the finish to phosphine oxide ($\geqslant$P=O). This process is idealized as shown in Scheme 10.I.

Continued modifications of the original THPC process showed that the treated cloth could be cured in the presence of gaseous ammonia (NH_3) at room temperature. This curing at low temperature preserves the original strength of the fabric. The excess ammonia

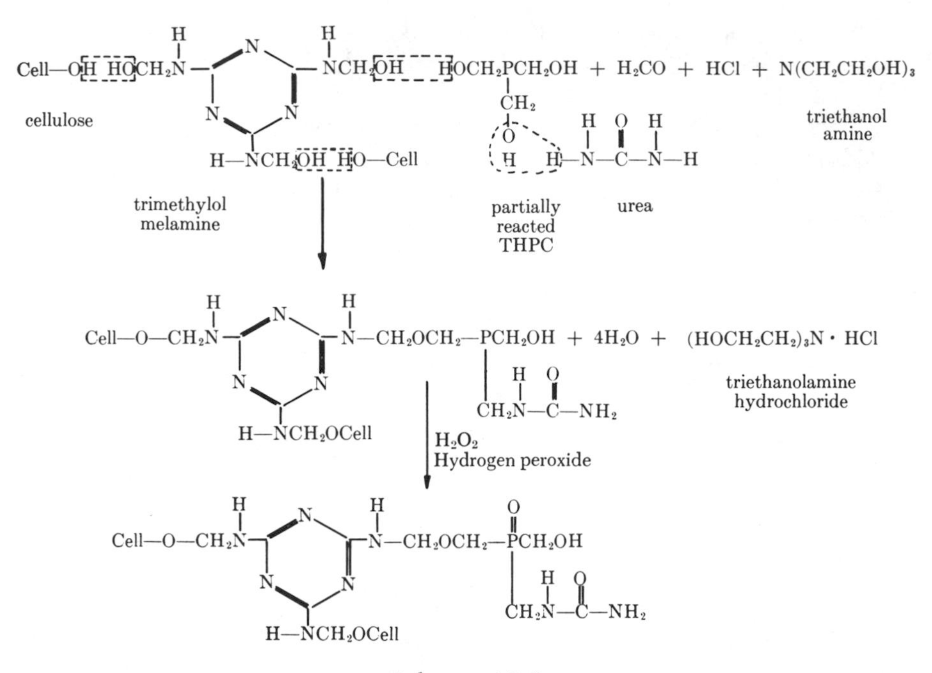

Scheme 10.1

also takes up the liberated hydrogen chloride and converts it to ammonium chloride. Curing with ammonia joins the methylol groups through the carbon-to-nitrogen-to-carbon linkage (–C–N–C–) in place of the carbon-to-oxygen-to-carbon linkage (–C–O–C–) shown previously. This development is an improvement over the original process as shown at the top of p 147.

Recent U.S. and British laws require certain wearing apparel, especially children's nightwear, to be flame-resistant. The modified THPC process was used extensively for this purpose. British regulations are particularly strict because many of their houses are still heated by fireplaces that burn "solid fuel" (a British term usually referring to coal). On cold mornings, children in long flannel nightgowns snuggle close to the fireplace, and their nightgowns often catch fire. Many children have been seriously injured or burned to death.

Even though the THPC flameproofing process is good, it has drawbacks; the treated cloth tends to be stiff, it crinkles like paper, and the treatment is expensive. According to Dr. J. W. Weaver, former research manager for Cone Mills Corp., the 1971 cost of material for children's sleepwear was about 27¢/yard. The flame-retardant process added another 20¢/yard. A garment made of flame-retardant material might retail for $3.98, for example, while the same garment,

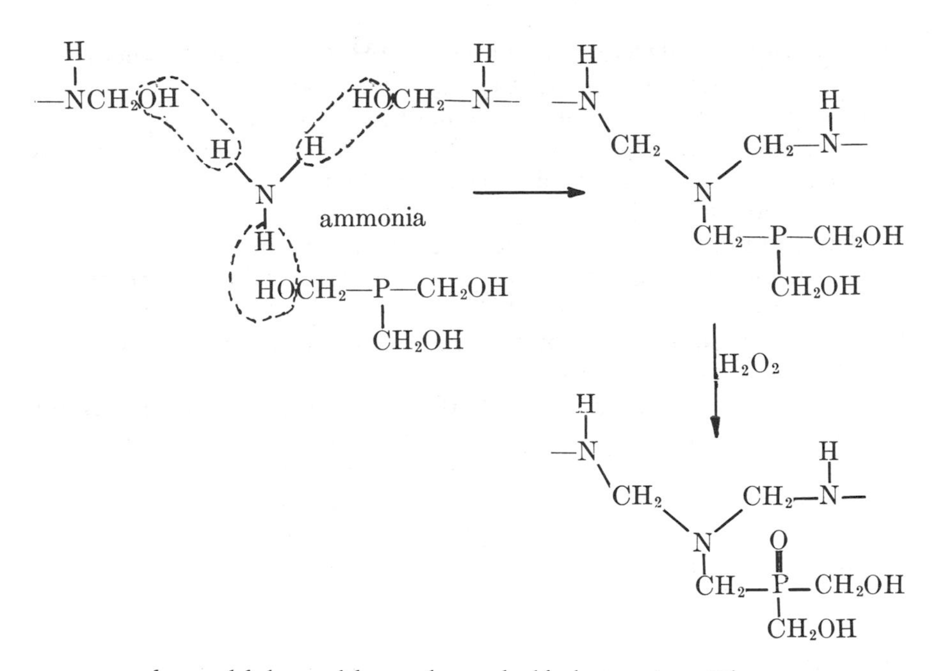

untreated, could be sold at about half that price. The same cost relationship is probably true today.

Some parents believe that fire accidents happen only to other people's children. I have been told that because by law only flame-retarded ready-made night garments are available for purchase in the stores, British parents try to save money by buying untreated flannelette and making the garments themselves.

Children's sleepwear made from polyester fabric has been able to pass the safety standards set by the government in recent years. Polyester sleepwear will burn. However, polyester, being thermoplastic, will also melt when heated. So the flaming portion usually falls off as hot drips; thus, the garment "passes" the government test. People still can suffer serious burn injuries from those flaming drips.

TETRAKIS(HYDROXYMETHYL)PHOSPHONIUM HYDROXIDE (THPOH). Research in this area is continuing. In 1967, Wilson Reeves and his co-workers announced the THPOH process. [The formula for THPOH, or tetrakis(hydroxymethyl)phosphonium hydroxide, is $(HOCH_2)_4POH$.] In this process, THPC is replaced by THPOH. THPOH is prepared from THPC by reaction with sodium hydroxide under carefully controlled conditions. This reaction is

$$\underset{\text{THPC}}{(HOCH_2)_4PCl} + NaOH \longrightarrow \underset{\text{THPOH}}{(HOCH_2)_4POH} + NaCl$$

Except for the replacement of THPC by THPOH, the basic chemistry of the two processes is essentially the same. Ammonia is used to cure the polymer onto the fabric. The new process is claimed to give a flame-resistant cloth that retains most of its original strength and is about as soft as the untreated fabric. In 1974, this process had reached commercial production.

The compound THPC has been replaced, in some instances, with the corresponding sulfate: $[(HOCH_2)_4P^+]_2SO_4^{2-}$ (THPS). The reason for this replacement is that THPC forms bis(chloromethyl) ether, $ClCH_2OCH_2Cl$, a rather potent carcinogen, under certain conditions during manufacture or use.

The present production of the THPC–THPS type of flame retardant is about 3 million pounds a year with most of it going into the flame retarding of garments for the military and for factory workers. Workers in steel mills who are constantly exposed to flying sparks and molten metals prefer the flame-retarded garments' stiffness, crinkle, higher cost, and all.

Organic Phosphonates

Pyrovatex CP. A flame-retardant process that had reached commercial-scale production is called Pyrovatex CP. According to the patent literature, the phosphorus-containing component in this product is

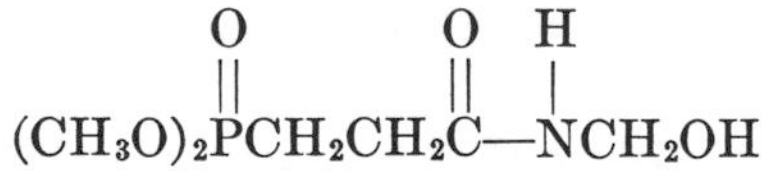

$$(CH_3O)_2\overset{\overset{\displaystyle O}{\|}}{P}CH_2CH_2\overset{\overset{\displaystyle O}{\|}}{C}-\overset{\overset{\displaystyle H}{|}}{N}CH_2OH$$

[3-[(hydroxymethyl)amino]-3-oxopropyl]phosphonic acid, dimethyl ester

The chemistry involved in attaching this molecule to cellulose is the same as that described for the THPC process—that is, through the methylol group, $-CH_2OH$. This group can react directly with the $-OH$ groups of the cellulose, or it can bond indirectly to the cellulose via such compounds as methylol urea or methylol melamine. Cotton fabric properly treated with Pyrovatex has a very good "hand" or softness. Its flame retardance withstands more than 50 machine launderings with phosphate detergents. However, laundering with nonphosphate detergents completely destroys its flame retardance. This destruction could be caused by the high alkalinity of the nonphosphorus detergents, which tend to decompose the phosphorus compound. The destruction may also be caused by the fact that nonphosphate detergents soften water by precipitating Ca^{2+} and Mg^{2+} cations as salts, whereas phosphate detergents remove Ca^{2+} and Mg^{2+} as water-soluble phosphate complexes. Residual precipitates of calcium and magnesium salts on the fabric, upon pyrolysis, would form non-flame-retardant phosphate salts rather than flame-retardant phosphoric acids. At this time, Pyrovatex is no longer of commercial importance.

Fyrol 76. Another flame-retardant process that reached commercial status in 1974 is called Fyrol 76 after the composition of the same name that contains a vinylphosphorus [$CH_2{=}CH{-}P({=}O)$] group. This compound bonds to cellulose also via a methylol group, but in this case, it does so indirectly via the methylol group of methylol acrylamide. A free radical catalyst, potassium persulfate, $K_2S_2O_8$ is used to join the double bonds of the vinylphosphonate with that of the acrylamide. The methylol group is attached to the cellulose when the treated fabric is heated at 300–350 °F:

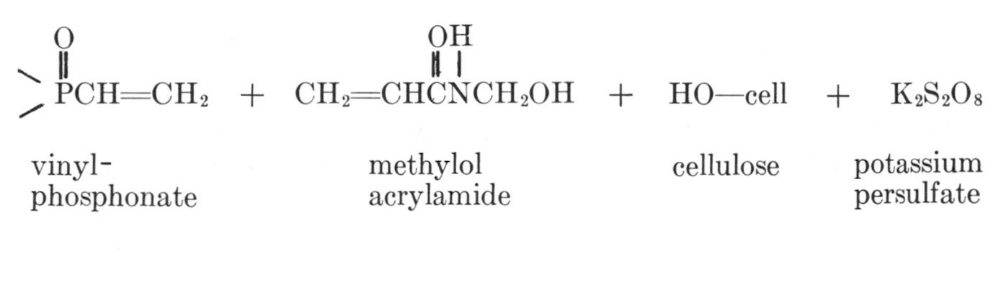

$$\longrightarrow \text{—CHCH}_2\text{CH}_2\text{CH}_2\overset{\text{O}}{\overset{\|}{\text{C}}}\text{—}\overset{\text{H}}{\overset{|}{\text{N}}}\text{CH}_2\text{OCell} \quad (\text{CH attached to } {>}\text{P=O})$$

Flame retardance from Fyrol 76 with a 20–25% add on by weight on lightweight cotton flannel withstands more than 50 launderings. The treated fabric exhibits soft hand and good strength. However, like the Pyrovatex and THPC treatments, the treated fabrics should not be laundered in nonphosphate detergents built with sodium carbonate or in soap in hard water.

The surprising softness (good hand) of the cotton fabric treated with Fyrol 76 was a welcome though unexpected property. We had discussed earlier the various modifications needed in order to obtain a good hand on the THPC–THPS-treated fabrics. Further research using a scanning electron microscope on the fabric treated with Fyrol 76 showed that the phosphorus compound penetrated deeply into the fiber rather than depositing only on the fiber surface (K. Kiss, Stauffer Chemical, personal communication) (Figure 10.1).

The phosphorus concentration was determined by X-ray diffraction of various points on the cotton fiber cross section. We believe this deep penetration of phosphorus into the cotton fiber is one of the reasons for the good bond. In general, surface coating alone results in a stiff fabric.

Fyrol 6. This compound, a phosphonate, was designed specifically for use as a flame retardant for rigid polyurethane foam. Its preparation involved the use of diethyl phosphonate as the intermediate. Diethyl phosphonate is prepared by the reaction of phosphorus trichloride with ethyl alcohol:

$$\underset{}{PCl_3} + \underset{\text{ethyl alcohol}}{3C_2H_5OH} \longrightarrow \underset{\text{diethyl phosphonate}}{(C_2H_5O)_2\overset{\overset{O}{\|}}{P}H} + \underset{\text{ethyl chloride}}{C_2H_5Cl} + 2HCl$$

Although diethyl phosphonate has been patented as a solvent in paint removers, its largest use is as an intermediate in preparing a reactive flame retardant for rigid polyurethane foam. Fyrol 6 is made by the reaction of diethyl phosphonate, $(C_2H_5O)_2P(=O)H$, with formaldehyde, CH_2O, and diethanolamine, $HN(CH_2CH_2OH)_2$.

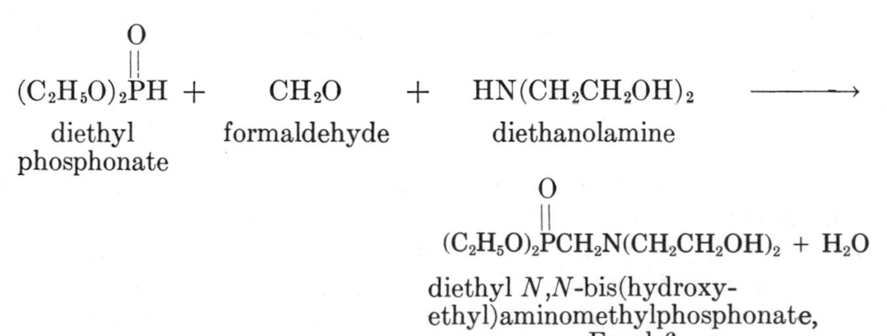

Rigid polyurethane foams are used extensively for packaging and insulation, particularly in household refrigerators, refrigerated railway cars ("reefers"), refrigerated trucks, oil tanks, and wall board

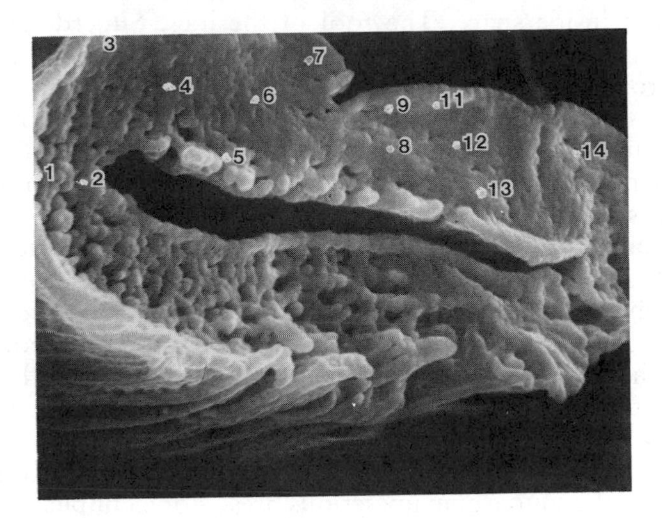

1 µm

Point No.	X-ray Counts	Evaluation
1	2050	
2	2005	average 2223
3	2485	
4	2368	standard
5	2396	deviation 225
6	2385	
7	2006	% relative
8	2110	error at ±22.4
9	1929	95% confidence
10	1917	level
11	1952	
12	2458	
13	2709	
14	2356	

Figure 10.1. Relative concentration of phosphorus at various points on cotton fiber cross section, as expressed by X-ray counts at each point.

insulation. As one of the newer insulating materials, polyurethane foam is so efficient that new refrigerators have thinner walls and thus more inside space. The roof of the huge Superdome in New Orleans is insulated with rigid polyurethane foam flame-retarded with Fyrol 6. When Fyrol 6 is added to the polyurethane formulation, it reacts with the diisocyanate during the foam formation stage to become an integral part of the foam. Such foams are usually applied by spraying. The resulting foam, with approximately 1.0–1.5% phosphorus, is flame-retarded.

Although Fyrol 6 is an effective flame retardant for rigid urethane foam, it is not suitable for flexible urethane foam because it imparts undesirable physical properties to the urethane. However, flexible urethane foams are used extensively as cushions in automobiles, furniture, and mattresses. Without proper flame retardants, these foams burn quite readily. As more cotton batting cushioning materials are replaced by flexible urethane foams, this problem becomes more serious. Even cotton batting cushions, when not properly flame-retarded, are the source of many serious fires—for example, the burning of mattresses by people who fall asleep while smoking in bed.

Phosphazene Derivative: Propyl Ester of Cyclic Phosphazene

A phosphorus compound that has received much attention over the years is phosphonitrilic chloride or phosphazene. It offers much promise as an intermediate for many commercial applications, including flame retardants. Phosphonitrilic chloride is the old name, and phosphazene is used in more recent publications. It is prepared by the reaction of PCl_5 with NH_4Cl in an organic solvent such as tetrachloroethane or chlorobenzene.

$$PCl_5 + NH_4Cl \longrightarrow (N{=}PCl_2)_{3-4} \xrightarrow{C_3H_7ONa} [N{=}P(OC_3H_7)_2]_{3-4}$$

The propyl ester is a high-boiling liquid. At one time, it was promoted in the United States for the flame retarding of rayon. The liquid propyl ester of phosphazene is mixed in with the soda cellulose dope just prior to the regenerating step when the soda cellulose is spun into the rayon fiber. The propyl ester of phosphazene is actually dispersed in the form of microglobules inside the fiber. The rayon fibers thus obtained are of high quality and quite flame-resistant. The major disadvantage of this flame retardant is its high manufacturing cost. As you know, rayon fiber is basically a low-cost substitute for cotton.

Organic Phosphates

The burning of plastics is also the result of the thermal degradation of the components of the plastic into highly combustible volatile materials. Because noncellulosic plastics do not possess the basic cellulose structure unit ($C_6H_{12}O_6$), the degradation does not go through the levoglucosan intermediate. However, phosphorus-based flame retardants also inhibit the decomposition of plastics into combustible volatiles. During the burning, the phosphorus atoms in organic phosphorus flame retardants oxidize to form phosphoric acid, which, upon further heating, condenses to polyphosphoric acid. Polyphosphoric acid then promotes the formation of chars, which, in turn, act as a physical barrier to the formation of gases and the release of heat. In general, a substantial fraction of the phosphorus is retained in the char. The polyphosphoric acid may also act as a barrier layer on the burning polymer. Also, in the case of polystyrene plastic, evidence indicates that the phosphorus-containing flame-retarding additive acts as a catalyzing agent in the thermal breakdown of the polymer melt. Polyphosphoric acid reduces the viscosity of the melt and thus favors the flow or drip from the combustion zone (*3*).

Model experiments using triphenylphosphine oxide and triphenyl phosphate as the flame retardants indicated that phosphorus compounds also inhibit burning at the vapor phase. This hypothesis is based on mass spectroscopic work that detected small molecular species such as PO, HPO_2, PO_2, and P_2 in the flame. Spectroscopic studies also showed that the hydrogen atom concentration in the flame is reduced in the presence of these phosphorus-containing species. The hydrogen atom concentration in the flame is rate-controlling in the burning process. However, one of the phosphorus model compounds, triphenyl phosphine oxide, is not a typical commercial phosphorus-based flame retardant (*3*).

Organic halogen compounds have been found also to be good flame retardants. The combustion product of aliphatic organic halogen compounds is believed to inhibit the combustion process in the vapor phase. As a consequence, as we shall see later, many flame-retardant synthesis chemists designed compounds containing both phosphorus and halogen atoms. In compounds with such combinations, flame retardancy in the solid phase is handled mostly by the phosphorus moiety, and the inhibition of the combustion process in the flame or vapor phase is handled by the aliphatic halogen portion of the molecule. Life for research chemists working in this area would have been very simple if having compounds containing phosphorus and aliphatic halogens were the only requirements. Experimental data have shown that many other factors are also involved. Certain properties in such synthesized compounds affect the properties of the finished products. Many cases occur in which the new flame retardant affects the properties of the finished products negatively. For example, so that a good flame retardant could be found for flexible polyurethane foam, hundreds of compounds were synthesized and evaluated before a promising candidate was found. Besides flame retardancy, factors such as toxicology, economics, and marketing factors have a large impact. Thus, developing a new flame retardant is not as straightforward as it might seem.

Firemaster T23P and Fyrol HB–32. A particularly effective flame retardant that is derived from phosphorus oxychloride is tris(dibromopropyl) phosphate (TDBPP). It is known by such trade names as T23P and HB–32.

$$POCl_3 + 3CH_2BrCHBrCH_2OH \longrightarrow OP(OCH_2CHBrCH_2Br)_3 + 3HCl$$

The effectiveness of this compound as a flame retardant is enhanced by the presence of bromine, a known efficient flame retardant. As shown in the reaction, this compound is formed by the direct esterification of the bromine-containing alcohol with phosphorus oxychloride. TDBPP is used in flexible and rigid polyurethane foams. It is also incorporated as an additive in solutions of cellulose polymers or acrylic polymers before they are extruded into fibers. These fibers are then used to make flame-retardant clothing, draperies, curtains, carpets, and even doll's hair. In a very recent application, TDBPP is used to impart some flame retardance to polyester fabric. In one technique, a water emulsion of TDBPP is added on the polyester fabric. After drying, the treated fabric is heated to about 400 °F for 1 min. Under these conditions, some TDBPP is absorbed into the fabric and is stable enough to resist a certain amount of laundering.

TDBPP, although an excellent flame retardant that has been used for several decades, is now banned. Health-hazard evaluation on TDBPP showed that it cannot pass the Ames test on mutagenicity. The Ames test is a screening procedure to detect potential carcinogens.

The Ames test was developed by Dr. Bruce N. Ames and his colleagues at the University of California, Berkley. The test offers a quick way to determine if a chemical causes mutations (i.e., changes in genes). Many but not all chemicals that cause mutations are known or suspected carcinogens (that are believed capable of initiating the cancer process). Because a high correlation had been shown between known carcinogens and a positive Ames test, the test is considered as a good first screen of suspected carcinogens. The test involves several mutant strains of the bacterium *Salmonella typhimurium*. None of these mutants can synthesize the essential amino acid histidine and therefore cannot sustain growth in its absence. If exposed to a mutagen, the mutation may correct this defect, enabling the bacteria to grow without added histidine. The count of bacteria acquiring this correction gives a measure of the mutagenic activity of the test chemical.

This and several similar screening tests based on microorganisms have attained considerable value as first-stage screens for carcinogenicity. Although it is far from absolute in its predictive power, a positive test gives warning that protective measures should be taken until further tests confirm or deny the activity of the chemical agent.

Fyrol CEF and Union Carbide 3CF. Chlorine is not as efficient a flame retardant as bromine when it is incorporated as a part of a flame-retardant molecule. However, chlorine is sufficiently effective and chlorine-containing compounds are usually much lower in cost than bromine-containing compounds. The compound tris(β-chloroethyl) phosphate, known by its trade names of Fyrol CEF and 3CF, is one of the flame retardants developed many years ago that contains both the phosphorus and chlorine moieties. This compound is still widely used. The chemistry for its preparation is based on the opening of the epoxide ring with phosphorus oxychloride using a catalyst, such as titanium tetrachloride:

$$POCl_3 + 3\,\overset{\diagup O \diagdown}{CH_2{-}CH_2} \xrightarrow{TiCl_4} OP(OCH_2CH_2Cl)_3$$

tris(β-chloroethyl) phosphate

Tris(β-chloroethyl) phosphate is used as a versatile flame retardant for flexible as well as rigid polyurethane foams. Even though tris(β-chloroethyl) phosphate is an additive type of flame retardant, that is, not incorporated as a part of the substrate by chemcial bonding, it still offers good retention except under very hot and humid conditions. Fyrol CEF is used in rigid urethane foam insulations, flexible foam carpet backing, and flame-laminated foams. It is also used in other polymer systems such as flame-retardant paints and lacquers, epoxy resins, phenolic resins, melamine resins, cast acrylic sheets, and polyester resins.

An important application of Fyrol CEF is as a secondary plasticizer in polyvinyl chloride to suppress the flammability resulting from the use of the flammable primary plasticizers, such as the phthalate esters.

Fyrol PCF. This compound is prepared by the reaction of phosphorus oxychloride with propylene oxide.

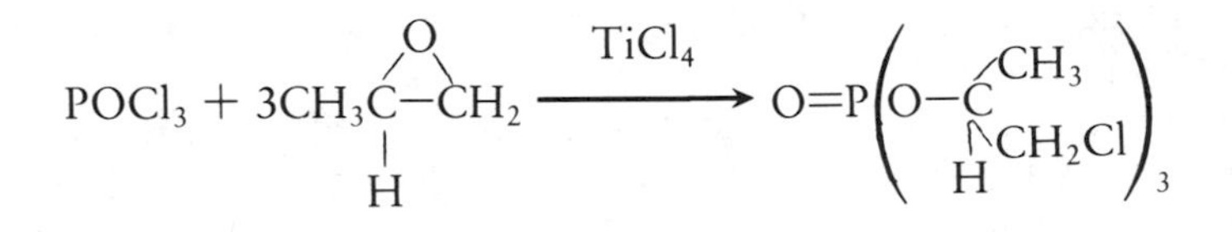

The chemical name for the product is tris(1-chloro-2-propyl) phosphate.

The formula as shown represents the principal component in the product even though it also contains some 2-chloro-1-propyl phosphate:

$$O{=}P\left(-OCH_2\underset{H}{\overset{Cl}{C}}CH_3\right)_3$$

The reason for the two structures is that, in opening the epoxy ring by the P–Cl groups, the cleavage can take place on either the left or the right side of the oxygen atom. The properties of this compound are rather close to those of tris(β-chloroethyl) phosphate. However, this compound does have lower reactivity to water and bases. Tris(1-chloro-2-propyl) phosphate is therefore quite useful in rigid polyurethane foam systems where good storage stability is important. This compound can be used either in the isocyanate component or in the polyol–catalyst mixture component. Rigid polyurethane foams that contain both Fyrol PCF and the methyl ester of terephthalic acid as flame-retardant components will pass the 25-ft tunnel test (this test is specially designed). The use of the terephthalic acid ester causes more char to form upon burning. A low smoke evolution formulation has been developed for sheathing foam using Fyrol PCF as the flame retardant. The compound is also useful in flexible molded foam.

Fyrol FR-2. Tris(dichloroisopropyl) phosphate, or Fyrol FR–2, is another important flame retardant for flexible urethane foam. This compound is prepared by the reaction of phosphorus oxychloride with epichlorohydrin.

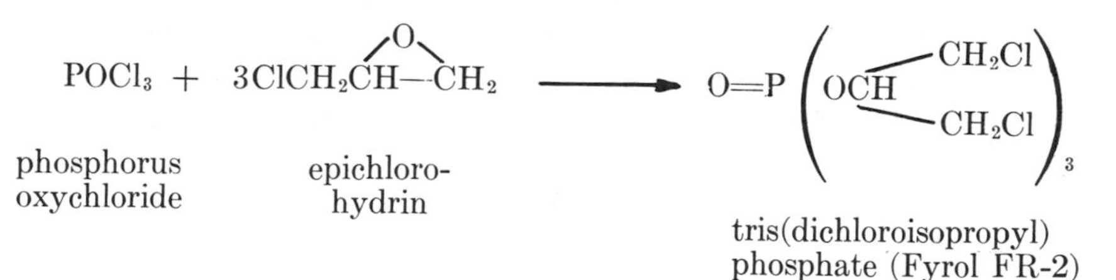

The product is another phosphate ester that is formed by the opening of the epoxide ring of the epichlorohydrin by the PCl group of phosphorus oxychloride.

As can be seen from its formula, Fyrol FR–2 is a larger molecule than either Fyrol CEF or Fyrol PCF. Therefore, it has much less tendency to volatize. It has also lower water solubility. Like Fyrol PCF, it is also quite stable to hydrolysis in water and resists attack by bases. This stability enchances its usefulness as an additive flame retardant for flexible polyurethane foams. With this stability, it does not interfere with the catalyst system used for the foam formulation. It does not cause discoloration or affect the tensile strength of the foam. Because foams containing Fyrol FR–2 as the flame retardant meet the Motor Vehicle Safety Standard 302 requirements for automotive seatings, it is widely used in automobiles.

Fyrol FR–2 has had numerous toxicology tests conducted by the Consumer Product Safety Commission; the National Institute of Health Science; and Stauffer Chemical Co., the manufacturer; and it has been classified as "not toxic". It has not produced birth defects, reproductive effects, or nerve damage. It is rapidly excreted by test animals and does not accumulate in their bodies. However, the manufacturer does not recommend the use of the product to treat clothing or similar materials intended for direct contact with the skin.

Diphosphates

These series of compounds are chemically related to tris(β-chloroethyl) phosphate. The members all have the group

$$(ClCH_2CH_2O)_2P(=O)—$$

present in tris(β-chloroethyl) phosphate. As can be seen later, many of the diphosphates are formed by having 2 mol of a reactive intermediate containing this group tied together chemically. Obviously, the diphosphates, with their higher molecular weight and lower volatility, have better permanency when used as an additive-type flame retardant. Besides lower volatility, these compounds have low water solubility and good-to-fair thermal stability. Foams containing these compounds show good stability in dry and humid aging. Commercially important members of this series are as follows.

Thermolin 101. The synthesis for this compound as described in the patent literature (*5*) involves the following reactions:

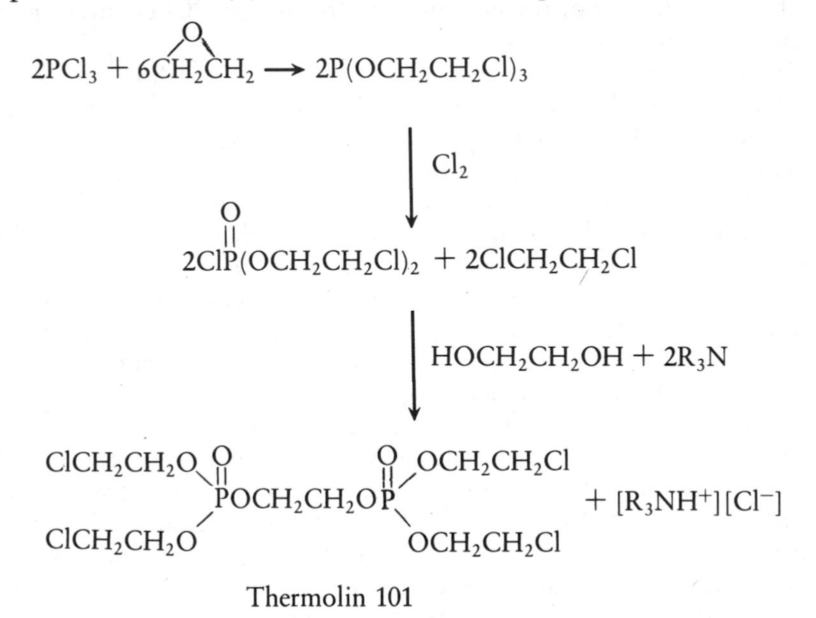

Thermolin 101

Phosgard 1227. This compound has the structure

$$(ClCH_2CH_2O)_2P(O)OCH_2CH_2OCH_2CH_2OP(O)(OCH_2CH_2Cl)_2$$

Phosgard 1227

Phosgard 2XC20. One phosphorus-containing flame retardant used commercially in flexible urethane foams—as well as in rigid urethane foams and other polymers—goes by the trade name of Phosgard 2XC20 (*6*). This compound too is made from phosphorus trichloride:

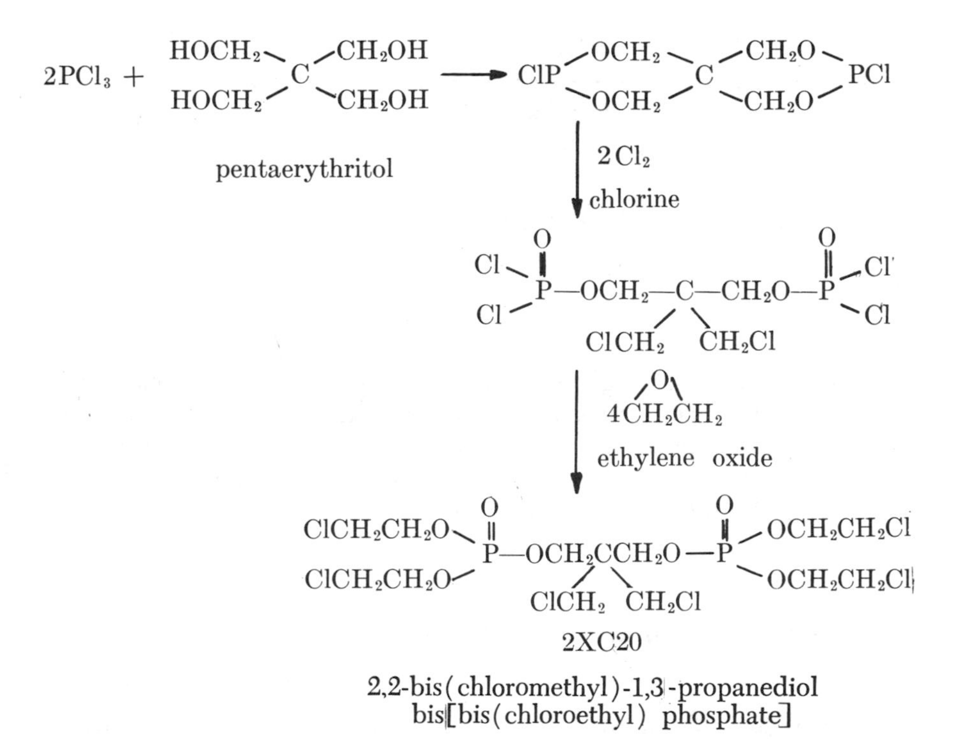

Even though the resulting chemical structure seems fairly complicated, the starting materials are quite common. The tetrahydric alcohol used—pentaerythritol—is an ingredient used to prepare alkyd resins for enamels applied to stoves, refrigerators, and washing machines.

Phosgard 2XC20, unlike Fyrol 6, is not a reactive flame retardant. It is incorporated into the system as an additive. Like all additive flame retardants in general, when subjected to elevated temperatures such as those encountered in automobile cushions while under a hot sun, a certain amount of additive volatilizes. Some additives volatilize more than others. These volatilized additives, when condensed on the colder windshield and windows, cause fogging, which obscures vision. To be acceptable for automotive foams, the volatility has to be low enough to meet the requirements of the automobile manufacturers.

Oligomeric Phosphates: Fyrol 99

This polymeric phosphate also contains the important $(ClCH_2CH_2O)_2P(=O)-$ group of tris(β-chloroethyl) phosphate. In fact, Fyrol 99 is prepared from tris(β-chloroethyl) phosphate (*7*, *8*).

$$(ClCH_2CH_2O)_3P{=}O \xrightarrow[\text{heat, basic catalyst}]{-ClCH_2CH_2Cl}$$

$$(ClCH_2CH_2O)_2\overset{O}{\overset{\|}{P}}{-}O{-}[CH_2CH_2O\underset{ClCH_2CH_2O}{\underset{|}{\overset{O}{\overset{\|}{P}}}}{-}O]_n{-}CH_2CH_2Cl$$

This low molecular weight polymeric phosphate has low volatility and has advantageous resistance to water and to solvents. It is used industrially to impart flame retardancy to resin-impregnated paper for use in automotive air filters, to flexible polyurethane foam, to rebonded polyurethane foam, and to structural foams.

Oligomeric Phosphate–Phosphonate: Fyrol 51

This product is prepared by the condensation of dimethyl methylphosphonate (a flame retardant produced commercially) with tris(β-chloroethyl) phosphate (9). In this condensation reaction, all the chlorine in tris(β-chloroethyl) phosphate is eliminated as methyl chloride. So the finished product is chlorine-free.

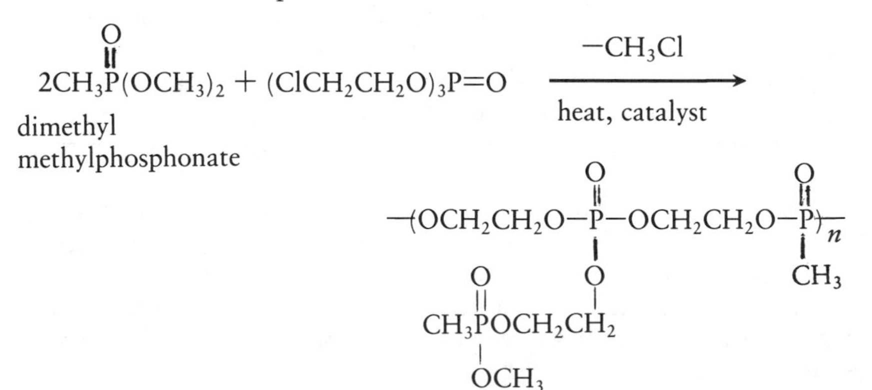

This liquid is water-soluble and is compatible with thermosetting resin. Fyrol 51 is also formulated with a small percentage of alcohol to reduce its viscosity. The new formulation is called Fyrol 58 and is used for resin-impregnated automobile air filters.

Triaryl Phosphates

A typical triaryl phosphate is tricresyl phosphate. It is prepared by heating phosphorus oxychloride with cresol:

$$\underset{\text{phosphorus oxychloride}}{POCl_3} + \underset{\text{cresol}}{3CH_3C_6H_4OH} \longrightarrow \underset{\text{tricresyl phosphate}}{O{=}P(OC_6H_4CH_3)_3} + \underset{\text{hydrogen chloride}}{3HCl}$$

Historically speaking, the first important phosphate esters introduced commercially were triphenyl and tricresyl phosphate. These compounds were used to make better quality "celluloid". Many years ago, billiard balls were made of ivory. The sport became so popular that an ivory shortage resulted. The first substitute for ivory was celluloid, a partially nitrated cellulose plasticized with camphor. Celluloid was also found to be excellent for motion picture film and toys.

Because the partially nitrated cellulose was made by the controlled reaction of cellulose (e.g., cotton) with nitric acid in the presence of sulfuric acid, it was a close relative of gun cotton, a more fully nitrated cellulose. When blended with camphor as the plasticizer, celluloid was extremely flammable. Movie films and some toys made with it practically exploded when ignited accidentally. Because some phosphorus compounds were known as flame retardants, triphenyl and tricresyl phosphates were checked and found to be effective replacements for camphor as plasticizers. This substitution took place in the 1910–1920 era. These phosphates do reduce the flammability of cellulose nitrate, but nothing can really make cellulose nitrate flameproof. Eventually, for movie films, cellulose nitrate was replaced with the much slower burning cellulose acetate.

The discovery of triaryl phosphates as plasticizers for cellulose nitrate opened an area for their application as plasticizers for other polymers. Triaryl phosphates perform very well in cellulose acetate and for vinyl polymers.

The discovery of tricesyl phosphate as a plasticizer for vinyl chloride polymer (PVC) was made in 1926 by Waldo L. Semon at BFGoodrich. In looking for an adhesive to bond rubber to metal, Semon found that boiling PVC in liquids, such as tricresyl phosphate or dibutyl phthalate, makes it highly elastic. Nowadays triaryl phosphate is commonly used in admixture with lower cost non-phosphate plasticizers, such as the phthalate esters. The triaryl phosphate imparts flame retardancy to the phthalate esters.

When triaryl phosphates are used as plasticizers, they not only impart good flame retardancy to the system but also provide faster processing and gelation. Other desirable properties imparted are good oil and gasoline extraction resistance, good microbial resistance, and good high-frequency heating (sealing) characteristics.

Tricresyl phosphate was originally prepared by using cresylic acids, isolated from petroleum and coal tar; the ortho isomer was avoided because it will produce the highly toxic *o*-cresyl phosphate. More recently, C_3 and C_4 alkylphenols, made synthetically by the alkylation of phenols, have been used as cresylic acid substitutes. Thus, commercial products such as isopropylphenyl diphenyl phosphate and *tert*-butylphenyl diphenyl phosphate are available. In the market, mixed alkyl aryl phosphates are also available. The mixed alkyl phenyl phosphates, when used as plasticizers for PVC, impart improved low-temperature flexibility to the vinyl films. Vinyl films plasticized with 2-ethylhexyl diphenyl phosphate have Food and Drug Administration approval for use in food-packaging applications.

Besides their use as plasticizers for PVC, triaryl phosphates are now also used extensively as flame retardants in engineering plastics such as General Electric's Noryl, a blend of polyphenylene oxide and polystyrene.

Research to find better flame retardants is a continuing effort in many industrial and governmental laboratories. Our concern with flame retardance is so great because studies show that every year 3000–5000 deaths and 150,000–250,000 injuries are caused by burns in connection with flammable fabrics in garments worn by millions of people who do not realize their potential hazard. The infamous torch sweaters brought about the passage of the Flammable Fabric Act of 1953; this Act was amended in 1967. The Department of Commerce has banned the sales of flammable sleepwear for children after July 1973. This department has also issued regulations covering carpets and mattresses. The Department of Transportation has issued flammability rules for automobile interiors and has upgraded regulations for passenger airplane interiors. The former Department of Health, Education, and Welfare (HEW) has set flammability standards on interior furnishings of those hospital and nursing homes built under the provisions of the Hill Barton Act. The governmental regulations are attempting to protect us from ourselves. Many sources of fires can be eliminated if we are careful. For example, cigarettes are a good source of fires. Fires started from butts, left unnoticed and smoldering on sofas, chairs, sheets, blankets, and mattresses, caused fully 35% of all residential fire deaths in 1982. Cigarettes were the culprits in 564,000 residential fires where 1,730 Americans were killed and 6,400 were seriously injured.

Flame-retarding processes for many systems are available, but not without shortcomings. For example, for clothing materials we still need a process that is easy to apply, is economical, will not adversely affect the strength, feel, and other properties of the fabric, and is permanent to laundering in phosphate and non-phosphate detergents. In other words, the development of a really good flame-retardant process for fabrics is still a challenge for scientists. Even more challenging is the development of a flame-retardant process for the widely used polyester-cotton blends. Pure polyester fabrics that are not flame-retarded burn by melting, and the flame drips off with the melt. The burns from this flame and hot melt are painful, but such injuries are not as severe as those that result when the whole garment catches on fire. Extensive studies by HEW showed that a remarkable correlation exists between the extent of the injury to the victim's body and the distribution of the burn on the clothing.

For pure cotton fabrics, at least, several flame retarding treatments do exist (with their merits and drawbacks). For polyester–cotton blends, however, even if the cotton portion is made flame-retardant, the blend still burns; in fact, the blend burns more dangerously than pure polyester because the char from the burned cotton supports the burning polyester portion and thus prevents it from dripping away. Some flame retardants for polyester–cotton fabrics are under development in various laboratories. These retardants still have enough drawbacks to preclude their large commercial use.

Literature Cited

1. Shen, C. Y.; Stahlheber, N. E.; Dyroff, D. R. *J. Am. Chem. Soc.* **1969,** *91*, 62.
2. Gay-Lussac, J. L. *Ann. Chim.* **1821,** *18*, 211.
3. Weil, E. D. *Kirk-Othmer's Encyclopedia of Chemical Technology*, 3rd ed.; Wiley: New York, 1981; Vol. 10, pp 396–419.
4. Lowe, E. J.; Ridgeway, F. A. U.S. Patent 3 371 994, March 5, 1968 (to Hooker Chemical; Neth. Applic. 301 712, Dec. 12, 1963 (to Albright and Wilson).
5. Turley, R. J. U.S. Patent 3 707 586, Dec. 26, 1972 (to Olin).
6. Birum, G. H. U.S. Patent 3 192 242, June 29, 1965 (to Monsanto Chemical).
7. Weil, E. D. U.S. Patent 3 896 187, July 22, 1975 (to Stauffer Chemical).
8. Weil, E. D. U.S. Patent 3 513 644, May 26, 1970 (to Stauffer Chemical).
9. Weil, E. D. U.S. Patent 3 891 727, June 24, 1975 (to Stauffer Chemical).

Chapter Eleven

Organic Phosphorus Polymers

MOST OF US HAVE LIVED IN A HOUSE or building that consisted to a greater or lesser extent of combustible materials. Even our clothing and furniture are made up mostly of combustible materials. The risk from fire is exacerbated as more and more of the traditional materials of construction such as metals, stone, and ceramics have been increasingly replaced by synthetic organic polymers. (Polymers are large molecules built by joining smaller ones into repeating chains.) With the rapid increase in population growth, more people are now living in areas of high population. The risk of injury from fire has also increased. In the United States, over the last couple of decades, the average number of deaths from fires has been about 12,000 per year. The cost in property damage from fire is in the order of several billion dollars a year.

In Chapter 10, we discussed the use of phosphorus-containing flame retardants. Tricresyl phosphate was one of the earlier phosphorus compounds used as a flame-retardant plasticizer for some polymers. With the advent of many new organic polymers, beginning in the 1930s and 1940s, the chemist tried not only to develop flame retardants based on phosphorus for application in organic polymers but also to prepare polymers with phosphorus as an integral part of the molecule, in other words, to prepare polymers that by themselves are flame-resistant.

Condensation Polymers

One of the first syntheses of a phosphorus-containing polymer we attempted was that of a polyaryl phosphate. The reaction involved

1002–0/87/0165$06.00/1

two bifunctional reactants, each with two reactive sites. The reactants used were phenyl phosphorodichloridate and hydroquinone:

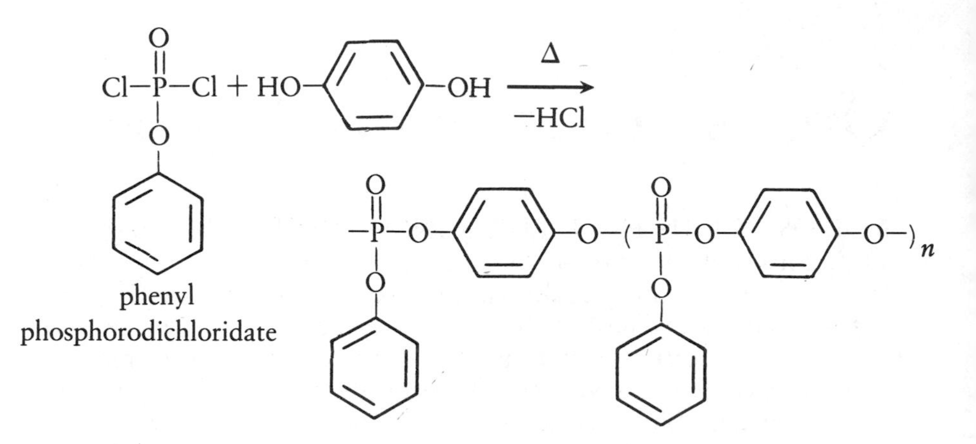

When these reactants were heated together for a long time under conditions suitable for producing a very high molecular weight polyester, the mixture became an infusible and insoluble gel. This gel formation occurred before the reaction went to completion. We found the reason for this formation was the phenyl phosphorodichloridate used. Phenyl phosphorodichloridate underwent a partial disproportionation reaction in the temperature range of 150–300 °C used for the polymerization:

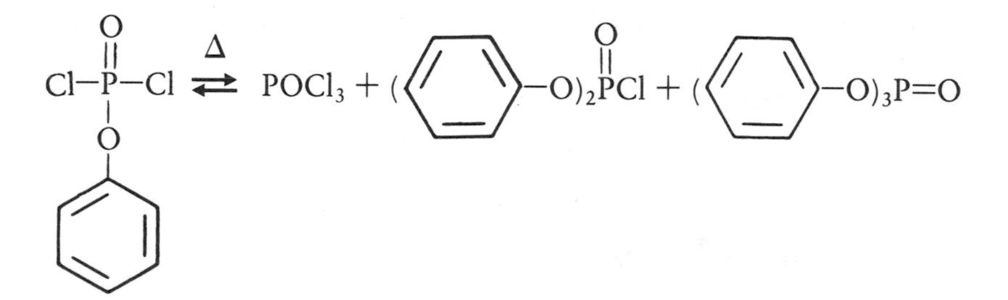

Phosphorus oxychloride, $POCl_3$, one of the disproportionation products, is a trifunctional reactant, that is, it has three reactive sites. It undergoes reaction with the difunctional hydroquinone to form a cross-link between polymer chains:

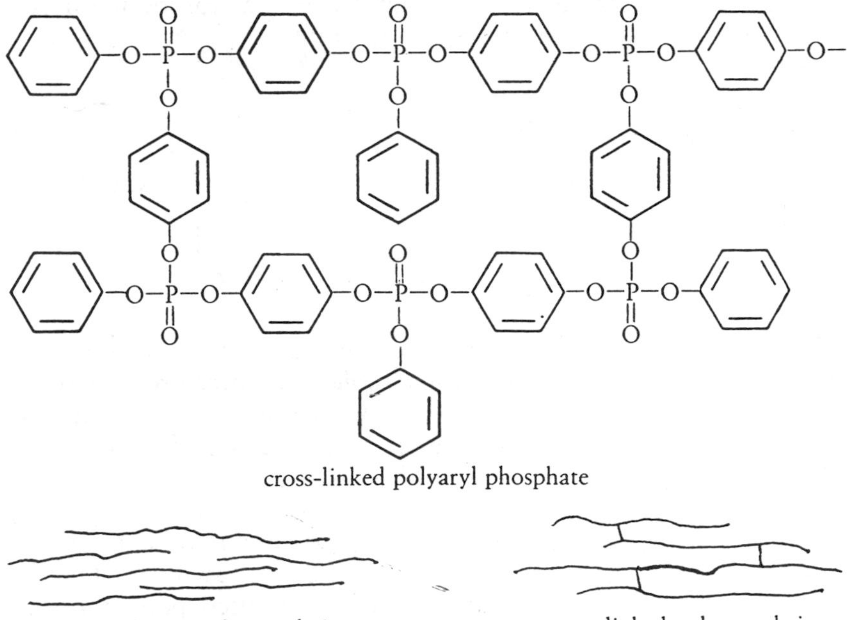

cross-linked polyaryl phosphate

oriented linear polymer chains

cross-linked polymer chains

Only a small quantity of trifunctional reactants is required to cause the whole reaction product to become a gel.

So that a truly thermoplastic phosphorus-containing polyester that remained fusible and soluble in organic solvents could be prepared, the use of bifunctional reactants that remain bifunctional under the reaction conditions used until they react completely was essential. In our case, the reactants would have to be stable up to about 300 °C in an inert atmosphere. One of the bifunctional organic phosphorus reactants meeting this requirement is phenylphosphonic dichloride.

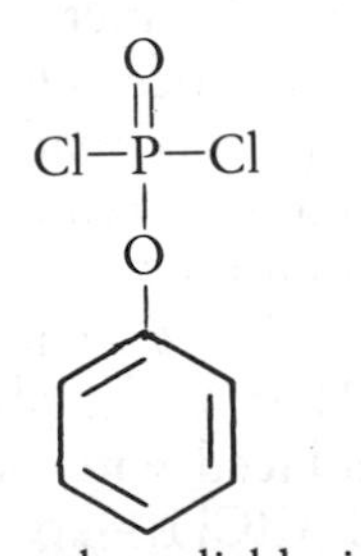

phenyl phosphorodichloridate

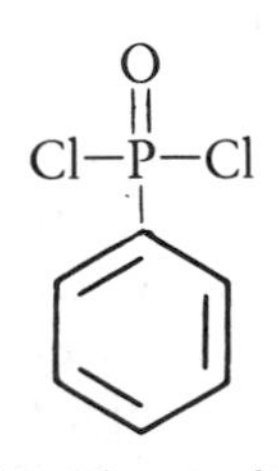

phenylphosphonic dichloride

In phenylphosphonic dichloride, the phenyl ring is attached directly to the phosphorus atom through a strong carbon-to-phosphorus bond. By contrast, the phenyl ring is connected to the phosphorus atom through weaker carbon-to-oxygen-to-phosphorus bonds in phenyl phosphorodichloridate. The polymerization reaction between phenylphosphonic dichloride and hydroquinone is

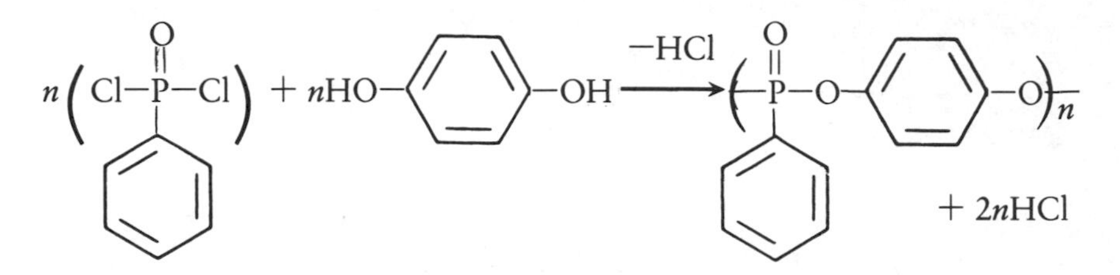

The polymer formed in this reaction is a true thermoplastic (*1*). It remains permanently fusible and soluble even after prolonged heating. It is very tough and hornlike. Upon melting, it can be pulled out into very long silklike fibers. The tensile strength of these fibers may be increased by the process of cold drawing.

In cold drawing of the fiber, that is, stretching the fiber at room temperature, the polymer molecules are oriented parallel to the fiber axis. The X-ray diffraction patterns of the cold-drawn-oriented fibers more closely resemble the pattern of an oriented polystrene fiber rather than the pattern of the highly crystalline polyamide or nylon fiber.

We found that the polymerization reaction must be carried out under meticulously anhydrous conditions. Phenylphosphonic dichloride is an acid chloride. It is very sensitive to hydrolysis when exposed to atmospheric moisture, and any inadvertent exposure to moisture causes partial hydrolysis. Some of the P–Cl groups become P–OH groups, and the P–OH groups do not condense with the hydroquinone. The polymer obtained when this happens has a lower molecular weight. It is a brittle rosinlike product.

Even though the tough hornlike polymer obtained in this reaction can be drawn out into fibers, it is not a suitable polymer for industrial applications. It lacks the other required properties. However, this phosphorus-containing polymer does not burn very readily. When it is ignited by a flame and the source of flame is removed, it self-extinguishes. This, of course, was our original goal.

Other polymers of this class may be obtained by varying the nature of the bifunctional phosphorus reactants. That is, the phenyl group in phenylphosphonic dichloride may be replaced with other substituents such as methyl (CH_3–), chloromethyl ($ClCH_2$–), or substituted phenyl groups such as chlorophenyl (ClC_6H_4–). The nature of the polymer may also be varied by replacing the hydroquinone with other dihydroxy aromatic compounds. For example, the melting point of the polymer is greatly increased when hydroquinone is replaced with tetrachlorobisphenol A in the condensation with phenylphosphonic dichloride.

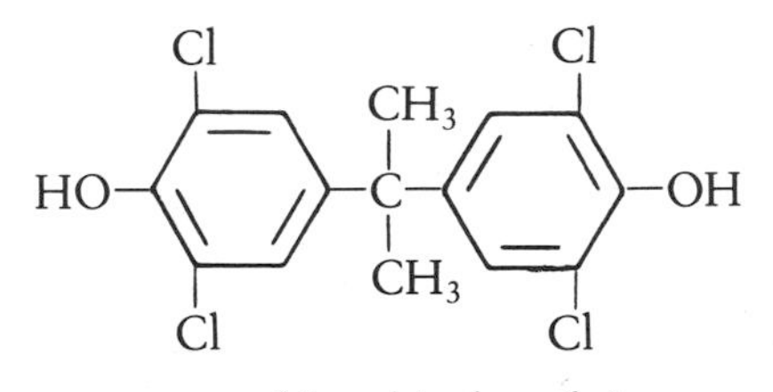

tetrachlorobisphenol A

A polymer of this polyester class was considered at one time for commercialization by Toyobo of Japan (2). The polymer was prepared by the reaction of phenylphosphonic dichloride and sulfonyl bisphenol:

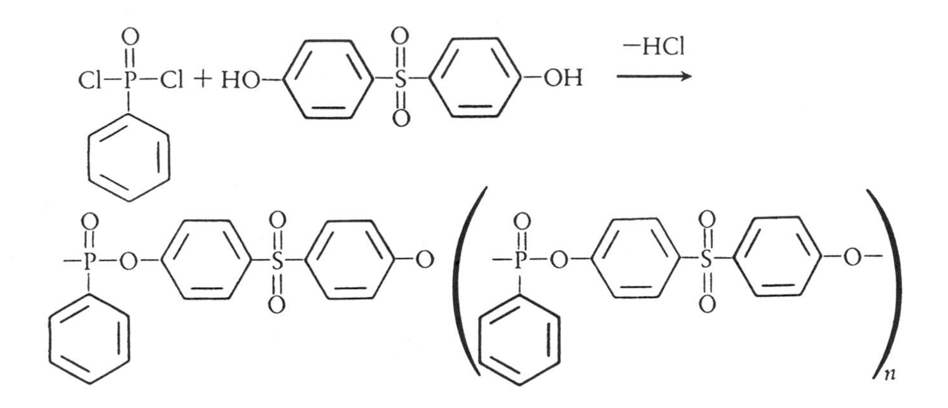

The polymer was to be used as an additive type of flame retardant for polyester fibers and fabrics. Such a polymeric flame retardant with its extremely low vapor pressure would have the advantage of being permanently incorporated with the polyester. This polymer does not vaporize off at the high temperature (250 °C) used during the processing of the polyester into fibers.

Organic phosphorus-containing polymers are also obtained in which a reactive phosphorus moiety is chemically incorporated into an organic polymer chain through covalent bonds. An example of this reaction is

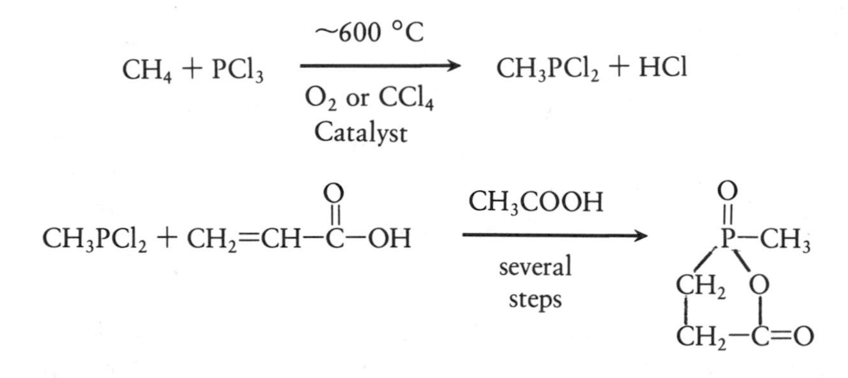

The product of these reactions is a reactive monomer. When it is added to the reaction mixture of ethylene glycol and dimethyl terephthalate in making polyester polymer, the phosphorus–oxygen–carbon acid anhydride bond opens up. The compound then enters into the polymerization reaction to make the polyester.

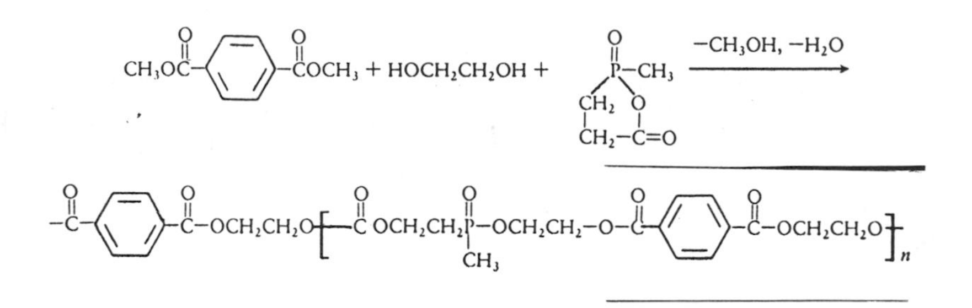

These reactions showed how the phosphorus monomer is incorporated into the polyester chain. Because the phosphorus reactant is chemically bonded into the polyester chain, it stays put during the high-temperature processing of the polyester into fibers. Such fibers, when woven into fabrics for clothing, have flame-retarding properties. Such clothing is used for children's sleepwear and for work clothing where a danger of exposure to fire exists.

Because only 1–2% phosphorus is needed to impart flame retardancy to the fiber, the ratio of the phosphorus monomer to the reactants for making the polyester polymer is much less than that shown in the equations. Flame-retardant polyester prepared in this manner was developed by Hoeschst Chemische Gesellschaft of Germany. This polyester has the trade name of Trevira 271 (*3*).

Another phosphorus-containing monomer that serves the same purpose of imparting flame retardancy to polyester fibers is Toyobo's GH (*4*). The chemistry for its preparation is

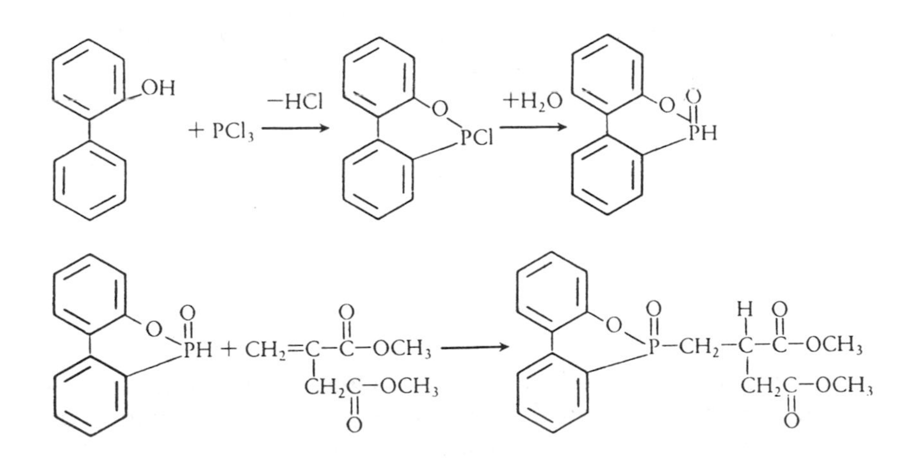

As can be seen by the formula of this phosphorus-containing monomer, it is the methyl ester of a substituted dicarboxylic acid and difunctional compound. Such an ester can cocondense with the ethylene glycol reactant to make the polyester polymer (*see* p 172). As with Trevira 271, this polyester needs only 1–2% of phosphorus based on the weight of the total polymer to impart flame retardancy, so the amount of Toyobo's GH monomer needed to make a flame-retardant polymer is relatively small. One of the advantages of this monomer is its high thermal and hydrolytic stability. Fabric made from this polymer has been offered for children's sleepwear in the United States and for use in curtains for public buildings in Japan.

Addition Polymers

A totally different class of polymers is formed by the polymerization of organic phosphorus compound monomers containing carbon–carbon double bonds. The polymerization occurs through the carbon–

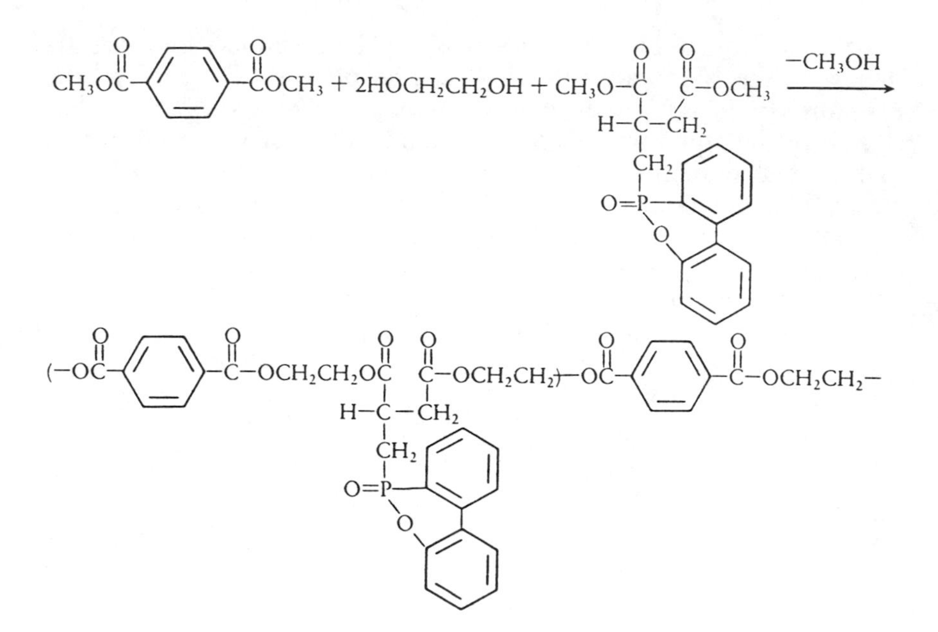

carbon double bonds of the monomers. One example of this class of monomer is bis(β-chloroethyl) vinylphosphonate. The vinyl compound is prepared by the following reactions:

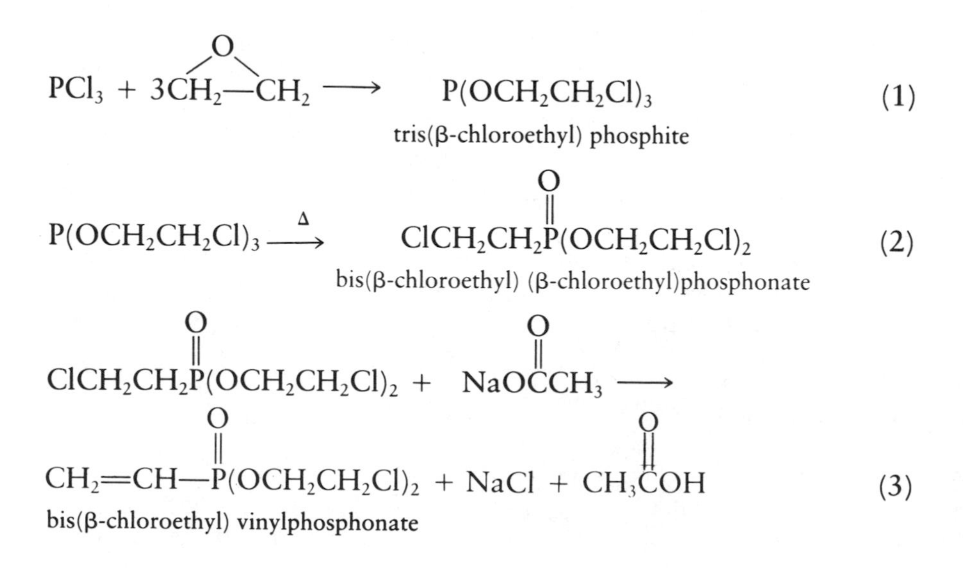

In equation 2, the phosphite has three carbon–oxygen–phosphorus bonds. When the compound is heated, one of the carbon–oxygen–phosphorus bonds undergoes rearrangement into a carbon-to-phosphorus bond. The phosphorus atom is converted from trivalent to pentavalent. This rearrangement is called the Michaelis–Arbuzov rearrangement. Michaelis was the German chemist who discovered this rearrangement reaction more than 100 years ago. Arbuzov was the Russian chemist who did an extensive study of this reaction during the early part of the century. Interestingly, the rearranged product, bis(β-chloroethyl) (β-chloroethyl)phosphonate is also the intermediate for the plant hormone ethephon, discussed in Chapter 21.

In equation 3, the β-chloroethyl group connected directly to the phosphorus atom through the carbon–phosphorus bond is dehydrochlorinated into a vinyl group by sodium acetate, a weak base. If a strong base, such as sodium hydroxide, were to be used, it would cause partial hydrolysis of β-chloroethyl ester groups with the carbon–oxygen–phosphorus bonds. A strong base would also cause some dehydrochlorination of the β-chloroethyl ester groups. Bis(β-chloroethyl) vinylphosphonate was mentioned in Chapter 10 where it was used as a component in Fyrol 76 for flame-retarding cotton fabric.

Bis(β-chloroethyl) vinylphosphonate now has a trade name of Fyrol Bis-Beta. It forms only very low molecular weight polymers when it polymerizes with itself to form homopolymers. However, it readily forms copolymers with vinyl chloride. We know, of course, that vinyl chloride can readily polymerize with itself to form very high molecular weight homopolymers:

$$(n+1)CH_2{=}CH{-}Cl \longrightarrow {-}CH_2\underset{\displaystyle Cl}{\underset{|}{C}}H(CH_2\underset{\displaystyle Cl}{\underset{|}{C}}H)_n{-}$$

Such polyvinyl chloride is hard, tough, and difficult to work with. It must be made more flexible to process it into sheets and other useful objects. This change is usually done by the addition of a plasticizer. The most commonly used plasticizers are the phthalate esters. Vinyl chloride polymer itself is not flammable. However, when it is plasticized with the flammable phthalate esters, the resultant plastic burns readily. If a self-extinguishing type of vinyl plastic is desired, flame retardant is added. Alternately, phosphate esters can be used as the plasticizers. Vinyl chloride polymer, when plasticized by the addition of such external plasticizers, has another deficiency. The added plasticizer has a tendency to migrate to the surface of the plastic sheet. In hot weather, some of the added plasticizer volatilizes. When an automobile with vinyl-covered seats is parked under the hot sun during a summer day, usually an oily film forms on the windshield and windows. The oily film is from the condensation of the vaporized plasticizer. Over a period of time, with more evaporation of the plasticizer, the vinyl sheet becomes hard and brittle. One way to decrease such evaporation is to use high molecular weight plasticizers with low vapor pressure. The preferred way is to have the plasticizer bonded chemically to the vinyl chloride polymer so that it will not evaporate at all. Fyrol Bis-Beta forms a copolymer with vinyl chloride in the presence of a free-radical catalyst such as AIBN (Vazo 64 from Du Pont). Such a chemically bonded Fyrol Bis-Beta acts as a chemically bonded internal plasticizer (*5*).

$$n\mathrm{CH_2{=}\underset{\displaystyle Cl}{\underset{|}{C}}H} + \mathrm{CH_2{=}\underset{\displaystyle O{=}P(OCH_2CH_2Cl)_2}{\underset{|}{C}}H} \xrightarrow[\text{catalyst}]{\text{Vazo 64}}$$

$$-\!\!\left[(\mathrm{CH_2{-}\underset{\displaystyle Cl}{\underset{|}{C}}H})_n\mathrm{CH_2\underset{\displaystyle O{=}P(OCH_2CH_2Cl)_2}{\underset{|}{C}}H}\right]\!\!-$$

A copolymer containing 7–12% of Fyrol Bis-Beta has properties desirable for producing flexible vinyl plastic sheets. These sheets, which are internally plasticized by the bonded phosphonate ester, remain flexible on aging. Such a copolymer also has favorable flame-retarding properties. The film prepared from the copolymer is commercialized with a trademark of Betaflex. Such film was found to have outstanding conformability characteristics when used as a covering over an irregular surface. An adhesive-backed sheet of this copolymer even hugs and adheres to surfaces with rivets and corrugations. It also adheres well to embossed metals and fiberglass-reinforced panels. The internal plasticizing allows this tough plastic sheet to withstand exposure to sun, rain, heat, and cold without loss of the plasticizer. Such properties make it ideal for durable films for use for signs over truck bodies.

Another class of organic phosphorus polymer is formed by the additional polymerization of diallyl organophosphonates (6). An example of such a monomer is diallyl phenylphosphonate:

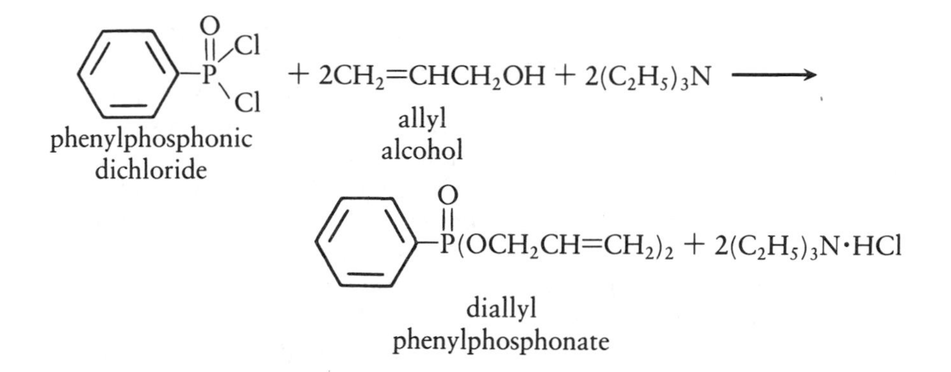

As shown in the equation, the phosphonate is formed by reaction of the organic phosphonic dichloride, such as phenylphosphonic dichloride, with allyl alcohol in the presence of a tertiary amine hydrogen chloride acceptor. Another method for the preparation of diallyl organophosphonate is through the addition of the P–H group of diallyl phosphonate to the double bonds of acrylic compounds. This reaction proceeds in the presence of the catalyst, sodium allylate.

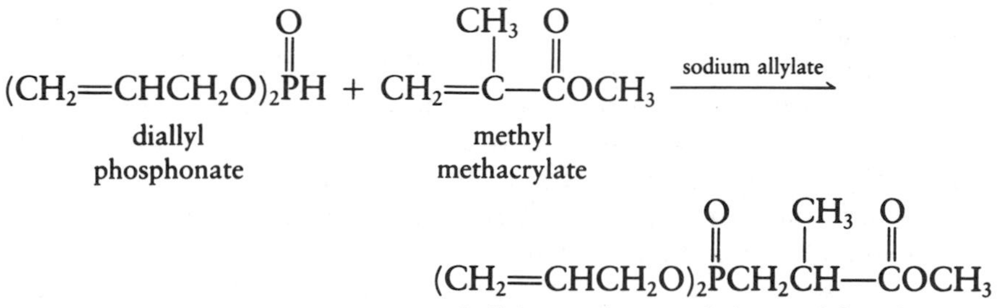

These allyl organophosphonates polymerize when heated in an oxygen-free atmosphere in the presence of benzoyl peroxide catalyst. The polymers are hard, glassy, transparent, flame-resistant solids. Because the polymer is formed by the addition reaction through two double bonds (tetrafunctional) in each molecule, it is highly cross-linked. The polymer is not soluble and does not melt on heating. When heated at high temperature, it decomposes.

Diallyl phenylphosphonate is also capable of forming solid copolymers with such monomers as methyl methacrylate, vinyl acetate, diallyl phthalate, and unsaturated polymers, but not with styrene. The copolymer with methyl methacrylate is rather interesting (7). Diallyl phenylphosphonate polymer has a high index of refraction while polymethyl methacrylate has a rather low index of refraction. The index of refraction of the copolymer increases with increasing percentage of diallyl phenylphosphonate. At a ratio of 2 parts of diallyl phenylphosphonate to 1 part of methyl methacrylate, the copolymer has an index of refraction, n^{25}_{D} of 1.544. This value matches the index of refraction of a heat-cleaned fiberglass fabric, that is, glass fabric in which the sizing was burned off. Glass fabric itself is opaque. However, when a plastic laminate is prepared from several layers of glass fabric and the methyl methacrylate–diallyl phenylphosphonate copolymer with a matching index of refraction, the resultant laminate is almost transparent. One reason for this clarity is that this copolymer adheres very tightly to the individual fibers of the glass fabric.

No microscopic air spaces occur between the polymer and the fibers. Air bubbles would show up as opaqueness. The copolymer with 60% vinyl acetate is also interesting. It is a very hard, glassy solid with a fair degree of flame resistance. The polymer of 100% vinyl acetate is a rather soft semisolid. So, in the case of this copolymer, the diallyl phenylphosphonate acts as a cross-linking agent for the vinyl acetate.

The most important copolymer of diallyl phenylphosphonate is the one with an unsaturated polyester. Copolymerization of the unsaturated polyester with 10–30% of diallyl phenylphosphonate will render the copolymer self-extinguishing when it is removed from a source of flame. For some time, diallyl phenylphosphonate was under commercial development, especially for use with unsaturated polyesters. However, it is not manufactured on a large scale at the present time.

Phosphazene Polymers

In Chapter 10, we discussed the propyl ester of cyclic chlorophosphazene as a flame retardant for cellulose rayon. Since its discovery by J. von Liebig in 1834, the unusual properties of chlorophosphazene itself have intrigued several generations of chemists. In 1897, Stokes (*8*) found that the reaction of ammonium chloride with phosphorus pentachloride produced various distillable chlorophosphazenes ranging from the cyclic trimer to the cyclic hexamer and an impure heptamer plus a polymeric residue. The cyclic trimer was the major product. Another interesting finding reported by Stokes was that when the cyclic trimer was heated in a sealed tube at 250–300 °C, it first melted and then it polymerized into a insoluble rubbery material. Because this rubbery material contains no organic moiety, that is, no carbon-containing groups, it has been called "inorganic rubber":

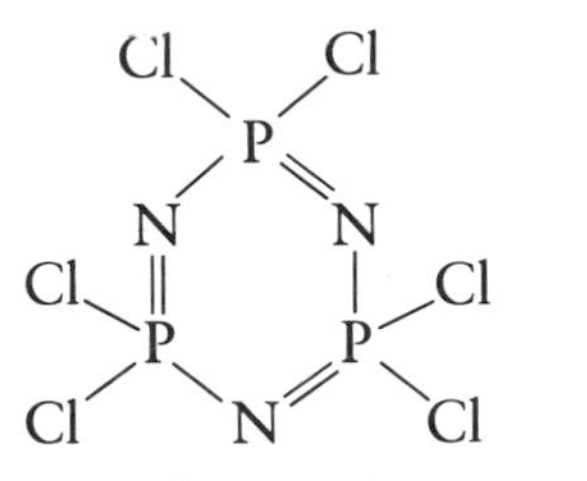

cyclic trimeric chlorophosphazene

$$\xrightarrow{250\ ^\circ\mathrm{C}} \quad -\overset{Cl\ \ Cl}{P}{=}N{+}\overset{Cl\ \ Cl}{P}{=}N{+}_n$$

poly(chlorophosphazene)

However, this material, even though it is a polymer, is also an acid chloride. So, when the material is allowed to stand and is exposed to atmospheric moisture, the P–Cl groups can gradually hydrolyze to P–OH groups and liberate HCl. The polymer also has a tendency to depolymerize back to the cyclic form. This depolymerization is rapid above 350 °C. Many chemists have sucessfully replaced the reactive acid chloride atoms in the cyclic trimer with such substituents as CH_3–, C_2H_5–, C_6H_5–, and various ArO–, RO–, and amino groups. These substituents are not sensitive to hydrolysis from atmospheric moisture. Unfortunately, once the acid chloride atoms are replaced with organic substituents, the substituted cyclic trimers can no longer be polymerized into high molecular weight polymers. At least nobody has found a method to do it to date.

The breakthrough in this area comes as the result of the research carried out by Allcock of Pennsylvania State University (*9*, *10*). He found that the rubbery, solvent-insoluble nature of the chlorophosphazene polymer reported in the literature was due to the presence of impurities (such as those resulting from atmospheric moisture), which caused cross-linking among the polymeric phosphazene chains:

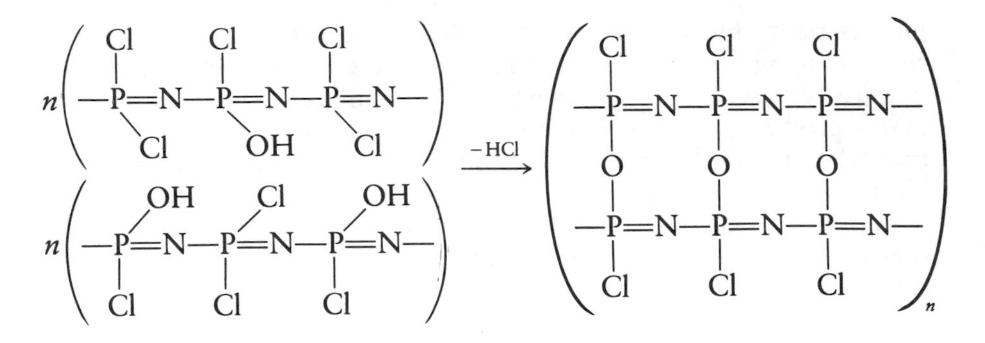

When many polymer chains are thus cross-linked, the product becomes a polymeric rubber. Of course, if too much cross-linking occurs, the polymer simply becomes a hard solid.

Professor Allcock found that by careful purification of the cyclic trimer and by avoiding exposure to atmospheric moisture, he was able to convert the cyclic trimer to a linear non-cross-linked elastomer. This elastomer has a length in excess of 15,000 $N{=}PCl_2$ units. It is soluble in some solvents such as benzene and toluene. When dissolved in the proper solvent, the acid chloride groups, $-PCl_2$, are quite reactive to reagents that can cleave the phosphorus–chloride bonds. Therefore, this linear polymer can be used as a reactant for reactions with organic and inorganic nucleophiles.

Even though the polychlorophosphazene is unstable in air because of the reactivity of the chloride ions to atmospheric moisture, it can be converted into derivatives that possess no phosphorus–chlorine groups. These derivatives are highly stable under a wide variety of conditions.

The nature of the substituents replacing the chlorine in polychlorophosphazene impart different physical and chemical characteristics to the resultant products. Thus, polymers can be "tailor made" with specific properties within certain limits.

One of the goals of preparing a polymer from phosphazene is to obtain a product that is exceedingly stable to hydrolysis, does not burn, does not dissolve in oil or cleaning solvent, and also does not change after years of outdoor exposure to strong sunlight. The polymer should also be capable of forming flexible fibers or films or be convertible into an elastomer. Such products should remain flexible even at extremely low temperatures. Professor Allcock and his coworkers accomplished almost all these goals in their research on phosphazene polymers.

For stability to hydrolysis, Professor Allcock employed a very hydrophobic group to replace the hydrolyzable chlorine atoms.

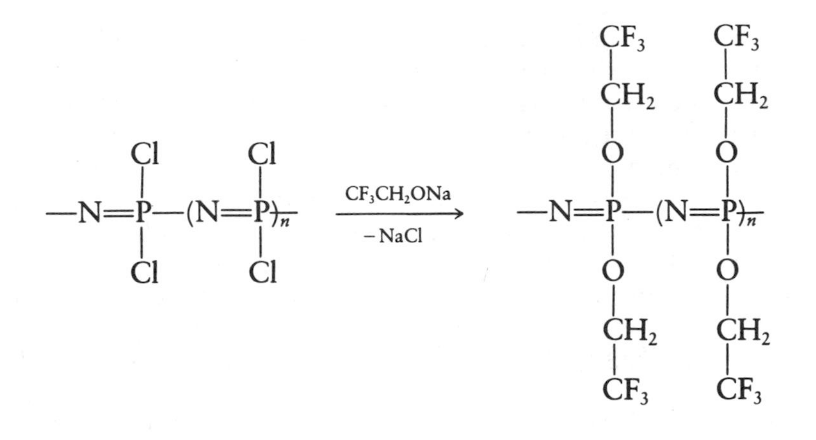

As shown, one of the hydrophobic groups he chose was the trifluoroethoxy group (CF_3CH_2O–).

With such a water-repellent group, water cannot gain a "foothold" on the molecule to cause any hydrolysis. The property of resistance to burning is already built into the polymer because it is a fundamental property of phosphazene with its phosphorus and nitrogen atoms. These two elements in a compound are known to have a synergistic effect for flame retardancy. Inertness to oil and most organic solvents is a characteristic of the fluorocarbon type of structure as represented by the trifluoroethoxy group. As to resistance to sunlight, even though the phosphazene structure appears to have a pseudounsaturated electronic structure, it does not absorb light at a wavelength longer than 2500 Å. The trifluoroethoxy group also does not absorb at wavelengths longer than 2500 Å.

According to Professor Allcock, the polymer's flexibility depends on the freedom of torsional movement of the bonds in the phosphazene polymer backbone. He has calculated that the nature of the substituents attached to the phosphorus exerts a powerful influence on the torsional mobility of the phosphazene backbone. The mobile substituents such as ethoxy (CH_3CH_2O–) and methoxy (CH_3O–) allow the skeletal bond to assume a wide range of torsional conformations. This flexibility, in turn, can be correlated with the low-temperature elasticity of the polyphosphazene with those particular substituents. Polyphosphazenes and phosphazene elastomers containing mixed fluoroalkoxy substituents such as trifluoroethoxy and polyfluoroalkoxy [$HCF_2(CF_2)_nCH_2O$–] groups are produced commercially. Elastomers containing these phosphazene compounds are reportedly used in the components of high-performance brake systems for military aircraft, where they must withstand high altitudes and temperatures as low as −70 °F. They can also withstand the high brake temperatures that are encountered on landing, up to 350 °F. They are used in O-rings in helicopters where they must maintain their flexibility even in the extremely low temperatures encountered in the Arctic. They are also used in battle tanks in ring seals for turbine engine dust filters. Here high abrasion resistance is necessary. The elastomer, when used as a damper for the vibration in the engine, provides constant performance over a wide temperature range. The elastomer was found to be compatible with tissue, and it bonds to denture base, where it is stable and remains resilient and soft. Thus far, its use has been limited to military equipment and similar specialized applications because of its high cost (around $140/lb in 1984).

The hydroquinone and resorcinol derivatives of the trimeric and tetrameric chlorophosphazene were also under commercial development at one time:

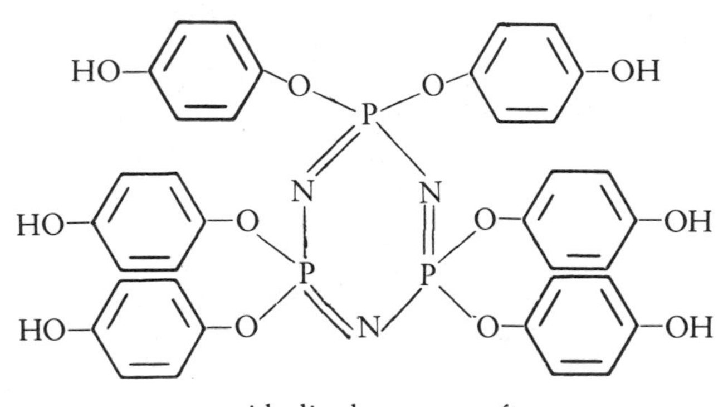

idealized structure of hydroquinone ester of trimeric phosphazene

The hydroquinone ester moieties of the compound undergo a condensation reaction with formaldehyde (H_2CO) in a reaction similar to the condensation of phenol and formaldehyde to form a resin. In the case of the condensation of formaldehyde with phenol, a low molecular weight polymer that is fusible and organic solvent soluble is first formed. When heated further, this polymer polymerizes into the cross-linked phenolic resin that is no longer soluble or fusible. Analogously, the hydroquinone ester of phosphazene also condenses with formaldehyde to form first a soluble and fusible polymer. When heated further, this soluble and fusible polymer also polymerizes into a highly cross-linked insoluble and infusible resin. Because this resin contains phosphorus and nitrogen, it is highly resistant to burning. However, the compound is expensive to prepare. The resultant polymer does not possess sufficient outstanding properties to justify the high cost. As a consequence, it was never produced commercially on a large scale.

Literature Cited

1. Toy, A. D. F. U.S. Patent 2 435 252, Feb. 3, 1948 (to Victor Chemical Works).
2. Masai, Y.; Kato, Y.; Fukui, N. U.S. Patent 3 719 727, March 6, 1973 (to Toyo Spinning).
3. Bollert, U.; Lohmar, E.; Ohorodnik, A. U.S. Patent 4 033 936, 1977 (to Hoechst A.G.).
4. Endo, S.; Kashihara, T.; Osako, A.; Shizuki, T.; Ikegami, T. U.S. Patent 4 127 590, 1978 (to Toyo Bosekikk).
5. Gallagher, R.; Hwa, J. C. H. *J. Polym. Sci., Polym. Symp.* **1978,** *64*, 329–337.
6. Toy, A. D. F. *J. Am. Chem. Soc.* **1948,** *70*, 186.
7. Toy, A. D. F.; Brown, L. V. *Ind. Eng. Chem.* **1948,** *40*, 2276.
8. Stokes, H. N. *Am. Chem. J.* **1897,** *19*, 782.
9. Allcock, H. R. *Phosphorus-Nitrogen Compounds*; Academic: New York, 1972.
10. Allcock, H. R. *Chem. Rev.* **1972,** *72*, 315.

Chapter Twelve

Water Treatment with Phosphorus Compounds

GROUND WATER FROM DEEP WELLS that is used for drinking, laundry, bathing, or generating steam in a power plant is called hard water because it contains an appreciable amount of calcium and magnesium ions, Ca^{2+} and Mg^{2+} (also iron, Fe^{2+}). These ions are usually present as the soluble calcium and magnesium hydrogen carbonates [$Ca(HCO_3)_2$ and $Mg(HCO_3)_2$] and magnesium sulfate ($MgSO_4$). They react with soap to form insoluble salts that are deposited as a bathtub or sink ring. When some hard waters are heated, a precipitate will appear as scale inside the heating container. Why does this scale form? The answer depends on the salts in the water. For example, when a metal hydrogen carbonate is heated, it decomposes to liberate CO_2 and forms the metal carbonate:

$$M(HCO_3)_2 \xrightarrow{\text{heat}} MCO_3 + CO_2 + H_2O$$

Thus, in hot-water heaters or in steam-generating plants, the soluble calcium and magnesium bicarbonates decompose into the insoluble calcium and magnesium carbonates, which appear as a hard scale. Some of the water evaporates, and calcium sulfate may also come out of solution. The scales thus formed decrease the efficiency of the heater because they are good heat insulators.

Softening Water with Trisodium Phosphate

Trisodium phosphate reacts with calcium and magnesium ions in water to form insoluble tricalcium and trimagnesium phosphates. When these ions are removed, the water is softened.

1002–0/87/0183$06.00/1

In practice, trisodium phosphate is used only in conjunction with other, cheaper water-softening processes, such as the soda-lime process, to precipitate residual calcium and magnesium ions as the very insoluble hydroxyapatite and trimagnesium phosphate. The precipitate is dispersed and does not form a hard scale. Therefore, the precipitate is easily purged. The reactions for removing calcium and magnesium from boiler water are

$$5Ca(HCO_3)_2 + 4Na_3PO_4 \xrightarrow{H_2O} \underset{\text{hydroxyapatite}}{Ca_5(OH)(PO_4)_3} + 10NaHCO_3 + Na_2HPO_4$$

$$3MgSO_4 + 2Na_3PO_4 \xrightarrow{H_2O} \underset{\text{trimagnesium phosphate}}{Mg_3(PO_4)_2} + 3Na_2SO_4$$

Metal hydrogen carbonate also is converted to the metal carbonate when neutralized with a base. In the soda-lime process, hard water is first treated with sodium carbonate and then with lime to convert the soluble calcium and magnesium bicarbonates into the relatively insoluble calcium and magnesium carbonates.

Classes of Sodium Metaphosphates (Sodium Polyphosphates)

Sodium metaphosphates represent one of the most important and interesting classes of phosphates. All three sodium metaphosphates described here are made from the same raw material. Because their chemical compositions are nearly identical, their properties are expected to be similar, but this situation is not the case. For example, one compound is extremely soluble in water and is a very good water softener. Another compound is insoluble in water and until recently was used as a polishing agent in toothpastes (Chapter 5). The third compound has an interesting ring structure and is used to modify starch employed in making clear, thick soups as described in Chapter 4. How can the same raw material be converted to products with such contrasting properties? Why are they useful? What research was done to put these compounds to work?

The term metaphosphate refers to compounds with one or more $NaOPO_2$ units. Because the polyphosphates are chains built up from metaphosphate units, they are sometimes called polymetaphosphates or simply polyphosphates. The term sodium hexametaphosphate is also used. It implies that the compound is a polyphosphate chain containing six (hexa-) metaphosphate ($NaOPO_2$) units. However, research has shown this notion to be incorrect. Commercial sodium hexametaphosphate actually contains an average of 13–18 metaphosphate units.

The three forms of sodium metaphosphates that have reached commercial importance are glassy sodium polyphosphate, cyclic sodium trimetaphosphate, and insoluble sodium metaphosphate. Today, the first two remain commercially important. As a result of changes in the technology of dental polishing agents (*see* Chapter 5), insoluble sodium metaphosphate is no longer produced at the commercial level. However, because this form of sodium metaphosphate has proved to be of such interest, we will continue to describe its preparation and properties.

As stated earlier, all three forms are prepared from the same material, monosodium phosphate, and the reaction conditions determine which product is formed. The preparation of cyclic sodium trimetaphosphate is described in detail in Chapter 4.

Glassy Sodium Polymetaphosphate. As described in Chapter 4, sodium acid pyrophosphate is prepared by heating monosodium phosphate at about 225–250 °C; thus, 2 mol of monosodium phosphate is condensed together, and 1 mol of water is eliminated:

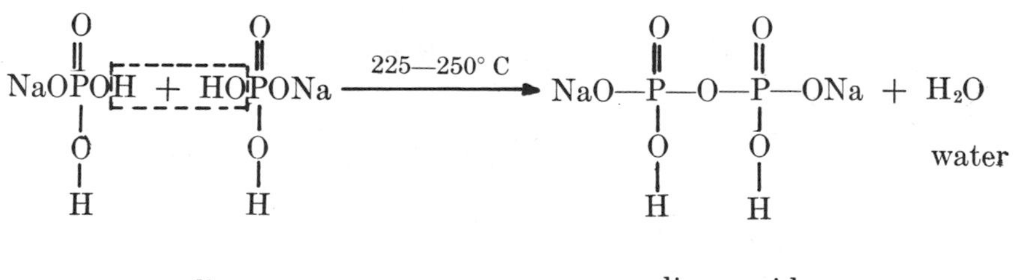

If monosodium phosphate is heated at 800–900 °C rather than at 225–250 °C, the reaction mixture becomes a hot molten liquid, and almost all water derivable from the P–OH groups is eliminated. If the molten liquid is cooled quickly (e.g., by pouring it over a cold steel sheet), the material solidifies into a thin layer of colorless glass. This glassy product is sodium polymetaphosphate, or Graham's salt (after Thomas Graham, who first described it in 1833). The size of each molecule depends on the ratio of sodium to phosphorus used. The reaction is

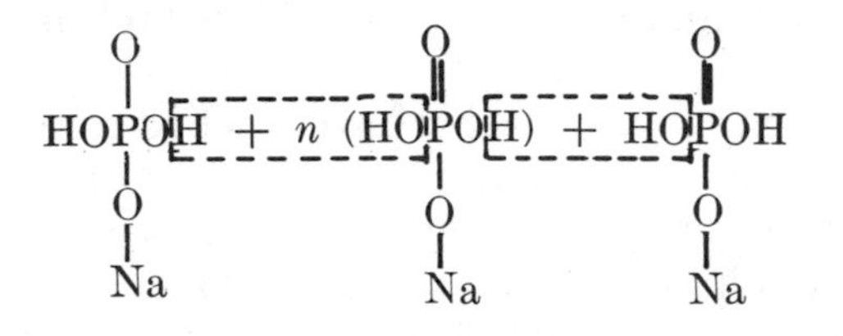

monosodium phosphate

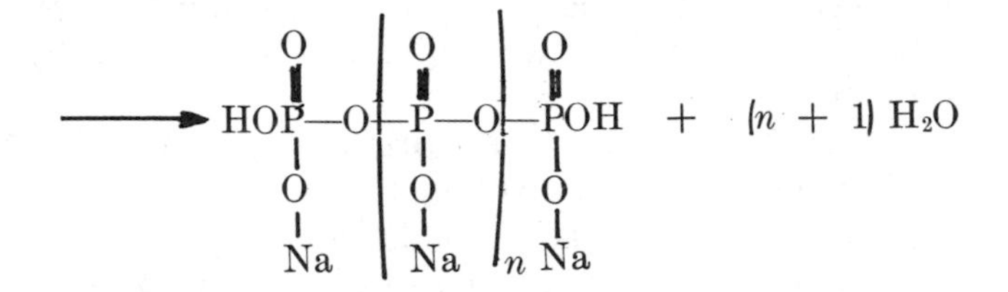

glassy sodium polymetaphosphate

The *n* in the equation represents the number of metaphosphate units in the chain other than the end units. Obviously, the larger the value of *n*, the longer the chain.

Cyclic Sodium Trimetaphosphate. If monosodium phosphate is heated to about 530–600 °C, it will not go into a molten form. Three moles of monosodium phosphate will join head-to-tail and lose 3 mol of water to form a ring. The compound formed is called cyclic trimetaphosphate and is a white crystalline product. Various other methods for the preparation of cyclic sodium trimetaphosphate are described in Chapter 4.

Insoluble Sodium Metaphosphate (IMP). IMP is sometimes known as Maddrell's salt, named after the German chemist R. Maddrell. Several forms of IMP exist, but the industrially important form is generally made by heating monosodium phosphate at about 475–500 °C. The crystalline product, which is almost insoluble in water, forms readily in small quantities even under lower temperature conditions. In fact, when cyclic trimetaphosphate or sodium acid pyrophosphate is prepared, by using monosodium phosphate as the raw material, IMP forms as a byproduct if reaction conditions are not controlled. When the soluble sodium phosphates, such as sodium acid pyrophosphate, are made, the presence of even 0.1% IMP as a byproduct renders water solutions of the desired water-soluble compound cloudy and turbid. Sometimes IMP even shows up as the undesirable byproduct in sodium tripolyphosphate.

Why is IMP insoluble in water when other sodium metaphosphates are very soluble? X-ray diffraction studies show that an IMP crystal is two long metaphosphate chains that spiral in opposite directions around the skew axes of the crystal. These chains are held together so tightly that a water molecule cannot easily slip into the chains. However, if some of the sodium ions in IMP are exchanged with larger potassium ions, the compactness is destroyed and IMP gradually dissolves. An interesting experiment is to add a mixture of IMP and potassium metaphosphate to distilled water. Neither compound alone is water-soluble, but the mixture gradually dissolves because some sodium and potassium ions have exchanged from one compound to the other.

IMP must be manufactured under static conditions or some soluble metaphosphate forms. The reason may be that because IMP's insolubility is the result of a compact crystal structure, this structure must be allowed to grow undisturbed or the proper structure will not form.

Softening Water with Glassy Sodium Metaphosphate

The most popular and well-known trade name for glassy sodium polymetaphosphate is Calgon, now a household word. Although this compound has been known since 1833 when Graham first described it, not until 1929 did Ralph E. Hall discover its unusual property—the ability to sequester many metallic ions as soluble complexes. Hall found this property useful for softening water, and his discovery led to many other applications that formed the basis of a large industry. The name Calgon was coined by Dr. Hall when he observed that glassy sodium polymetaphosphate could sequester calcium ions in water and make it seem as if they were not there—that is, gone, or calcium gone, Calgon.

The theoretical value of P_4O_{10} in a sodium metaphosphate unit ($NaOPO_2$) is 69.9%. If the theoretical value for both the P_4O_{10} content and the sodium oxide content—that is, 100% pure monosodium phosphate as the raw material—is used, the glassy polymetaphosphate obtained would be a long chain of infinite length. However, this situation is not practical. For industrial use, Calgon generally has a P_4O_{10} content of around 67.0–67.8%, which is adequate for most applications. This amount is 2.1–2.9% below the theoretical value for sodium metaphosphate. This finding also means that the sodium oxide value is above the theoretical value. In other words, the starting material is not pure monosodium phosphate, but monosodium phosphate with a little disodium phosphate.

Disodium phosphate [$(NaO)_2PO(OH)$] acts as a chain terminator—that is, this phosphate becomes part of the end of the chain. Thus, when a polymetaphosphate chain reacts with disodium phosphate, the chain is terminated:

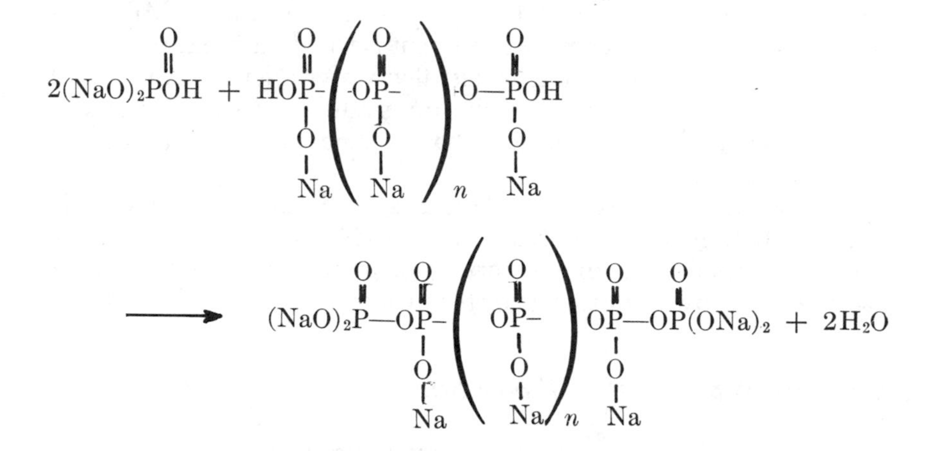

Of course, the more chain terminators, the shorter the chains. For commercial Calgon with 67.0–67.8% P_4O_{10}, the chain length is between 13 and 18 metaphosphate units. If more chain terminators were present (e.g., if the total P_4O_{10} used were lower), shorter chains would be possible. For instance, if the P_4O_{10} content were down to 52.8%, the chain length would be only the two end units, or tetrasodium pyrophosphate. The chain length of commercial glassy sodium polymetaphosphate is chosen to obtain good water solubility and low hygroscopicity.

Sodium polymetaphosphate was first used to soften water. This compound forms a soluble complex with the calcium and magnesium ions of hard water and renders these ions unable to react with soap to form an insoluble scum. Even if an insoluble soap should form, glassy sodium polymetaphosphate can redissolve it. Some soap is always wasted by reaction with magnesium and calcium ions to form useless insoluble soaps, but if glassy sodium polymetaphosphate is added, no soap is wasted. Of course, if the water has already been softened, sodium polymetaphosphate is not needed.

Another important use for this compound is in the so-called "threshold" treatment. As described earlier, when hard water is heated in a boiler, the soluble calcium and magnesium salts originally present begin to precipitate as calcium sulfate and calcium and magnesium carbonates. These salts deposit as crystals (scales) on the inside surfaces of boilers. If these crystals are to form, nuclei must be present upon which more molecules can build in a regular and orderly fashion. Through this process, the nuclei grow into the large crystals known as scale. If anything is added to prevent this orderly growth, the crystal will not grow.

A small amount of glassy sodium polymetaphosphate interferes with crystal growth; this interference prevents scale formation. The addition of an amount of sodium polymetaphosphate that will sequester all calcium ions and magnesium ions into soluble complexes is not necessary—just enough to prevent crystal growth must be added. This addition is called the "threshold" treatment. If glassy sodium polymetaphosphate were added to sequester all the calcium, a glassy sodium polymetaphosphate to calcium ratio of 2.5:1 would be required. In many threshold treatments, 1 ppm (part per million) of glassy sodium polymetaphosphate prevents the precipitation of calcium carbonate from a solution containing 200 ppm of calcium bicarbonate, $Ca(HCO_3)_2$.

Threshold treatment prevents the precipitation of scale in boilers. This fact allows the boiler to maintain high heat-transfer efficiency, but the treatment is not without drawbacks. A clean surface is also more prone to chemical attack and corrosion. Fortunately, the threshold treatment that prevents scale formation also protects against corrosion. A submicroscopic protective film is formed by the reaction of the metal and metallic oxides on the inside surface of the boiler with the glassy sodium polymetaphosphate. This film insulates the clean metal surface from attack by oxygen and other corrosive elements in water. The film is so thin that it is visible only as a multicolored iridescence on a polished metal surface. In cold water, 2 ppm of Calgon will control corrosion. In hot water, 40–60 ppm is needed.

The threshold treatment also prevents iron or manganese compounds from precipitating when water is exposed to air or chlorine. Precipitation of ferric oxides gives red water. Precipitation of manganese compounds gives black water. In general, for 1 ppm of iron or manganese, only 2–4 ppm of glassy sodium polymetaphosphate is needed. Threshold treatment is used in almost all industrial water supplies to prevent lime scales. This treatment is useful in boilers, pipes, screens, washers, heat-exchange equipment, and condensers.

Treatment of Hard Water with Tetrasodium Pyrophosphate

In addition to calcium and magnesium ions, tetrasodium pyrophosphate (TSPP) is also able to sequester other heavy metal ions such as iron and vanadium. In the textile industry, many cottons are bleached by peroxy compounds such as hydrogen peroxide. However, peroxides are easily decomposed in the presence of even traces of heavy metal ions. Therefore, TSPP and sodium silicate must be used to sequester these metal ions; thus, the metal ions cannot catalytically decompose the peroxides.

Scale Inhibition and Removal with Organophosphorus Compounds

The water-softening effect and the threshold effect provided by the soluble phosphate are important in preventing scale formations. Also, the sequestering properties of the inorganic phosphates are important in scale removal. However, the effectiveness of the soluble phosphate is related in a large part to its chain length. The longer chained polyphosphates are much more effective in all three properties than are the shorter chained polyphosphates and the orthophosphates. Also, orthophosphates, even at relatively low concentrations, can react with metal ions and precipitate as metal phosphates; this reaction contributes to the scale formation. Unfortunately, the longer chained polyphosphates are hydrolytically unstable. They slowly hydrolyze to form shorter chained polyphosphates and orthophosphates. In situations where the longer chained polyphosphates are used to treat water streams where the water is held only for a short time, such as in fertilizer solutions, this hydrolysis is not a serious problem, because the water is used or transferred before a significant amount of hydrolysis of the polyphosphate can take place. However, when the water is to be retained, as in the case of a boiler or a storage tank, hydrolysis can become a serious problem.

Recently, a number of organophosphorus compounds have been found to provide the scale-inhibiting properties of the inorganic polyphosphates (*1*). These compounds are as effective as the inorganic polyphosphates with respect to scale inhibition, yet they provide hydrolytic stability. Thus, organophosphorus compounds are of particular value in treating waters such as those that must be heated for long periods of time, for example, boilers.

These organophosphorus compounds are the polyphosphonic acids, such as (aminomethyl)phosphonic acids (i.e., NMPA) (*2–4*) and (hydroxyalkylidene)phosphonic acids (HEDPA). These compounds are described in Chapter 7. These materials are widely used as scale preventers in recycled cooling water systems and in oil-well drilling. Although they are more expensive than the inorganic phosphates, when durability of performance is important, these organophosphorus compounds are preferred.

Literature Cited

1. Cowan, J. S.; Weintritt, D. J. *Water-Formed Scale Deposits*; Gulf: Houston, TX, 1976; pp 273–285.
2. Ralston, P. H. U.S. Patent 3 434 969, March 25, 1969 (to Calgon).
3. Ralston, P. H. *JPT, J. Pet. Technol.* **1969**, *21*, 1029.
4. Irani, R. R.; Lyons, J. W. U.S. Patent 3 346 487, 1967 (to Monsanto).

Chapter Thirteen

Electroless Plating and Electroplating

Electroless Plating

Many chrome-plated parts on new automobiles are not metal at all but chrome-plated plastic. Sodium hypophosphite is the compound usually responsible for this successful chrome plating. This compound is also used to plate corrosion-resistant nickel coatings onto industrial machinery. Because all plating using sodium hypophosphite is done without an electric current, it is called electroless plating.

In sodium hypophosphite, phosphorus has a valence state of +1 in contrast to the +3 in phosphite. The compound is made by the reaction of yellow phosphorus with a boiling water solution of a base such as lime [calcium hydroxide, $Ca(OH)_2$] or sodium hydroxide (NaOH). When lime is used, the products, in addition to water-soluble calcium hypophosphite, are insoluble calcium phosphite (removed by filtration) and the gaseous byproducts, phosphine (PH_3) and hydrogen. The mechanism for this reaction is quite complicated and not perfectly understood. The following reaction shows only the overall process:

$$\underset{\text{phosphorus}}{2P_4} + \underset{\text{calcium hydroxide}}{4Ca(OH)_2} + \underset{\text{water}}{6H_2O} \longrightarrow$$

$$\underset{\text{calcium hypophosphite}}{2Ca(H_2PO_2)_2} + \underset{\text{calcium phosphite precipitate}}{2CaHPO_3} + \underset{\text{hydrogen gas}}{2H_2} + \underset{\text{phosphine gas}}{2PH_3}$$

The water solution of calcium hypophosphite is converted to sodium hypophosphite by addition of sodium sulfate (Na_2SO_4). The byprod-

1002–0/87/0193$06.00/1

uct from this reaction—insoluble calcium sulfate—is removed by filtration. The solution is then concentrated to recover sodium hypophosphite crystals as the monohydrate ($NaH_2PO_2 \cdot H_2O$):

$$\underset{\text{calcium hypophosphite}}{Ca(H_2PO_2)_2} + \underset{\text{sodium sulfate}}{Na_2SO_4} \xrightarrow{H_2O} \underset{\text{sodium hypophosphite}}{2NaH_2PO_2 \cdot H_2O} + \underset{\text{calcium sulfate precipitate}}{CaSO_4}$$

Sodium hypophosphite electroless plating was first developed by General American Transportation Corp. (*1*) under the trade name Kanigen for plating odd-shaped metal parts or the interiors of railroad tank cars with corrosion-resistant nickel coatings (*2, 3*). The nickel coating produced also contains 8–10% phosphorus, apparently as phosphide. So that this coating can be applied, a bath consisting principally of sodium hypophosphite and nickel sulfate ($NiSO_4 \cdot 6H_2O$) is used; the bath also contains a complexing agent to prevent the precipitation of nickel phosphite, accelerators to enhance the rate of nickel deposition, and stabilizers to prevent decomposition of the bath.

Many proprietary formulations on the market contain different auxiliary agents. The exact operating conditions for each bath depend on the nature of these added reagents. However, the main reaction for nickel plating involves the reduction of the nickel ion, Ni^{2+}, from nickel sulfate to nickel metal or Ni^0:

$$H_2PO_2^- + H_2O \longrightarrow H_2PO_3^- + 2H^+ + 2e^-$$

$$Ni^{2+} + 2e^- \longrightarrow Ni^0$$

$$\underset{\text{hypophosphite ion}}{H_2PO_2^-} + H_2O + Ni^{2+} \xrightarrow[\text{catalyst}]{Ni} \underset{\text{phosphite ion}}{H_2PO_3^-} + 2H^+ + \underset{\text{nickel metal}}{Ni^0}$$

Theoretically, electroless nickel plating with hypophosphite involves the same reaction as does plating with an electric current. That is, electroless nickel plating requires the action of two electrons (e^-) to reduce the nickel ion of +2 valence (Ni^{2+}) to nickel metal (Ni^0) of 0 valence. In electroplating, the two electrons are supplied by electric current. In electroless plating using sodium hypophosphite, the two electrons come from the oxidation of the hypophosphite ($H_2PO_2^-$) to the phosphite ($H_2PO_3^-$). In other words, in the oxidation of the hypophosphite ion to the phosphite ion, the phosphorus atom is changed from a valence state of +1 to +3 with a loss of two electrons. This reaction is catalyzed by freshly deposited nickel and by such metals as palladium. Substrates to be plated such as iron or aluminum are placed in the plating solution; because these metals are more electropositive than nickel, some metallic nickel deposits, and simultaneously some iron or aluminum dissolves. This thin layer of deposited nickel then catalyzes the electroless plating of nickel on the iron or aluminum surface.

In many respects, electroless plating is more versatile than electroplating. Objects of complicated design, with many narrow recesses, crevices, and small holes cannot be completely coated in electroplating. The electric current jumps across narrow gaps, leaving the crevices and holes unplated and unprotected. In electroless plating, coverage occurs wherever the solution can reach. This results in complete, protective coverage.

Small objects such as valve and pump parts made of steel become quite corrosion-resistant when nickel-plated. The interiors of large tank cars are very difficult to coat by electroplating. Electroless nickel plating, on the other hand, works quite well; the entire tank car is used as the plating bath. The ultimate goal of electroless plating is the formation of a corrosion-resistant coating on metals such as iron and aluminum. In the chemical industry, corrosion is an important problem.

Chrome-Plated Parts

Most automobile trim that is not made of bright-dip aluminum is fabricated of chrome-plated zinc die-cast pieces. The chrome is electroplated over a metal surface that has previously been plated (also electrically) with copper and nickel. In fact, on some older automobiles—including very expensive models—some of the shiny chrome plates are marred by blisters. These blisters are the result of imperfect plating, which leads to corrosion of the zinc die-cast metal base.

Plastic automobile parts resist the ravages of weather and salts used on roads in winter. A chrome plate applied to a smooth plastic surface imparts the desired decorative effect and yet will not blister from corrosion. In addition, plastic parts are easy to mold and for this purpose are cheaper than zinc die-cast metals; one reason is that, plastic is much lighter than metal. One pound of plastic makes more parts than 1 lb of metal. As a result, many chrome-plated plastic parts have replaced the commonly used zinc die-cast metal parts.

Plating of Plastics

One night many years ago while waiting to be served in a restaurant, I was idly playing with the salt shaker. The cap looked like any chrome-plated metal top, except that it was quite lightweight. Then I realized that it was made of chrome-plated plastic. Although I knew that many of the shiny parts in new automobiles (knobs, handles, trim, and housings for the lights) are chrome-plated plastic, I did not realize that this technology had also invaded the household and restaurant markets.

Objects for electroplating must conduct electricity. Because plastics are nonconductors, they can be made conducting by coating with an electroless plating of a nickel underlayer using a sodium hypophosphite-nickel sulfate bath. Electroless nickel plating of plastics is the result of much research. For many years, only a specially developed plastic, known as ABS resin (acrylonitrile–butadiene–styrene), was suited to electroless plating. Now other plastics such as polypropylene have been reported to be platable. For successful plating, the coating must adhere tightly to the surface of an object. Because plastics are generally smooth, they present quite a challenge. Thus, a special type of plastic was developed whose surface could be treated to anchor the metal coating without seriously affecting the shine of the final chrome plate.

Here is how ABS is plated. The plastic piece is first etched with a sulfuric acid–chromic acid solution. The butadiene component in the plastic is attacked by this solution, and small pits are etched randomly on the surface. These pits are about 0.5 μm deep (1 μm = 0.001 mm) and have porous bases and sides. (To the naked eye, the plastic surface is still quite smooth.) The object is then placed in a bath containing stannous chloride ($SnCl_2$) and palladium chloride ($PdCl_2$). When these chemicals react, palladium chloride is reduced to palladium metal (Pd^0), and stannous chloride is oxidized to stannic chloride ($SnCl_4$). Palladium metal is deposited on the porous surface of the etched pits. This oxidation–reduction reaction is as follows:

$$\underset{\text{stannous chloride}}{SnCl_2} + \underset{\text{palladium chloride}}{PdCl_2} \longrightarrow \underset{\text{palladium metal}}{Pd^0} + \underset{\text{stannic chloride}}{SnCl_4}$$

Because the palladium metal deposits only on the inside of the random pits, the plastic surface is still not conductive enough for electroplating. The plastic piece is then immersed in an electroless nickel bath, and the palladium metal spots catalyze the reduction of the nickel ions to nickel metal by the hypophosphite ions.

Nickel metal deposits first on the palladium spots and then the deposited metal in turn catalyzes the deposition of more nickel. The nickel-covered spots then begin to spread, covering the entire plastic surface with a continuous nickel coating (containing a little phosphorus). This nickel-coated surface is very thin but conducts enough electricity so that the plastic piece can then be plated electrically in a standard electrobath.

In common practice, the chemical nickel coating is plated over with a layer of electrocopper, then a layer of electrobright nickel, and finally a layer of chrome. The electrocopper layer fills in the uneven dull surface of the electroless nickel–plastic surface to provide a smooth foundation for subsequent coatings. Also, the ductility of copper acts as a cushion—it absorbs the difference in thermal expansion between the plastic base and the metal coatings and thus prevents peeling. However, because copper is not corrosion-resistant, it is protected by a bright nickel coating that is corrosion-resistant. Why must the nickel coating be bright? Because the final chrome layer is simply decorative, it is so thin that it is transparent. The total thickness of the metal coatings is only about 0.0015 mm. Thus, in a chrome-plated plastic piece, the top layer is chrome, below that is bright nickel, then copper, and finally the electroless nickel plate, which is anchored to the plastic surface by small dots of palladium anchored to the porous surface of the little pits etched on the surface of the plastic. When properly done, this metal coating adheres tenaciously to the plastic surface. Research is in progress to plate glass, wood, cellophane, steel, and even plastic cloths with electroless nickel as an undercoating.

Electroplating

Electroplating entails using an electric current to reduce metal ions from solution on the surface of a cathode. As an example

$$\underset{\text{cupric ion}}{Cu^{2+}} + \underset{\text{electrons}}{2e^-} \longrightarrow \underset{\text{copper metal}}{Cu}$$

The electroplating bath has two electrodes, the cathode, where electrons enter and at which point the reduction occurs, and anode, where oxidation occurs. If the anode is made of metal capable of being oxidized under the plating bath conditions, the reaction can be the opposite of that at the cathode. Namely, in the case of copper

$$Cu^0 \longrightarrow Cu^{2+} + 2e^-$$

The piece to be plated is immersed in an electrically conductive electroplating bath containing a solution of the metal ions of the metal to be plated on the piece. The piece is connected to an outside source of electric current so that it becomes the cathode. During the plating process, the metal cations in solution, such as copper ions, migrate to the cathode and are reduced to metallic copper and plate out on the piece. Under ideal conditions, when the anode is made of the same metal that is being plated on the cathode, oxidation at the anode releases the equivalent amount of metal ions into the solution. Thus, the metal ion concentration in solution remains constant. In actual operation, side reactions occur, such as the reduction of protons or the oxidation of anions, which lead to variations in the composition of the baths. Therefore, make-up materials must be added periodically.

A typical example of commercial electroplating is the decorative plating of some automobile parts, such as the metal trim and the bumper. Here steel is plated, first with a layer of copper, then with a layer of nickel, and finally with a thin layer of chromium. Most electrolytic plating baths of commerical importance contain little, if any, phosphorus compounds. However, one bath, the copper pyrophosphate bath, has gained prominence in recent years. This bath is used particularly in the plating of printed circuit boards for the electronics industry (*4*). Such pyrophosphate baths allow high electrode efficiency, a good plating rate, thick copper deposits, and "throwing power". Throwing power is the ability to deposit copper in crevices or holes. Of these properties, good throwing power and high current efficiency are perhaps the most important reasons for the popularity of this plating bath.

A typical copper pyrophosphate electroplating bath contains 30 g/L of copper ion, 200 g/L of pyrophosphate ion, and small amounts of nitrate ion and ammonia. The pyrophosphate is added as potassium pyrophosphate; the copper is added as the soluble copper pyrophosphate. These reagents form a soluble complex:

$$\underset{\text{tetrapotassium pyrophosphate}}{3K_4P_2O_7} + \underset{\text{cupric ion}}{Cu_2P_2O_7} \longrightarrow \underset{\text{potassium copper pyrophosphate complex}}{2K_6Cu(P_2O_7)_2}$$

This complex, which is very soluble in water, can be precipitated as a hexahydrate [$K_6Cu(P_2O_7)_2 \cdot 6H_2O$]. The pyrophosphate bath is operated at 122–140 °F, a pH of 8.0–8.5, and a cathode current density of 10–70 A/ft^2. The ratio of $P_2O_7^{4-}$ to Cu^{2+} is held in the range of 7–8 to 1. The high concentration of pyrophosphate salts increases the conductivity of the bath and aids in the dissolution of the copper anode. Proprietary brighteners are added to promote smooth deposits. These compounds are often organic sulfur or organic nitrogen compounds. The ammonia is added to assist in effecting efficient dissolution of the copper anode. The nitrate ion inhibits a side reaction at the cathode leading to the formation of hydrogen when operating at high current densities.

Some hydrolysis of the pyrophosphate occurs during operation. When hydrolysis is held to certain limits, it is beneficial, because the orthophosphate formed aids in anode dissolution. However, at above about 70 g/L, the orthophosphate leads to dull and banded deposits. Normally, plating efficiency is close to 100%; however, the pH must be carefully controlled to obtain high efficiency and good bath performance. Too high a pH leads to reduced anode efficiency; too low a pH results in loss of throwing power. Also, high bath temperatures result in too rapid a hydrolysis of the pyrophosphate. Air agitation of the bath is beneficial, particularly if some brighteners are used. Copper deposits are fine-grained, semibright, and easy to polish.

Because this bath is more expensive than the cyanide or acid plating baths, it is used for special applications. Examples are situations where the plater wishes to take advantage of the throwing power, such as for complex designs that have crevices or holes. An important example is the "through-hole" plating of printed circuit boards. These circuit boards are plated and etched to leave conductive copper circuits that allow the interconnecting of the various devices on the circuit board, such as the capacitors, resistors, and integrated circuit chips. Holes are drilled in these circuit boards prior to plating with copper to allow electrical connections to be made from the backside through the holes. The throwing power of the copper pyrophosphate bath allows the walls of these holes to become conductive. Other examples of the use of this bath are where thick deposits, from 0.3 to 1 mil (1 mil = 0.001 in.), are desired or when the electroplater wishes to avoid the disposal problem that arises when using copper cyanide baths.

One defect of the conventional copper pyrophosphate baths is that the copper deposits on steel, aluminum, magnesium, or zinc have poor adhesion. This problem is solved by first using a copper "strike" coating. This coating is very thin, adherent copper that provides bonding to these metals. The strike is then followed by the conventional copper pyrophosphate bath. The copper "strike" bath is usually a dilute cyanide or pyrophosphate bath. The dilution (e.g., 7–10 g/L of copper and 25 g/L of pyrophosphate) results in lower electrical conductivity with consequent loss of current efficiency; efficiency is reduced to as low as 30%. However, because only thin films, on the order of 0.01 mil, are required, this low efficiency can be tolerated. If a copper cyanide strike is used, the part must be thoroughly rinsed before immersing in the copper pyrophosphate bath. Although copper pyrophosphate baths are relatively insensitive to contaminants, cyanide and lead do affect performance adversely. Trace amounts of cyanide can be removed with hydrogen peroxide; lead is removed by electrodeposition.

An analogous zinc pyrophosphate electrolytic plating bath exists that is based on an alkaline solution of $K_6Zn(P_2O_7)_2$ (*5*). Again, excess potassium pyrophosphate is used. The bath operates at 110–125 °F at cathode current densities of 2–10 A/ft^2. This plating bath can result in bright zinc finishes with a minimum of hydrogen embrittlement. The latter property is associated with high cathode efficiency of the bath—for example, little reduction of protons to hydrogen. This bath too has the advantage over zinc cyanide baths that the operator does not have the problem of decontaminating waste streams containing cyanide.

Literature Cited

1. Gutzeit, G. U.S. Patents 2 658 841 and 2 658 842, Nov. 10, 1953 (to General American Transport).
2. *Sodium Hypophosphite: Its Use in Electroless Plating*; Stauffer Chemical: Westport, CT, 1970.
3. Feldstein, N. *Met. Finish.* **1981,** *79* (*1A*), 504–508.
4. *Metals Handbook*; American Society for Metals: 1964; Vol. 2, pp 425–431.
5. Kosmos, J. *Met. Finish.* **1981,** *79* (*1A*), 348–350.

Chapter Fourteen

Miscellaneous Industrial Uses for Inorganic Phosphorus Compounds

Polyphosphoric Acid

Polyphosphoric acid is used by petroleum refiners to make cheaper and better gasoline and to make intermediates for detergents and plastics from materials that previously were considered to be refinery waste and were destroyed. Polyphosphoric acid is a thick, viscous liquid that consists of molecules of different chain lengths. It is produced by heating orthophosphoric acid to drive off its water until a P_4O_{10} content of 82–84% is reached. The resulting polyphosphoric acid is used extensively as a catalyst in the chemical and petroleum industries for the polymerization, alkylation, dehydration, condensation, and isomerization processes.

Because of its high viscosity, polyphosphoric acid is most easily handled when it adheres to and partially reacts with some porous material such as diatomaceous earth. The formation of a supported polyphosphoric acid catalyst, usually in the form of pellets or beads, results. The catalytic activity of the supported polyphosphoric acid depends largely on its hydrogen ions; this activity decreases when the acid is severely dehydrated to metaphosphoric acid (a long-chain phosphoric acid with only one hydrogen per phosphorus atom). Because dehydration begins at 425 °F (232 °C), most reactions using this catalyst are carried out below this temperature.

So that dehydration is suppressed, wherever conditions permit, 2–10% water is usually introduced as steam with the reactants. In the

1002–0/87/0201$06.00/1

petroleum industry, spent supported polyphosphoric catalyst is often regenerated by passing air over or through it to burn off the carbonaceous residue. The resulting metaphosphoric acid is then broken down to lower polyphosphoric acids by reaction with steam at about 500 °F (260 °C).

As discussed in Chapter 11, polymerization is the process of building large molecules by joining smaller ones into repeating chains. A polymerization that results in joining only a few of the smaller molecules to form a moderately sized larger molecule is called oligomerization, and the products are called oligomers. *Oligos* is a Greek word meaning few or little. In petroleum refining, large molecules in crude oil are cracked (or broken) into smaller molecules of different sizes. In the early days of gasoline production, much larger quantities of small molecules, such as propylene (C_3H_6), were produced than could be sold, and these leftovers were disposed of by burning. Later, it was found that oligomers of propylene could be prepared by using polyphosphoric acid as a catalyst. These oligomers are branched olefins having 9, 12, and higher numbers of carbons per molecule.

An example of such oligomerization is shown. These oligomers are large enough to be added to the gasoline fraction in concentrations up to several percent. Also, as will be mentioned later, their branched structure contributes to the octane rating of the gasoline.

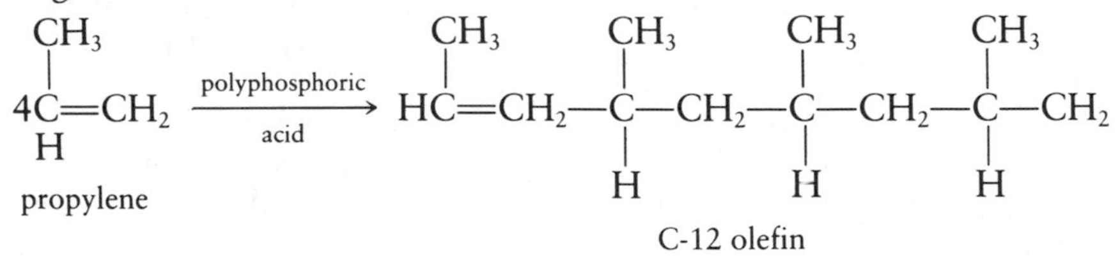

Although the activity of the polyphosphoric acid catalyst results only in the linkage between a few propylene units, today modern polymerization catalysts exist that are based on compounds of transition metals, such as titanium and chromium, that can join thousands of propylene units to form the solid plastic known as polypropylene.

An olefin, such as the one described above containing 12 carbons, can react with a benzene ring to form an alkylbenzene. This alkylation reaction is also catalyzed by polyphosphoric acid. Such an alkylbenzene, when sulfonated by sulfuric acid, has been used extensively as the surface-active agent in detergent formulations. These surface-active agents, which contain branched alkyl groups, are not as readily degraded by bacteria as products having linear alkyl groups. Therefore, because of the demand for surfactants that are more readily digested by microbes, these branched alkylbenzene sulfonates are being replaced by linear alkylbenzene sulfonates, compounds made from the alkylation of benzene with linear olefins. An alkylation reaction using a linear olefin is as follows:

$$CH_3CH_2CH_2CH_2CH_2CH{=}CHCH_2CH_2CH_2CH_2CH_3 + C_6H_6$$

linear C-12 olefin benzene

$$\xrightarrow[\text{acid}]{\text{polyphosphoric}} CH_3CH_2CH_2CH_2CH_2CH(C_6H_5)CH_2CH_2CH_2CH_2CH_2CH_3$$

linear alkylbenzene

The alkylation of benzene with propylene to form cumene (isopropylbenzene) has become an important industrial process. Cumene is an intermediate in a process that simultaneously produces phenol and acetone—both useful industrial chemicals. In this process, the cumene is oxidized to form cumene hydroperoxide. This intermediate is then cleaved by acid to form phenol and acetone. The oxidation is usually done with air, using several stages or steps, starting at 11 °C and ending at 90 °C. The reaction is halted when the cumene peroxide concentration is about 35%. Most of the unreacted cumene is distilled and recycled. The residue is then held at 60–100 °C and treated with a nonoxidizing inorganic acid to effect the cleavage. This process is outlined in Scheme 14.I. Polyphosphoric acid is used in the first step. Less expensive acids, such as sulfuric acid, are used for the second step. Overall yields are in the range of 95%. In the United States, more than 98% of the 1.6 billion pounds of phenol produced is made this way.

Molecular dehydration is the removal of the elements of water from one molecule or between two molecules. An example of the first reaction is the dehydration of cyclohexanol to form cyclohexene.

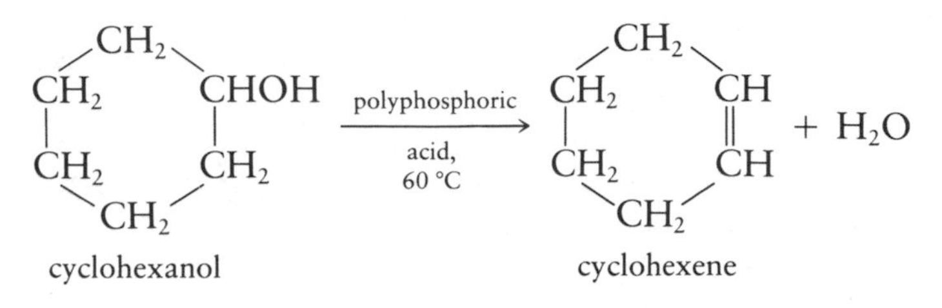

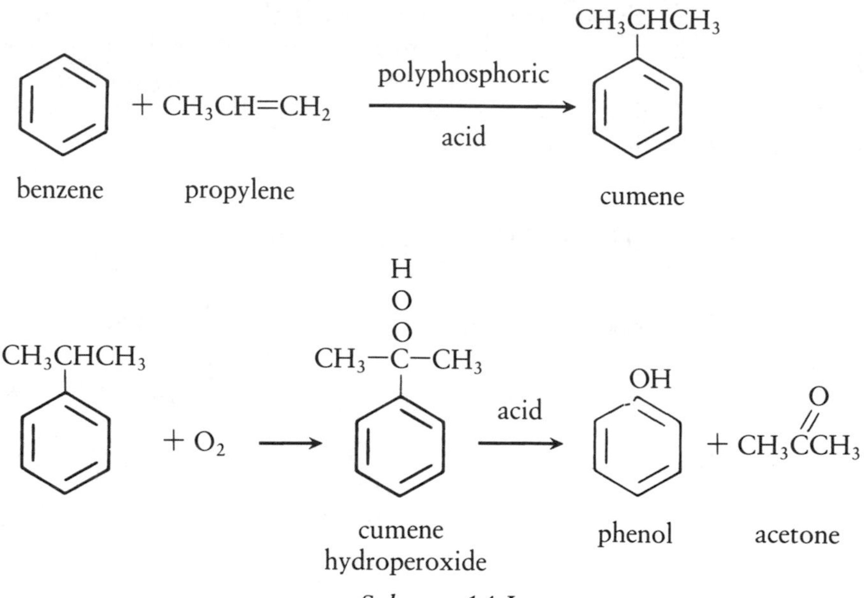

Scheme 14.I

Under proper conditions, polyphosphoric acid can also catalyze the reverse reaction, the hydration of an olefin. Polyphosphoric acid, through its ability to remove the elements of water, is an effective catalyst for condensation reactions. An example is the cyclization of phenylbutyric acid to tetralone:

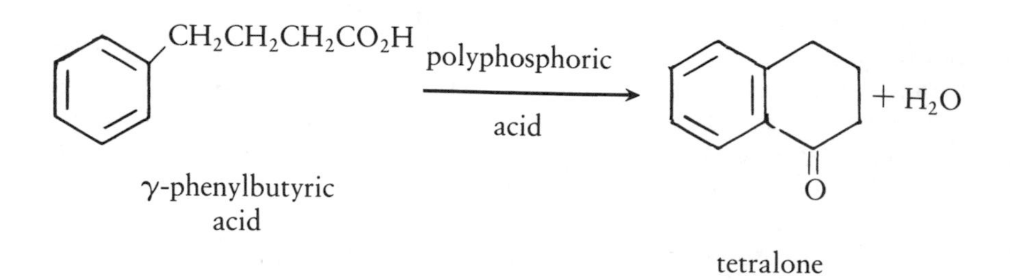

Isomerization converts straight-chain hydrocarbons into branched-chain compounds. This reaction is important in the petroleum industry. Of the hydrocarbons suitable for gasoline, those hydrocarbons with branched chains have higher octane ratings than those with straight chains. Thus, branched-chain hydrocarbons give less knock in an automobile engine.

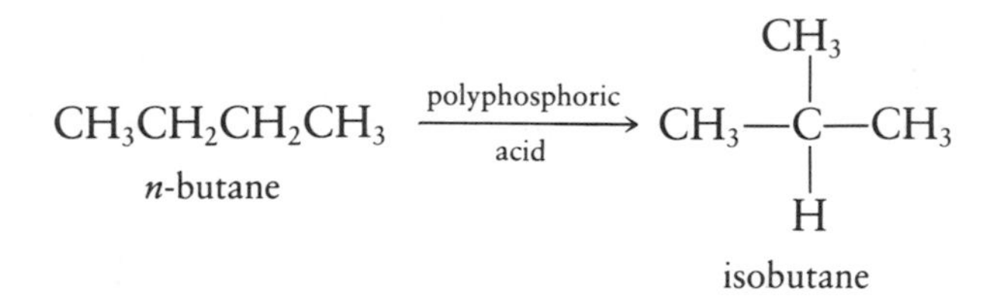

Although polyphosphoric acid has been used as a catalyst for many other organic reactions, few of these reactions have reached the commercial scale.

Monocalcium Phosphate Monohydrate

Effervescent tablets containing monocalcium phosphate monohydrate [$Ca(H_2PO_4)_2 \cdot H_2O$] and sodium bicarbonate give off bubbles when added to water. This bubbling is cause by the quick release of carbon dioxide gas that results from the action of acidic monocalcium phosphate on sodium biocarbonate. These tablets contain the same components found in baking powder. Children use this rapidly generated gas to power toy rockets. If they use baking powder and do not replace the container lid tightly, moisture seeps in. When this powder is used to bake a cake, the batter will rise to much less than its normal height while the baker's ire rises to a much greater height!

Phosphates in Ceramics, Cement, and Glass

Ceramic Binders: Aluminum Acid Phosphate and Other Acid Phosphate Salts. Phosphoric acid and a number of its salts are used as bonding agents for various ceramic powdered materials such as alumina to form refractory products. Such refractories can withstand the high temperatures encountered in the steel industry. Phosphoric acid can also bind silica or magnesia to form refractory molds for metal casting.

The use of phosphates as ceramic binders was described in the early 1930s (*2*), and the first systematic study of phosphate binding in refractories was reported in 1950 (*3*). This study showed that metal ions such aluminum, magnesium, iron, and beryllium greatly increase the bonding power of phosphoric acid, whereas calcium, barium, and thorium ions decreased its effectiveness. Kingery found the acidic phosphates were essential to form bonds. So that strong bonds can be obtained, the cation in the acid phosphate binder should be of small ionic radius and either weakly acidic or amphoteric. Such cations provide the disordered structures that result in ease of formation of the desired amorphous binders. Aluminum is such a cation. Aluminum acid phosphate provides a very strong bond. Not surprisingly, binders based on aluminum attained the greatest commercial importance. Aluminum acid phosphate binders can be made in situ by the addition of phosphoric acid to a blend of the filler and hydrated alumina, or separately, by prereacting the hydrated alumina with the phosphoric acid.

A number of commercial aluminum acid phosphate solutions are available. These solutions in phosphoric acid have mole ratios of aluminum to phosphorus ranging from 1:2.3 to 1:4.1. These commercial solutions are dense (1.4–1.6 g/cm^3), viscous (35–90 cP) liquids containing about 40% free water. They are acidic, having a pH of about 2.5, and they are miscible with water. The excess phosphoric acid is desirable for obtaining the highest binding strength. Also, a number of commercially available water-soluble, solid aluminum acid phosphates exist, including one that is an aluminum chlorophosphate.

Aluminum acid phosphate binders have been shown effective as binders for alumina, silica, kaolin, zirconia, mullite, and silicon carbide. One of the most important uses for aluminum acid phosphate binders, and for phosphoric acid itself, is the preparation of so-called unfired refractory brick. The advantage is these bricks can be cold set, set in place as liners for a furnace, and fired in place. Such bricks are used for furnace liners, electric arc furnace roofs, and incinerator liners. A second important advantage for the use of aluminum phosphate binders in high-temperature furnace liners is their high resistance to attack by slag (*4*).

Aluminum phosphate binders have been used in ramming, troweling, and pouring and in castable formulations. These formulations are used in such applications as patching furnace liners and preparing refractory molds (*5*). Aluminum acid phosphates have also been used for the preparation of protective coatings for metals, such as steel (*6*). At one time, an important use of aluminum acid phosphate was to bind asbestos for toaster boards and brake linings. Because asbestos has been found to be a hazardous substance, this use no longer exists.

Acid phosphate salts of magnesium, iron, and chromium have also been used commercially as binders, as have phosphate binders having mixtures of these metal salts.

Ceramic Binders: Glassy Sodium Polyphosphates. In the basic oxygen furnace for steel production, the furnace liner consists of bonded magnesium oxide, a basic material. Chromium oxide was the original binder and at one time the binder most commonly used. In 1966, bonding the magnesium oxide with long-chained glassy sodium polyphosphate and curing at 250 °F (121 °C) was found to lead to a brick of increased strength. Also, the greater the chain length of the glassy sodium polyphosphate, the greater the strength of the bonded brick. Probably, this glassy phosphate bonding is a result of degrading or cleavage of the glassy phosphates during the curing step and the degraded phosphate reacting with the magnesium oxide to form the bond. X-ray data confirm this belief; the data suggest that the bonding agent is an amorphous magnesium phosphate glass formed in situ. At 2200–2300 °F (1205–1260 °C), this glass begins to crystallize to form $Mg_3(PO_4)_2$, at which point the glass no longer acts as a binder.

A modification of this procedure entailed the incorporation of lime along with a glassy sodium polyphosphate as the binder. This reaction produced magnesium oxide bricks with improved strength up to 2850 °F (1565 °C) (*7*). Subsequent studies suggested, in this case, that the bonding agent was an $NaCaPO_4$ complex. This complex softens to a viscous liquid at 1500–1600 °C; however, before complete melting occurs, sodium oxide (Na_2O) is volatilized and the bond composition is shifted to $Ca_3(PO_4)_2$, whitlockite, a refractory material with a melting point of 1775 °C (*8, 9*).

Both of these bonding systems, the addition of glassy sodium polyphosphate and the additon of glassy sodium polyphosphate plus lime, improved the bond strength of chromium-bonded magnesium oxide refractory as well as straight magnesium oxide refractory.

Cements: Magnesium and Aluminum Phosphates. A number of phosphate-based cements have been described and used commercially. Perhaps the earliest such cements were the dental cements based on zinc oxide and phosphoric acid (*10–12*). These cements are relatively fast setting (3–10 min) cements in which the main component is dibasic zinc phosphate, $ZnHPO_4 \cdot 3H_2O$. A typical formulation entails mixing calcined zinc oxide and magnesium oxide. This blend is mixed with a phosphoric acid solution of monoaluminum phosphate. The powdered zinc and magnesium oxides in part react with the phosphoric acid to form salts, and the unreacted excess oxide acts as filler. Such cements are used for cavity linings and for temporary fillings.

Also, a number of industrial cementlike materials exist based on phosphorus compounds, usually using magnesium phosphate or aluminum phosphate (*13–15*). These materials are often used for anchoring or patching purposes. Recently, such cements have gained interest as patching materials for highways, because the present patching materials, asphalt and concrete, are slow setting and have poor long-term durability. The potential advantage of these phosphate-based cements is their ability to remain workable for sufficient time to apply and then to set quickly (within 8–10 min) and develop sufficient compressive strength that traffic can use the road within 2–3 h of the repair.

To fit within existing road-patching technology, these phosphate-based cements must be applied in a manner similar to that of Portland cement. The users want the product to be a dry powder to which they need only add water to convert it to a patching mortar. Also, because of the potential for injury, they do not want their employees to handle phosphoric acid or a solution of a metal acid phosphate. So that this problem, that is, to be able to provide the phosphate component in a form that will react with the metal oxides in the cementation step, can be solved, most of these formulations contain either mono- or diammonium phosphate as the source of phosphate ions. When water is added, the magnesium oxide or aluminum oxide reacts with the ammonium phosphate to form the metal phosphate and release ammonia.

Several such formulations are being tested on state and interstate highways, and so far, some formulations are performing quite well. Of course, many other competitors exist for this patching business, including formulations based on modified calcium sulfate, calcium aluminate, and Portland cement. Time will tell if the phosphate-based patching cements will provide sufficient benefit that they will become commercial on a very large scale.

Phosphate Glasses. Glasses are noncrystalline solids with extended three-dimensional random network structures. A number of inorganic oxides form glasses; these oxides include SiO_2, B_2O_3, GeO_2, Al_2O_3, P_4O_{10}, V_2O_5, Sb_2O_5, and ZrO_2.

Phosphate glasses are based on a random network structure of PO_4 tetrahedrons. We are all familiar with bone china. Few of us realize it actually is made from bone. It is made from calcined cattle bones, which consist mainly of calcium phosphate. When fired, the bones melt to form a glass with a high index of refraction. This high refractive index contributes to the characteristic translucency of the china. The opalescence is in part a result of the fluorine content in bones. The fluoride causes the crystallization of calcium fluorapatite in the glassy calcium phosphate matrix.

Many specialty types of phosphate glass exist. One type is heat-absorbing glass. Such glasses are made with a phosphate glass containing a few percent of iron oxide. The iron oxide in phosphate glasses absorbs more sharply the ultraviolet and infrared (heat) band of the spectrum than silicate glasses, yet iron-containing phosphate glasses are nearly transparent to visible light. Therefore, almost clear heat-absorbing glasses can be made.

Silicate glasses are readily etched when exposed to hydrogen fluoride (HF). Phosphate glasses are more resistant to fluoride attack.

Therefore, some commercial optical glasses containing phosphate are produced to take advantage of this property.

We have all seen the chromatic aberration caused by light passing through lenses. This deviation is caused by the dispersion of the components of the light by the lens (i.e., the colors are separated from each other). Such lenses in cameras result in less sharp pictures. Some glasses that contain both fluoride and phosphate have very low optical dispersion (i.e., these glasses show little change in refractive index with the wavelength of light). Therefore, these compounds are of great value for producing lenses of minumum chromatic aberration as desired for camera lenses. An example of such a composition is $LiF \cdot Al(PO_3)_3$ (*16*).

A serious unsolved problem of the nuclear energy industry is the disposal of radioactive wastes. Phosphate and silicophosphate glasses have been considered as media suitable for embedding high-level radioactive waste. Recently, a lead–iron–phosphate was reported as even more effective in encapsulating such radioactive wastes. This glass is structurally stable and extremely resistant to ground water leaching (*17*).

A number of other uses for phosphates in glasses are described in the literature. These applications include (1) colored glasses, based on the effect of the phosphate on transition metal ions, such as cobalt, nickel, and iron; (2) semiconducting glasses, based on vanadium phosphate; and (3) laser glasses, in which rare earths are embedded in a phosphate matrix. An example is $La_2O_3 \cdot BaO \cdot P_4O_{10} \cdot Nd_2O_3$, wherein the efficiency is related to the interaction of the phosphate with the neodymium (Nd).

Aluminum Phosphate Zeolites

Zeolite is the name given to a large family of crystalline hydrous aluminum silicates having open or caged structures. The name, coined in the 18th century, is derived from the Greek work *zein*, meaning to boil; boil refers to the location of the first discovered members, in the cavities of lavas where the stones appear to boil. Today not only have many naturally occurring members of this family been discovered but also many synthetic forms have been produced, forms not yet seen in nature. Important features of zeolites are their uniform open structures, with pores of specific dimensions. These pores permit them to act as hosts only for gases and organic molecules of specific sizes and shapes. Their ability to seperate molecules by shape and size led to the zeolites being called molecular sieves.

Thus, zeolites, with their unique properties, have found many uses, which include water removers from liquid and gas streams, separators of organic molecules of specific shapes or sizes, catalysts for organic reactions, catalyst supports, and, as a result of their ion-exchange capabilities, agents in water softeners and replacements for phosphates as detergent builders (*see* Chapter 7). Current U.S. sales of zeolites are more than 300 million pounds per year.

Recently, a new family of zeolite-like molecular sieves was discovered (*18*). This family is based on aluminum phosphate rather than on aluminum silicate. These molecular sieves are prepared by the hydrothermal (i.e., heating in an aqueous solution, often at temperatures requiring the system to be put under pressure) reaction of aluminum and phosphate ions in the presence of amines or quaternary ammonium ions and at temperatures ranging from 100 to 250 °C. The structure of the resulting zeolite is controlled by the choice of amine or quaternary ammonion ion, because these compounds act as templates by establishing the shape and dimensions of the interior cage or void space. After these zeolites crystallize, they are calcined at 600–1000 °C to destroy and remove the amines and to form the characteristic void space. At least 20 different three-dimensional, microporous framework structures of the aluminum phosphate type of zeolites have been prepared, 14 of which have the open structure and six of which are two-dimensional. The structure of one such member was labeled by the discoverers as $AlPO_4$–5.

The aluminum phosphate based zeolites have several important features that differ from the aluminum silicate based zeolites. The most prominent difference is that aluminum phosphate based zeolites are electrically neutral. The charges of positive trivalent aluminum ion and the negative trivalent phosphate ion cancel each other. These materials have no accompanying cation or anion; therefore they have no meaningful ion-exchange capacity. Pore diameters range from 3 to 10 Å; the actual value depends on the template used. The aluminum phosphate based zeolites are strongly hydrophilic and mildly acidic. Applications of immediate interest are as desiccants for drying gas and liquid streams, separators of gases and organic molecules by size and shape, catalysts, and catalyst supports.

Even more recently modified aluminum phosphate molecular sieves have been developed that provide ion-exchange properties and improve catalytic properties (*19*). These sieves are silicoaluminophosphate molecular sieves. They are prepared by the same hydrothermal process used for the aluminum phosphate based zeolites, including the use of an amine template, except that silicate is included in the reaction medium. At least 13 three-dimensional microporous framework structures of this class are known. Pore diameters range from 3 to 8 Å. Some of these silicoaluminophosphates have exchangeable countercations and Brønsted acid sites. Such silicoaluminophosphates have catalytic properties (i.e., for cracking of long-chained hydrocarbons) that are superior to the aluminum phosphate type zeolites.

Inorganic Phosphate Fibers

Kurrol's salt is a name given to a series of high molecular weight alkali metal metaphosphates that have fibrous structures. These metaphosphates can be made by heating the mono alkali metal salt of phosphoric acid under certain conditions to drive off water. They condense to form very long chained fibrous structures. The potassium form of Kurrol's salt, referred to earlier, is made by heating monopotassium phosphate above 150 °C. The sodium form requires heating monosodium phosphate to 660 °C to form a melt, seeding the melt at 580 °C, and tempering the supercooled melt at about 550 °C for 1–1.5 h. Both forms are only slightly soluble in water but slowly swell and eventually dissolve to give very viscous solutions. Their dissolution in water is greatly accelerated when soluble salts of other cations, cations that do not precipitate phosphate, are present. Incidentally, these long-chained alkali metal metaphosphates differ from the insoluble sodium metaphosphate (IMP), described previously as dentifrice polishing agents (Chapter 5) and cheese emulsifiers (Chapter 4), because IMP (known as Maddrel salt) is neither soluble in water nor fibrous in structure. IMP is made by carefully heating sodium monophosphate to about 400 °C; no melt is formed. Another important requirement in making IMP is that the reaction mixture should not be disturbed or only mildly mixed during the heating period.

The Kurrol-type salts had been of scientific interest only. Until recently, no significant commercial use had been identified. However, in 1981, scientists at Monsanto reported preparing two new types of fibrous inorganic Kurrol-type phosphates. These compounds are calcium polyphosphate fibers and sodium or lithium calcium polyphosphate fibers (*20–22*).

These fibers generally range in length from 5 μm to 1 mm, and some as long as 3 cm have been isolated. As an example of the preparation, the process for making fibrous calcium polyphosphate entails heating calcium carbonate and monosodium phosphate with phosphoric acid to 1050 °C. The resulting melt is cooled to 800 °C and seeded with fibers of the potassium form of Kurrol's salt. After crystallization, the mixture is cooled and extracted with boiling water. The insoluble fibers are isolated in yields as high as 50%. The fibers are highly crystalline with high tensile strength and high modulus of elasticity. The calcium form melts at 970 °C. Perhaps a unique property of these fibers is that they can be enzymatically degraded. Therefore, they are being considered as an alternate to asbestos for insulation.

Asbestos is a term used to describe a variety of naturally occurring forms of hydrated silicates that, upon mechanical processing, separate into mineral fibers. These fibers are unique minerals and until recently served many useful purposes as industrial and residential thermal and electrical insulation products. However, certain diseases have been associated with asbestos, particularly with long-term (20–30 years) inhalation of asbestos dust. These diseases include asbestosis, a nonmalignant fibrotic lung condition; bronchogenic (or lung) carcinoma; and mesothelioma, a cancer of the lining of the chest or abdominal cavities. Reduction of asbestos dust exposure is at present the only known method of preventing such asbestos-related diseases. As a result of these diseases, society has been forced to seek alternates to asbestos. The prevailing view is that the diseases associated with asbestos are a result of fiber size, morphology, and stability rather than a result of any specific chemical properties of the fiber. Therefore, any refractory material of similar morphology would probably instigate the same illnesses. Thus, the concept of a biodegradable fibrous material, such as these insoluble phosphate fibers, offers much hope toward identifying a truly effective, yet safe, replacement for asbestos. However, such phosphate fibrous material is expected to be much more expensive than asbestos.

The final approval of these phosphate fibers remains to be given. The biodegradation data are being reviewed by key medical specialists, both in industry and in government. If the biodegradation data are found sufficient, no doubt these phosphate fibers will be subject to long-term animal tests before they receive government approval for use for electrical and thermal insulation or for flame-retardant garments. However, this is another example of scientists addressing a serious need with meaningful solutions.

Phosphate-Based Catalysts

As already discussed, polyphosphoric acid and the silicoaluminophosphate molecular sieves have catalytic activity. Although polyphosphoric acid and phosphoric acid are the most important phosphorus-based catalysts used in the petroleum industry, a number of other inorganic phosphate materials that provide commercially important performance as catalysts exist. Examples are aluminum phosphates, bismuth phosphate, boron phosphate, tricalcium phosphate, vanadium phosphate and zirconium phosphate.

The catalytic properties of these phosphates depend on the method of preparation and the stoichiometry used in their preparation. Many of the more active phosphate catalysts are nonstoichiometric; also, the surface area and pore volumes available for catalytic activity are greatly affected by the method of preparation.

Studies indicate that the active sites on many of the phosphate catalysts are acidic, and many of the reactions catalyzed are those generally catalyzed by acids. Examples are the dehydration of alcohols, the isomerization of olefins, and the alkylation of phenols (*23*).

Dipotassium Phosphate and Tetrapotassium Pyrophosphate

Two moles of potassium hydroxide reacts with 1 mol of phosphoric acid to give dipotassium phosphate. The anhydrous form of this compound is very soluble in water (at 30 °C, about 70 g will dissolve per 100 g of water). So that the compound is obtained in crystalline form, water is evaporated under vacuum at around 60 °C.

If the water were boiled off at atmospheric pressure rather than under vacuum—for example, if the mixture were heated to too high a temperature—2 mol of dipotassium phosphate would combine, with the elimination of 1 mol of water, to form tetrapotassium pyrophosphate as a byproduct:

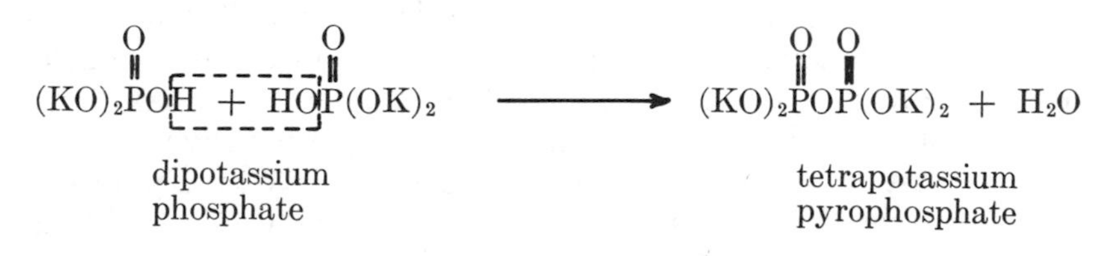

Dipotassium phosphate is used as a buffer in car radiators (a buffer is a compound used to maintain the pH of a solution within a desired range). Antifreeze solutions such as water-glycol mixtures become acidic (lower pH) with time through oxidative reactions. This acid mixture corrodes the radiator metal and leaks develop. Dipotassium phosphate buffer maintains a pH of around 9, which prevents corrosion.

The commercial process for tetrapotassium pyrophosphate (TKPP) entails deliberately heating dipotassium phosphate to above 300 °C. Yields are essentially quantitative. TKPP, like dipotassium phosphate, has a high solubility in water: 61 g/100 g of solution. Because TKPP has detergent builder properties similar to tetrasodium pyrophosphate (TSPP), TKPP is used as a builder in liquid heavy-duty detergents. TKPP is much more expensive than TSPP, and therefore TSPP is preferred in cases where high water solubility is not required.

Tripotassium Phosphate

Tripotassium phosphate is made by neutralizing the three acidic hydrogens in 1 mol of phosphoric acid with 3 mol of potassium hydroxide. The solution can be concentrated and then cooled to yield crystals. However, a much easier method is to heat the mixture to 300–400 °C and drive off the water quickly. The compound is not affected by high temperature.

Tripotassium phosphate is used to manufacture styrene–butadiene synthetic rubber by polymerization. Tripotassium phosphate acts as an electrolyte to regulate the polymerization rate and controls the stability of the resulting synthetic latex.

Tripotassium phosphate is also used to remove hydrogen sulfide gas (H_2S) from industrial vapors. Because H_2S is a weak acid, it is absorbed in the basic water solution of tripotassium phosphate. When the solution is heated later, H_2S gas is driven off, and the tripotassium phosphate can be used again. This absorption process is called scrubbing; it recovers H_2S as a valuable byproduct and prevents its rotten-egg odor from polluting the atmosphere.

Potassium Polymetaphosphate

Potassium polymetaphosphate can be made into a ball that bounces like rubber—an unusual property for an inorganic compound. Because its molecules are packed tightly together in long chains, potassium polymetaphosphate is practically insoluble in water. When a sodium salt is added, the sodium replaces some of the potassium and loosens the chain. If enough sodium is added, the insoluble salt can be made into a viscous gel that still consists of long-chained molecules. At this point, if a bivalent ion such as calcium ion is added, each calcium ion can replace two sodium or two potassium ions from two separate chains. The net result is that the chains are cross-linked by calcium ions.

A compound cross-linked in this way is no longer water-soluble. The many small holes in between the network of long chains can hold some water. The finished product is a rubbery gum that can be stretched into a film or squeezed into a round, bouncing ball. Unfortunately, the water cannot be held permanently. The water evaporates with time, and when the material is dry, it becomes a solid powder.

Diammonium Phosphate

The propensity of diammonium phosphate to decompose slowly to form ammonia and the acidic monoammonium phosphate makes it useful as a dye-leveling agent in woolen dye baths—that is, diammonium phosphate promotes even dyeing. Colloidal wool dyes that have a high affinity for wool yield color that sturdily resists washing and perspiration. However, the dye tends to precipitate too rapidly on the surface of the cloth; this property gives an uneven color distribution and causes streaks. Rapid deposition of the dye also prevents deep penetration of the color into the fabric. If the bath is kept alkaline, the colloidal dye remains dispersed in the bath and will not precipitate on the fabric.

Diammonium phosphate promotes and maintains an alkaline bath. The wool swells and allows the dye to penetrate. When the dye solution is boiled, the diammonium phosphates gradually decompose, evolving ammonia, and the bath gradually becomes acidic. A typical bath may contain from 0.1% to 0.5% diammonium phosphate; the pH may start at 7.85 and, after 2.5 h of heating, drop to 5.78. As the bath becomes more acidic, the dye starts to precipitate, but because the change in pH is gradual, dye precipitation is also gradual, and even dyeing with good penetration is obtained.

As indicated in Chapter 6, the ammonium phosphates are of great value as fertilizers, and as mentioned in Chapter 10, they are effective as nonpermanent flame retardants.

Tetrasodium Pyrophosphate

Tetrasodium pyrophosphate (TSPP) can sequester heavy metal ions such as iron and vanadium, an important factor in bleaching textiles. Many cotton textiles are bleached by peroxy compounds such as hydrogen peroxide, and peroxides are easily decomposed in the presence of even a trace of heavy metal ions. TSPP and sodium silicate are added to sequester these metal ions, preventing them from catalytically decomposing the peroxides.

Sodium Tripolyphosphate as Dispersing Agent

The tripolyphosphate anion in detergents, which has many negative charges, is an effective dispersing agent not only for dirt but also for many industrially important materials. For example, in drilling an oil well, liquid lubricant must be used to cool the drill bit as it bites through rock. In addition, a liquid is needed to suspend and carry the cuttings to the surface. For most oil-well operations, a water suspension of clay is used with a high-density compound such as barium sulfate. Sodium tripolyphosphate is often used as the dispersion agent to keep the drilling mud in proper suspension, although some of this usage is being replaced by the organophosphorus sequestering agents, such as nitrilotris(methylenephosphonic acid), NMPA (*see* Chapter 12).

Another example is found in the manufacture of cement and brick. In "wet-process" cement production, the raw materials are ground together in the form of a water slurry, and then the slurry is calcined to remove water. Advantageously, as little water as possible should be used to obtain the desired viscosity in the reactant mixture (as usual, evaporation of less water is less costly). Addition of a fraction of 1% sodium tripolyphosphate does this process very well. Thus, more solid per liquid volume can be handled, and the production rate of a cement plant can be increased as much as 10%.

Phosphine

Some uses for phosphine depend on its toxic properties—for example, as a fumigant against insects and rodents in stored products. For this purpose, phosphine is generated on the spot from aluminum phosphide (AlP). The commercial product is a tablet of aluminum phosphide and ammonium carbamate impregnated with paraffin. Upon reaction with water or atmospheric moisture, phosphine is released. The ammonium carbamate decomposes to carbon dioxide and ammonia, and these gases prevent the phosphine generated from igniting spontaneously.

Zinc phosphide (Zn_3P_2) has long been used as a rodent poison and in grain baits against field mice and gophers. The acute oral toxicity of zinc phosphide to rats is LD_{50} = 45.7 mg/kg of body weight. LD_{50} represents the minimum lethal dose required to kill 50% of the rats in a given test.

Phosphine is also used as an intermediate for the preparation of several flame retardants, as discussed in Chapter 10.

Phosphate Parting Agents: Tricalcium Phosphate

A major application of tricalcium phosphate is for flow conditioning of solids. An example, as shown in Chapter 4, is flowable table salt. The materials to be flow-conditioned are usually coarser than the flow conditioner. The conditioner, as a finely divided powder, adheres to the surfaces and alters the adhesion (i.e., surface activity) between adjacent particles of the originally poor flowing material. This action results in reduced friction between particles. In the case of materials that cake, particularly where caking is induced by moisture, the flow conditioner can physically separate the particles by coating the surfaces.

Tricalcium phosphate is a finely divided, insoluble solid that is unusually effective as a flow improver. It can readily coat particles of materials such as salt or sugar and prevent caking or bridging. As expected from the mechanism, the finer the tricalcium phosphate particles are, the more effective it is. As an example, 1% of tricalcium phosphate having an average diameter of 2 μm can prevent the salt in the salt cellar from caking.

Another application of this coating effect is in the suspension polymerization of styrene. Here polystyrene beads are prepared. During the preparation, beads or droplets of styrene are formed and suspended in an aqueous medium; the polymerization occurs within the preformed beads. Importantly, the beads must not coalesce, especially during the stages of partial polymerization, where the beads develop sticky surfaces. Fine particle size tricalcium phosphate has been found particularly useful for preventing such coalescence. During the polymerization, the tricalcium phosphate particles coat the polymerizing beads and thus prevent agglomeration. When the polymerization is complete and the polystyrene beads are no longer capable of coalescing, the residual tricalcium phosphate is dissolved with a mineral acid, such as hydrochloric acid. This application is one of the major uses for tricalcium phosphate.

One other closely related application of tricalcium phosphate is as an insecticide. Evidence exists that fine particles of tricalcium phosphate will protect grain or flour from insects. The mechanism is believed to be a result of the tricalcium phosphate particle either abrading or damaging the insect's exoskeleton, that is, embedding itself in joints or crevices of the insect and, through abrasion or irration, severly injuring the insect. As a result of the effect, large quantities of tricalcium phosphate have been sold to protect grain during shipment. This action was done under the U.S. Foods for Peace program.

Literature Cited

1. van Wazer, J. R. *Phosphorus and Its Compounds*; Interscience: New York, 1961; Vol. 2, pp 1947–1951.
2. Morgan, J. D.; Bjorkstedt, W. G.; Lowe, R. E. U.S. Patent 1 811 242, June 23, 1931.
3. Kingery, W. D. *J. Am. Chem. Soc.* **1950,** *33*, 239, 242.
4. Morris, J. H.; Perkins, P. G.; Rose, A. E. A.; Smith, W. E. *Chem. Soc. Rev.* **1977,** *6* (2), 173–194.
5. Gitzen, W. H.; Hart, L. B.; Mac Zura, G. *Am. Ceram. Soc. Bull.* **1966,** *35* (6), 217.
6. Ott, E.; Allen, E. R. WADD Technical Report 61-137, 1961; Aeronautical Systems Division, Air Force Systems Command, U.S. Air Force, Wright-Pattersen Air Force Base, Ohio.

7. Lyon, J. E.; Fox, T. V.; Lyons, J. W. *Am. Ceram. Soc. Bull.* **1966,** *45* (2), 1078.
8. Venables, C. L.; Treffner, W. S. *Am. Ceram. Soc. Bull.* **1970,** *49*, 660.
9. Foessel, A. H.; Treffner, W. S. *Am. Ceram. Soc. Bull.* **1970,** *49*, 652.
10. *Guide to Dental Materials*, 2nd ed.; American Dental Association: Chicago, IL, 1964; p 28.
11. Lu, K. W.; Porter, D. R. *J. Dent. Res.* **1962,** *41*, 1277.
12. Watts, C. H., Jr. U.S. Patent 2 675 322, April 13, 1954 (to Pre-Vest).
13. Horvitz, H. J.; Gray, A. P. U.S. Patent 4 152 167, May 1, 1979 (to Set Products).
14. Sherif, F. G.; Michaels, E. S. U.S. Patent 4 487 632, Dec. 11, 1984 (to Stauffer Chemical).
15. Sherif, F. G.: Michaels, E. S. U.S. Patent 4 505 752, March 19, 1985 (to Stauffer Chemical).
16. Westman, A. E. R. *Top. Phosphorus Chem.* **1985,** *9*, 326.
17. Sales, B. C.; Abraham, M. M.; Bates, J. B.; Boatner, L. A. *J. Non-Cryst. Solids* **1985,** *71*, 103.
18. Wilson, S. T.; Lok, B. M.; Messina, C. A.; Cannon, T. R.; Flanigen, E. M. *J. Am. Chem. Soc.* **1982,** *104*, 1146.
19. Lok, B. M.; Messina, C. A.; Patton, R. L.; Gajek, R. T.; Cannon, T. R.; Flanigen, E. M. *J. Am. Chem. Soc.* **1984,** *106*, 6092.
20. Griffith, E. J. In *Phosphorus Chemistry*; Quin, L. D.; Verkade, J. G., Eds.; ACS Symposium Series 171; American Chemical Society: Washington, DC, 1981; pp 361–365.
21. Griffith, E. J. U.S. Patent 4 360 625, Aug. 24, 1982 (to Monsanto).
22. Griffith, E. J. U.S. Patent 4 346 028, Aug. 24, 1982 (to Monsanto).
23. Moffat, J. B. *Top. Phosphorus Chem.* **1980,** *10*, 285–340.
24. *Chem. Eng. News* **1983,** *61* (*June 20*), 36.

Chapter Fifteen

Miscellaneous Industrial Uses for Organic Phosphorus Compounds

Phosphorus Chloride Derivatives: Miscellaneous Uses

Although we seldom encounter phosphorus chlorides in our everyday life, they are important chemical agents for making products we need—for example, mild synthetic bar soaps, clear nonyellowing plastic sheets, compounds for silver polishes and insecticides, and compounds for extracting uranium for the atomic bomb. The common phosphorus chlorides are

PCl_3	PCl_5	$POCl_3$	$PSCl_3$
phosphorus trichloride	phosphorus pentachloride	phosphorus oxychloride	phosphorus thiochloride

Phosphorus Pentachloride. PCl_5 is a free-flowing solid made by chlorinating elemental phosphorus. Liquid phosphorus trichloride (PCl_3) is first formed as the intermediate, and as additional chlorine is added the trichloride is immediately converted into the solid pentachloride.

$$\underset{\text{phosphorus}}{P_4} + \underset{\text{chlorine}}{6Cl_2} \longrightarrow \underset{\text{phosphorus trichloride}}{4PCl_3}$$

$$4PCl_3 + 4Cl_2 \longrightarrow \underset{\text{phosphorus pentachloride}}{4PCl_5}$$

1002–0/87/0219$10.00/1

Phosphorus pentachloride is a very reactive material. When exposed to moist air, it reacts spontaneously with the water to form phosphorus oxychloride ($POCl_3$) and releases hydrogen chloride (HCl). Phosphorus pentachloride reacts with most organic compounds having a free hydroxyl group to convert that group to a chloride group with the release of hydrogen chloride and the formation of $POCl_3$.

$$ROH + PCl_5 \longrightarrow RCl + POCl_3$$

One of the industrially interesting reactions of PCl_5 is the reaction with ammonium chloride to form phosphazenes. In older literature, phosphazenes are referred to as phosphonitrilic chlorides.

$$PCl_5 + NH_4Cl \longrightarrow \underset{\text{phosphazenes}}{(PNCl_2)_{3,4,\ldots n}} + HCl$$

This reaction, discovered in 1897, is the basis of the "inorganic rubbers" mentioned in Chapter 11. Aside from its use in this reaction, PCl_5 is not used in any commercially important applications. Because of its high reactivity, PCl_5 is difficult to produce and isolate. Therefore, for those few industrial uses requiring PCl_5, the compound is usually made in situ by dissolving PCl_3 in a solvent such as carbon tetrachloride and adding chlorine to convert the dissolved PCl_3 to a suspension of PCl_5. Usually, the PCl_5 can be used as such without need for separation from the suspending fluid.

Of the four phosphorus chlorides listed previously, only phosphorus trichloride and phosphorus oxychloride are industrially important.

Phosphorus Trichloride Derivatives. For the manufacture of phosphorus trichloride (PCl_3) industrially, elemental phosphorus is suspended in previously prepared phosphorus trichloride. Chlorine gas is bubbled into this suspension to convert the elemental phosphorus into more phosphorus trichloride. This product is then purified by distillation.

Phosphorus trichloride is highly reactive. When mixed with water, it erupts, producing hydrochloric acid and phosphorous acid:

$$\underset{\text{phosphorus trichloride}}{PCl_3} + 3H_2O \longrightarrow \underset{\text{phosphorous acid}}{H_3PO_3} + \underset{\text{hydrochloric acid}}{3HCl}$$

The most important use for phosphorus trichloride is as an intermediate or reagent in the manufacture of many industrial chemicals, ranging from insecticides to synthetic surfactants to ingredients for silver polish. For example, PCl_3 is heated with sulfur in the presence of catalysts such as aluminum chloride or activated charcoal to give phosphorus thiochloride ($PSCl_3$). Phosphorus trichloride is also used as an intermediate in synthesizing phosphorus oxychloride.

As an intermediate, phosphorus trichloride is almost unparalleled in its wide-ranging applications. It is used to make acyl chlorides from which we get synthetic surfactants, bar soaps, and detergents; it is used to make alkyl chlorides, the intermediate for mercaptans, which are components in silver polishes, synthetic rubber, and vinyl plastics; PCl_3 is also used to make dialkyl phosphonates, which are themselves used to make insecticides, wetting agents, and metal extractants. Miscellaneous reactions use phosphorus trichloride as an intermediate in nylon manufacture and resin stabilizers. These uses and reactions are discussed in detail in the following paragraphs.

PHOSPHOROUS ACID. Although the official name of this compound is phosphonic acid, it is known commercially by its older, and now trivial, name, phosphorous acid. Phosphorous acid (the reaction product of PCl_3 and water shown in the previous reaction), when converted to its lead salt (basic lead phosphite, $2PbO \cdot PbHPO_3$), is used in formulations to stabilize polyvinyl chloride (PVC) plastics—that is, to prevent PVC plastics from discoloring.

Another commercially important class of compounds derived from phosphorous acid is the organophosphorus sequestering agents, discussed in Chapter 7. An example is the preparation of nitrilotris(methylenephosphonic acid), NMPA. NMPA is prepared by the reaction of phosphorous acid with formaldehyde and ammonium chloride in the presence of hydrochloric acid:

$$\underset{\text{phosphorous acid}}{3H_3PO_3} + \underset{\text{ammonium chloride}}{NH_4Cl} + \underset{\text{formaldehyde}}{3CH_2O} \xrightarrow[H_2O]{HCl} \underset{\text{NMPA}}{N[CH_2\overset{\overset{\displaystyle O}{\|}}{P}(OH)_2]_3} + 3H_2O + HCl$$

NMPA behaves similarly to sodium pyrophosphate, sodium tripolyphosphate, and glassy sodium polyphosphate in that it is an efficient chelating agent for metal ions such as iron, calcium, and magnesium. Thus, NMPA is useful in detergent compositions for softening water, for cooling water-tower systems, and for treating boiler water to prevent scale formation by the threshold effect (effective at 1–10-ppm concentration). NMPA is also an effective corrosion inhibitor for metals when used in a water solution along with zinc chloride. Despite its higher cost, NMPA is used in place of sodium polyphosphates in many of the applications just named. It has the advantage of being stable under pH and temperature conditions where polyphosphates would lose their effectiveness by decomposing to orthophosphates.

Phosphorous acid can also be used as an intermediate for the herbicide glyphosate, discussed in Chapter 21.

As you will note in the next few pages, phosphorous acid is obtained as a byproduct from many of the processes that use PCl_3 to convert an organically bound OH group to an organically bound chloride. At one time, this byproduct phosphorous acid was made in quantities sufficient to meet all the industrial requirements for the acid. Companies would purchase the crude byproduct phosphorous acid, purify it, and then use it in their own processes. However, the commercial demand for phosphorous acid has now grown to an extent that it must be manufactured directly from PCl_3.

ACYL AND ALKYL CHLORIDES AND THEIR DERIVATIVES FOR SURFACTANTS AND MERCAPTANS. If you have ever worked with phosphorus trichloride and observed firsthand its violently reactive character and corrosive properties, it is difficult to believe that PCl_3 could be used to make any product that is mild to the skin. However, phosphorus trichloride is an effective chlorinating agent, which makes it an ideal intermediate in preparing surfactants for mild, bar-type synthetic detergents. PCl_3 converts fatty acids into fatty acid chlorides. For example, PCl_3 is used to convert lauric acid (derived from coconut oil) into lauroyl chloride:

$$\underset{\text{phosphorus trichloride}}{PCl_3} + \underset{\text{lauric acid}}{3CH_3(CH_2)_{10}COOH} \longrightarrow \underset{\text{lauroyl chloride}}{3CH_3(CH_2)_{10}COCl} + \underset{\text{phosphorous acid}}{H_3PO_3}$$

These fatty acid chlorides are intermediates for synthetic surfactants. When lauroyl chloride reacts with isethionic acid ($HOCH_2CH_2SO_3H$), lauroyl isethionic acid is formed. The sodium salt of this acid is the actual surfactant used in a popular bar detergent (in contrast to bar soap). When this detergent cleans the skin, it removes only a minimum of the skin's natural oil. Surfactants normally used in laundry detergents, such as sodium alkylbenzenesulfonate, discussed previously, are too efficient as cleansers; they remove everything from the skin, including natural, protective oils; this removal results in dry, itchy skin. Sodium lauroyl isethionate is prepared by the following reaction:

$$\underset{\text{lauroyl chloride}}{CH_3(CH_2)_{10}CO\boxed{Cl + H}O} CH_2CH_2SO_3H \xrightarrow{2NaOH} \underset{\text{sodium lauroyl isethionate}}{CH_3(CH_2)_{10}COOCH_2CH_2SO_3Na}$$

(isethionic acid: $HOCH_2CH_2SO_3H$)

The fatty acid from tallow—tallow acid—is converted to the tallow acid chloride by phosphorus trichloride. The sodium salt of the reaction product of tallow acid chloride with taurine ($H_2NCH_2CH_2SO_3H$), known as sodium tallow taurate, is also used as the surfactant for another type of gentle synthetic detergent bar. The reaction for preparing sodium tallow taurate is

$$\underset{\text{tallow acid chloride}}{\text{tallow–CO}\boxed{Cl + H}} \overset{H}{N}CH_2CH_2SO_3H \xrightarrow{2NaOH} \underset{\text{sodium tallow taurate}}{\text{tallow–CONHCH}_2CH_2SO_3Na} + NaCl$$

(taurine: $H_2NCH_2CH_2SO_3H$)

Phosphorus trichloride also chlorinates alkyl alcohols to their corresponding alkyl chlorides. For example, phosphorus trichloride reacts with octyl alcohol [$CH_3(CH_2)_6CH_2OH$] to give octyl chloride:

$$\underset{\text{phosphorus trichloride}}{PCl_3} + \underset{\text{octyl alcohol}}{3CH_3(CH_2)_6CH_2OH} \longrightarrow \underset{\text{octyl chloride}}{3CH_3(CH_2)_6CH_2Cl} + \underset{\text{phosphorous acid}}{H_3PO_3}$$

One of the most important uses of alkyl chloride is as an intermediate for preparing mercaptans by reaction with sodium hydrosulfide, NaSH:

$$\underset{\text{octyl chloride}}{CH_3(CH_2)_6CH_2Cl} + \underset{\substack{\text{sodium}\\ \text{hydrosulfide}}}{NaSH} \longrightarrow \underset{\text{octyl mercaptan}}{CH_3(CH_2)_6CH_2SH} + \underset{\substack{\text{sodium}\\ \text{chloride}}}{NaCl}$$

Octyl mercaptan is used in silver polish. This compound reacts with tarnish (silver sulfide) to form silver octyl mercaptide, a colorless product that is easily wiped off. Also, a very thin coating of silver octyl mercaptide is left on the surface of the silver. This coating protects the silver from the tarnishing action of hydrogen sulfide in the air.

Long-chain alkyl mercaptans are also used to control molecular weight in synthetic rubber manufacture. Some tin organic mercaptides are used as stabilizers for vinyl plastics.

DIALKYL PHOSPHONATES AS INTERMEDIATES FOR INSECTICIDES, WETTING AGENTS, METAL EXTRACTANTS, AND OIL ADDITIVES. Phosphorus trichloride is the intermediate used to prepare the following important classes of compounds: the dialkyl phosphonates $[(RO)_2P(=O)H]$, the trialkyl phosphites $[(RO)_3P]$, and the triaryl phosphites $[(ArO)_3P]$. (R represents an alkyl group, and Ar refers to an aryl group.)

The use of phosphorus trichloride as an intermediate in preparing dialkyl phosphonates, as exemplified by diisopropyl phosphonate $[(i\text{-}C_3H_7O)_2P(=O)H]$, for the nerve gas diisopropyl phosphorofluoridate (DFP), is discussed in Chapter 18. The series of reactions, starting with phosphorus trichloride and ending with DFP, is summarized in the following three equations:

$$\underset{\substack{\text{phosphorus}\\ \text{trichloride}}}{PCl_3} + \underset{\substack{\text{isopropyl}\\ \text{alcohol}}}{3\ i\text{-}C_3H_7OH} \rightarrow \underset{\substack{\text{diisopropyl}\\ \text{phosphonate}}}{(i\text{-}C_3H_7O)_2\overset{\overset{O}{\|}}{P}H} + \underset{\substack{\text{isopropyl}\\ \text{chloride}}}{i\text{-}C_3H_7Cl} + \underset{\substack{\text{hydrogen}\\ \text{chloride}}}{2\ HCl} \quad (1)$$

$$(i\text{-}C_3H_7O)_2\overset{\overset{O}{\|}}{P}H + \underset{\text{chlorine}}{Cl_2} \rightarrow \underset{\substack{\text{diisopropyl}\\ \text{phosphorochloridate}}}{(i\text{-}C_3H_7O_2)\overset{\overset{O}{\|}}{P}Cl} + HCl \quad (2)$$

$$(i\text{-}C_3H_7O)_2P(=O)Cl + NaF \rightarrow (i\text{-}C_3H_7O)_2P(=O)F + NaCl \qquad (3)$$

sodium fluoride; diisopropyl phosphorofluoridate (DFP)

Reaction 1 is not limited to isopropyl alcohol. Other alkyl alcohols react similarly with phosphorus trichloride to give the corresponding dialkyl phosphonate $[(RO)_2P(=O)H]$. For example, with methyl alcohol (CH_3OH), the product is dimethyl phosphonate:

$$PCl_3 + 3CH_3OH \longrightarrow (CH_3O)_2P(=O)H + CH_3Cl + 2HCl$$

methyl alcohol; dimethyl phosphonate; methyl chloride; hydrogen chloride

With octyl alcohol, $C_8H_{17}OH$, the product is dioctyl phosphonate:

$$PCl_3 + 3C_8H_{17}OH \longrightarrow (C_8H_{17}O)_2P(=O)H + C_8H_{17}Cl + 2HCl$$

octyl alcohol; dioctyl phosphonate; octyl chloride

If this reaction is conducted under mild conditions (low temperature) and the byproduct hydrogen chloride is removed and thereby not allowed to attack the ester group of the dialkyl phosphonate, excellent yields of dialkyl phosphonate are obtained. However, if the reaction is conducted under more severe conditions (high temperature) and the byproduct hydrogen chloride is not removed, primarily alkyl halide and phosphorous acid are obtained, as shown by the equation for the formation of octyl chloride:

$$3C_8H_{17}OH + PCl_3 \longrightarrow H_3PO_3 + 3C_8H_{17}Cl$$

Dimethyl phosphonate from the reaction of phosphorus trichloride with methyl alcohol is the intermediate used in making the insecticide Dipterex:

$$(CH_3O)_2P(=O)H + Cl_3CCHO \longrightarrow (CH_3O)_2P(=O)CH(OH)CCl_3$$

dimethyl phosphonate; chloral; Dipterex (dimethyl 1-hydroxy-2,2,2-trichloroethylphosphonate)

Dioctyl phosphonate [$(C_8H_{17}O)_2P(=O)H$] is obtained by the action of phosphorus trichloride on octyl alcohol under controlled temperature conditions. Dioctyl phosphonate can be chlorinated to give the dioctyl phosphorochloridate [$(C_8H_{17}O)_2P(=O)Cl$], which upon hydrolysis gives dioctyl phosphate [$(C_8H_{17}O)_2P(=O)OH$]:

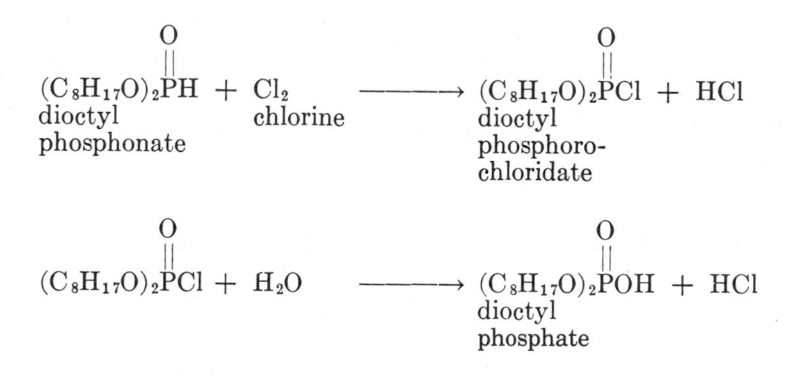

The sodium salt of dioctyl phosphate is a very good wetting agent and is used to some extent for this purpose. Dioctyl phosphate, in which the octyl alcohol is 2-ethylhexyl alcohol [the product is then properly called bis(2-ethylhexyl) phosphate], is important in extracting uranium from its ore. The ore is dissolved in 10% sulfuric acid, and the uranium solution is then extracted with kerosene containing 5–6% bis(2-ethylhexyl) phosphate. Other impurities are left in the original sulfuric acid solution, and uranium is extracted from the kerosene as a fairly pure sodium salt by using a 10% solution of soda ash.

Dibutyl phosphonate is an antiwear and extreme-pressure additive for petroleum oils that are components of industrial hydraulic fluids and industrial gear oils (*1*, *2*).

TRIALKYL AND TRIARYL PHOSPHITES FOR INSECTICIDES AND POLYMER STABILIZERS. Trialkyl phosphites are prepared by the reaction of phosphorus trichloride with alcohols in the presence of a base, such as a tertiary amine (R_3N), which absorbs the hydrogen chloride (HCl) evolved.

$$\underset{\text{phosphorus trichloride}}{PCl_3} + \underset{\text{alkyl alcohol}}{3ROH} + \underset{\text{tertiary amine}}{3R_3N} \longrightarrow \underset{\text{trialkyl phosphite}}{(RO)_3P} + \underset{\text{tertiary amine hydrochloride}}{3R_3N \cdot HCl}$$

If an absorbing agent were not used, HCl would attack the trialkyl phosphite formed and convert it into a dialkyl phosphonate.

$$(RO)_3P + HCl \longrightarrow (RO)_2P(=O)H + RCl$$

trialkyl phosphite	hydrogen chloride	alkyl phosphonate	alkyl chloride

If methyl alcohol is used, trimethyl phosphite $[(CH_3O)_3P]$, is formed. This product is probably the most important trialkyl phosphite produced today. Its main use is as the intermediate in synthesizing a series of insecticides such as DDVP and phosdrin. For example, the reaction to make DDVP involves the action of trimethyl phosphite on chloral (Cl_3CCHO) (*see* Chapter 19).

Other trialkyl phosphites are used industrially. For example, triisooctyl phosphite $[(i\text{-}C_8H_{17}O)_3P]$, which is prepared from isooctyl alcohol ($i\text{-}C_8H_{17}OH$), is a stabilizer for polyvinyl chloride plastics. When vinyl plastic is overheated or exposed to ultraviolet radiation, it discolors. One theory is that the action of heat and ultraviolet radiation causes polyvinyl chloride plastics to lose hydrogen chloride (HCl) and absorb oxygen. So that this degradation is prevented, selected trialkyl phosphites may be added. Phosphite has a trivalent phosphorus atom that can react with oxygen to form the stable pentavalent phosphate. Phosphite thus consumes the oxygen and prevents it from attacking the polymer. The organic groups in the phosphite make it soluble and compatible with the polymer system.

Triphenyl phosphite is made from phenol and PCl_3:

$$PCl_3 + 3C_6H_5OH \longrightarrow (C_6H_5O)_3P + 3HCl$$

phosphorus trichloride	phenol	triphenyl phosphite

In this reaction, phosphorus trichloride reacts directly with phenol; no base is required (as it is for synthesizing trialkyl phosphites) because the hydrogen chloride evolved does not attack triaryl phosphite, $(ArO)_3P$.

Obviously, triphenyl phosphite is easier to make than trialkyl phosphite. In fact, chemists often prefer to make triphenyl phosphite first and then convert it to the trialkyl phosphite by ester exchange. The following reaction illustrates the preparation of tridecyl phosphite by the ester exchange reaction; here, decyl alcohol replaces phenol:

$$(C_6H_6O)_3P + 3C_{10}H_{21}OH \longrightarrow (C_{10}H_{21}O)_3P + 3C_6H_5OH$$

triphenyl phosphite — decyl alcohol — tridecyl phosphite — phenol

Phenol can be recovered for reuse with phosphorus trichloride to make more triphenyl phosphite to start the reaction over again.

Triphenyl phosphite is used alone as a resin stabilizer. For example, in cooking, alkyd resins are used in enamel coatings for large appliances. A small amount of triphenyl phosphite in the reaction mixture results in a lighter colored resin. Triphenyl phosphite is also used in stabilizer formulations for vinyl plastics. A special triaryl phosphite, tris(nonylphenyl) phosphite [$(C_9H_{19}C_6H_4O)_3P$], is a very effective stabilizer for the government rubber–styrene (GRS, made during World War II) type of synthetic rubber and polyolefins. Tris(nonylphenyl) phosphite protects them from degradation by heat and light.

PHENYLPHOSPHONOUS DICHLORIDE AND DERIVATIVES USED IN NYLON AND INSECTICIDES. Phosphorus trichloride reacts at 600–700 °C with benzene (C_6H_6) to give phenylphosphonous dichloride:

$$PCl_3 + C_6H_6 \xrightarrow{600\text{–}700\ ^\circ C} C_6H_5PCl_2 + HCl$$

phosphorus trichloride — benzene — phenylphosphonous dichloride

Phenylphosphonous dichloride ($C_6H_5PCl_2$) reacts with water to give phenylphosphinic acid, $C_6H_5P({=}O)(OH)H$. This acid has been used as a stabilizer to prevent nylon from discoloring during process heating. Phenylphosphonous dichloride also reacts with sulfur to give phenylphosphonothionic dichloride [$C_6H_5P({=}S)Cl_2$]. This compound is the intermediate in preparing the insecticide EPN (Chapter 19).

Phosphorus Oxychloride Derivatives. Phosphorus oxychloride, a colorless corrosive liquid, is a versatile intermediate for manufacturing many important organic phosphates. This compound is prepared by one of two commercial methods. The most widely used process involves the direct oxidation of phosphorus trichloride with oxygen under carefully controlled conditions.

$$2PCl_3 + O_2 \longrightarrow 2POCl_3$$

phosphorus trichloride	oxygen	phosphorus oxychloride

The other method is the chlorination of phosphorus trichloride, PCl_3, in the presence of phosphoric anhydride (P_4O_{10}).

$$6PCl_3 + P_4O_{10} + 6Cl_2 \longrightarrow 10POCl_3$$

Phosphorus oxychloride's most important use is as a reactant in the manufacture of the triaryl and trialkyl phosphates, as well as the mixed alkyl aryl phosphates. The discussion of the chemistry and application of phosphorus oxychloride in this section, therefore, concerns the chemistry and application of the phosphate esters.

TRIARYL PHOSPHATES AND THEIR USES. A typical phosphate, tricresyl phosphate, is made by heating phosphorus oxychloride with cresol:

$$POCl_3 + 3CH_3C_6H_4OH \longrightarrow OP(OC_6H_4CH_3)_3 + 3HCl$$

phosphorus oxychloride	cresol	tricresyl phosphate	hydrogen chloride

The first important phosphate esters introduced commercially were triphenyl and tricresyl phosphates; they were used or plasticized to make a better quality "celluloid" (*see* Chapter 10).

The discovery of triaryl phosphates as plasticizers for cellulose nitrate opened an area for their application as plasticizers for other polymers. Triaryl phosphates perform very well in cellulose acetate and for vinyl polymers. The use of the triaryl phosphate esters as plasticizers not only permits plastics to be processed at a lower temperature and with less effort but also imparts some flame resistance to the plastics in which they are used.

Aryl phosphates are also used as gasoline additives. When gasoline containing tetraethyl lead burns in an automobile engine, lead deposits form on the spark plugs and on the inside of the cylinder head. These deposits catalytically ignite the gasoline–air mixture in the cylinder prior to ignition by the spark plugs. These misfires are heard as wild pings and reduce gasoline mileage.

Aryl phosphates such as tricresyl phosphate [$(CH_3C_6H_4O)_3P{=}O$] or cresyl diphenyl phosphates [$(CH_3C_6H_4O)(C_6H_5O)_2P{=}O$], when added in small percentages to the gasoline, combat misfires. The theory is that when triaryl phosphates are burned in the gasoline mixture, the phosphorus moiety combines with the lead decomposition product to form a phosphorus-containing lead compound, such as lead phosphate. This compound deposits on the spark plug and inside the cylinder heads, and unlike other lead deposits, it is no longer active catalytically—that is, it does not cause premature firing. In the 1960s and 1970s, this application consumed millions of pounds of triaryl phosphates annually. However, with the gradual elimination of lead from gasoline, this application has decreased substantially.

Because triaryl phosphates are less flammable than petroleum oil, they have gradually replaced the petroleum oil as lubricants, coolants, and hydraulic fluids in machinery where fire is a potential danger. For example, triaryl phosphates are used as lubricants and coolants in the large bearings in huge gas turbines that generate electricity and steam. These turbines are also used to pump oil and gas through large pipelines and to drive air compressors in the chemical-processing industries.

Triaryl phosphates are used as hydraulic fluids to control the opening and shutting of valves in the electrohydraulic control units of the large equipment. The big machinery in steel mills that control the size, shape, and thickness of steel sheets and bars during processing are also controlled hydraulically. Because the steel pieces are very hot, less flammable fluids must be used in case of accidental leakage in the hydraulic system.

Mining equipment is also hydraulically controlled. Because the threat of fire and explosion inside mines is also present, less flammable fluids such as those based on triaryl phosphates are recommended.

A minor use for tricresyl phosphate is as an additive in lubricating oils to reduce wear on bearing surfaces. Theoretically, under actual driving conditions, tricresyl phosphate gradually decomposes to cresyl acid phosphates that react with the iron surfaces of the bearings, forming a microscopic protective coating over them.

Although the tricresyl phosphates are used extensively, a particular mode of neurotoxicity was discovered in the United States during the prohibition period, in the early 1930s, when tricresyl phosphate was used to adulterate bootleg liquor. The toxicity, which caused a characteristic type of paralysis, was associated with the presence of *o*-cresol esters in the tricresyl phosphate (*3*). The details of the chemistry and the toxicology are described in Chapter 19. After the discovery of this cause of the toxicity, the use of cresylic acid feed stocks containing low concentrations of *o*-cresol became a standard industrial practice for the manufacture of tricresyl phosphate.

Recently, products equivalent in performance to tricresyl phosphate were obtained by using alkylated phenols in place of the cresylic acid. These alkylphenols were prepared synthetically from phenol and an olefin in the presence of an acid catalyst. An example is the preparation of isopropylphenol.

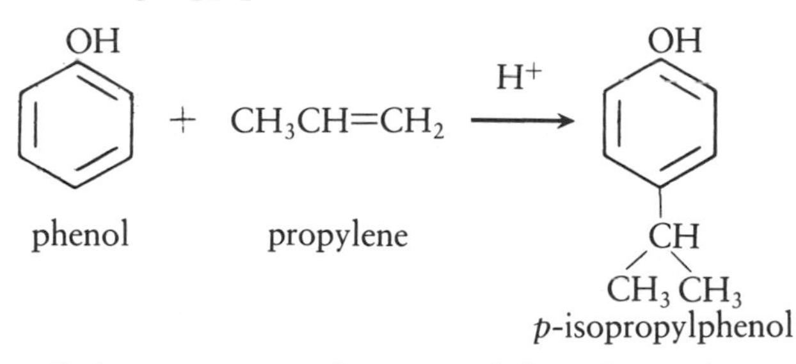

Actually, all three isomers of isopropylphenol are obtained in this reaction, but none of the phosphate esters, even those of the ortho isomer, causes the neurotoxicity associated with the phosphate esters of *o*-cresol.

Two such alkylphenols, isopropylphenol and *tert*-butylphenol, are gaining more and more use as replacements for cresylic acid for preparing triaryl phosphates. Compounds such as isopropylphenyl diphenyl phosphate (*4*) have replaced tricresyl phosphate in plasticizer applications, and because of their oxidative thermal stability, aryl phosphte esters of *tert*-butylphenol, such as tert-butylphenyl diphenyl phosphate (*5*), are replacing tricresyl phosphate in lubricants and hydraulic fluid applications.

TRIALKYL AND ALKYL ARYL PHOSPHATES AND THEIR USES. Other important phosphate esters prepared from phosphorus oxychloride are the trialkyl and mixed alkyl aryl phosphates. The trialkyl phosphates, as exemplified by tris(2-ethylhexyl) phosphate and tributyl phosphate, are prepared by the reaction of phosphorus oxychloride with the respective alcohols. The conditions for making these phosphates are much milder than those used for triaryl phosphates. Tributyl phosphate is prepared in the following way:

$$POCl_3 + 3C_4H_9OH \longrightarrow OP(OC_4H_9)_3 + 3HCl$$

phosphorus oxychloride	butyl alcohol	tributyl phosphate

When alcohols other than butyl are used, the corresponding trialkyl phosphates can be prepared by this general reaction.

Tributyl phosphate is used to purify uranium. As noted before, dioctyl phosphoric acid is used in the solvent extraction of uranium from its ore. Although the uranium compound obtained is fairly pure, it is not pure enough for the conversion to uranium metal for atomic reactors. In one purification process, the extracted uranium is dissolved in nitric acid. This solution is extracted further with tributyl phosphate in a kerosene solution. In this step, the tributyl phosphate actually forms a compound with the uranyl nitrate that is soluble in the kerosene solution: $UO_2(NO_3)_2[(C_4H_9O)_3P{=}O]_2$ (*6–8*). The uranium salt is then extracted from the kerosene solution with a 10% solution of sodium carbonate. The selectivity of tributyl phosphate in extracting one metal from another also makes it useful for separating normally difficult-to-separate elements such as hafnium from zirconium.

Tributyl phosphate is also used as a nonflammable component in the fluid used in the hydraulic system of large commercial aircraft. Another important application is as a defoamer in preparing latex paints and printing inks. Tributyl phosphate is also used in processing industries such as paper manufacturing, where agitation creates foam and results in spillage. Less than 1% of this compound depresses foaming by changing the surface tension of the liquid.

Another important trialkyl phosphate is tris(2-ethylhexyl) phosphate (*9*). As a plasticizer for vinyl polymer sheets, not only does tris(2-ethylhexyl) phosphate impart flame resistance to the sheets but also the plasticized sheets remain flexible at low temperatures. When such sheets are used as seat covers in automobiles and trucks, the covers will not stiffen or crack in cold weather.

Of the mixed alkyl aryl phosphates, the most important one is 2-ethylhexyl diphenyl phosphate, $C_8H_{17}OP({=}O)(OC_6H_5)_2$ (*10*). This nontoxic compound finds use as a plasticizer for food-wrapping films, for films for tubings for skinless sausages, and for other meat packaging. 2-Ethylhexyl diphenyl phosphate has been approved for this use by both the Bureau of Animal Industry and the Food and Drug Administration. This compound is also the main component in certain nonflammable hydraulic fluids used in large aircraft. Because 2-ethylhexyl diphenyl phosphate does not ignite spontaneously below 1000 °F and remains fluid at low temperatures, it is ideal for this purpose.

TRIALKYL TRITHIOPHOSPHATES AND TRIALKYL TRITHIOPHOSPHITES. A very interesting compound is the reaction product of phosphorus oxychloride with butyl mercaptan (an essence of skunk):

$$\underset{\text{phosphorus oxychloride}}{POCl_3} + \underset{\text{butyl mercaptan}}{3CH_3(CH_2)_2CH_2SH} \longrightarrow \underset{S,S,S\text{-tributyl trithiophosphate}}{OP[SCH_2(CH_2)_2CH_3]_3} + 3HCl$$

S,S,S-Tributyl trithiophosphate is a good defoliant; it removes leaves from cotton plants without killing them (*11, 12*). When these plants are machine-harvested, the green leaves normally stain the white cotton. If the plants are defoliated by a spray of a water emulsion of a compound such as *S,S,S*-tributyl trithiophosphate, the staining problem is eliminated. The compound *S,S,S*-tributyl phosphite, prepared from the reaction of PCl_3 with butyl mercaptan, serves the same function in defoliating cotton. This reaction is discussed further in Chapter 21.

Miscellaneous Uses of Phosphoric Anhydride Derivatives

Phosphoric Anhydride Reactions with Water. Phosphoric anhydride, P_4O_{10}, is generally referred to as P_2O_5. However, the correct molecular formula is P_4O_{10}. Phosphoric anhydride forms when elemental phophorus burns in air. Most P_4O_{10} manufactured is immediately converted into phosphoric acid by reaction with water. Phosphoric anhydride, however, is the water-free form of phosphoric acid. [When phosphorus burns in a limited amount of air, the lower oxide of phosphorus, phosphorous anhydride (P_4O_6) also forms. Because this oxide is not important commercially, it is not discussed in this chapter.]

As a free-flowing white powder, phosphoric anhydride has a strong affinity for water and will readily absorb moisture from the air or any place it can obtain moisture. This property makes phosphoric anhydride a natural candidate as a desiccant, and it is widely used for drying small amounts of compounds in chemical laboratories. As P_4O_{10} absorbs water, it becomes a gummy form of polyphosphoric acid. So that this gummy material does not coat unreacted P_4O_{10} and thus reduce its dehydrating capacity, P_4O_{10} is generally used as a porous mixture with activated carbon.

Pictorially, the tetrahedral structure of phosphorous anhydride (P_4O_6) resembles elemental phosphorus with oxygen atoms inserted between the phosphorus atoms. Phosphoric anhydride (P_4O_{10}) has a structure related to that of P_4O_6, with an additional oxygen atom on each phosphorus atom. When water reaches P_4O_{10}, it splits the POP bonds—that is, 1 mol of water splits one POP bond. Because six POP bonds are present in each P_4O_{10} tetrahedron, 6 mol of water is required to break all of the POP bonds; 4 mol of phosphoric acid [$(HO)_3P{=}O$ or H_3PO_4] is produced.

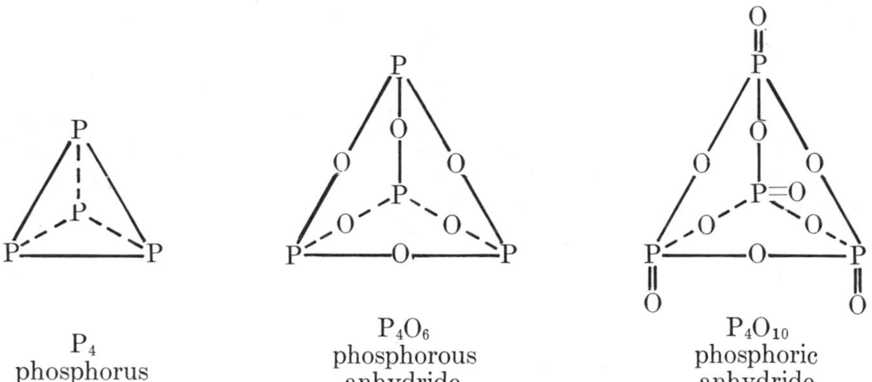

P_4
phosphorus

P_4O_6
phosphorous anhydride

P_4O_{10}
phosphoric anhydride

When only two bonds are selectively broken with 2 mol of water, cyclic tetrametaphosphoric acid is formed:

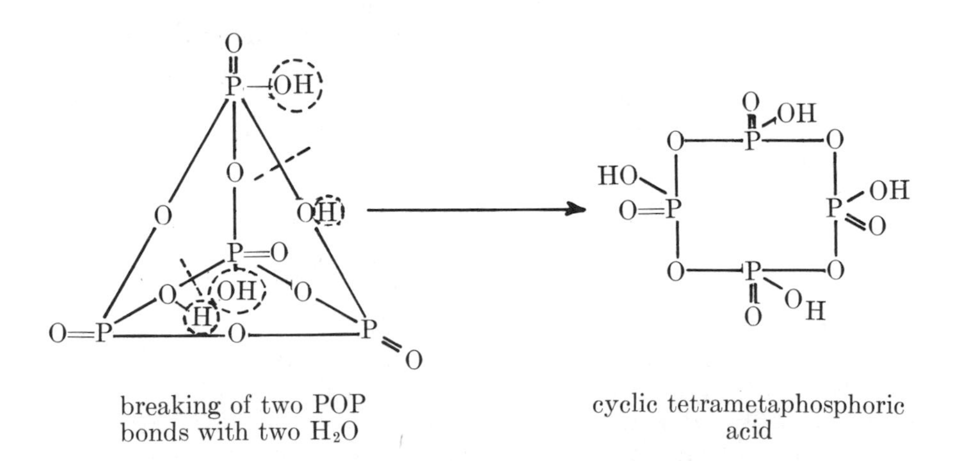

breaking of two POP bonds with two H_2O

cyclic tetrametaphosphoric acid

As a matter of fact, one method for preparing sodium cyclic tetrametaphosphate in fairly good yield is by adding P_4O_{10} to a water solution of sodium hydroxide at O °C. The product can be separated from other sodium phosphates by fractional crystallization.

When water reacts with P_4O_{10}, each POP bond splits into two POH groups. Above 150 °C, the two POH groups from 2 mol of phosphoric acid [$O{=}P(OH)_3$] can combine and eliminate 1 mol of water to form a new POP bond:

$$(HO)_2P(=O)\text{—}OH + HO\text{—}P(=O)(OH)_2 \rightleftharpoons (HO)_2P(=O)\text{—}O\text{—}P(=O)(OH)_2 + H_2O$$

In other words, in the proper temperature range, the splitting of a POP bond by a water molecule and the formation of a new POP bond with the elimination of 1 mol of water is a reversible, equilibrium reaction.

POH groups can also split another POP and at the same time create a new POP bond. This fact is illustrated by the reaction of orthophosphoric acid [$(HO)_3P{=}O$] with tripolyphosphoric acid to form 2 mol of pyrophosphoric acid.

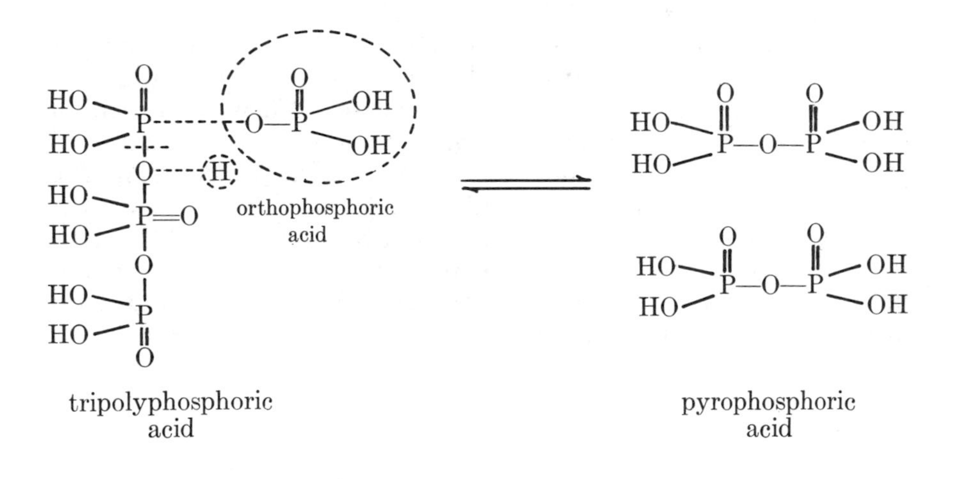

Bond splitting and new bond formation continue until equilibrium is reached. At that point, the product is a mixture of phosphoric acids of various chain lengths. As indicated in the chart in Chapter 3, the composition at equilibrium depends on the concentration of P_4O_{10} in the mixture.

Alkyl Acid Phosphates and Their Uses. The P_4O_{10} tetrahedron also reacts with alcohols, ROH (and phenols, ArOH); the reaction is generally performed at 60-70 °C; ROH reacts with the POP bond as follows:

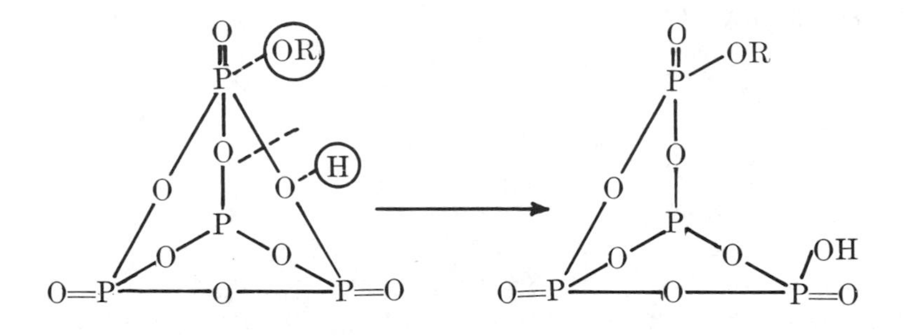

The reaction of ROH with the various POP bonds in the tetrahedron follows the laws of probability; sometimes the OR groups go to the P in one end of the POP bond that already has an RO group, and sometimes they go to the end of the POP bond without an RO group. When 1 mol of P_4O_{10} reacts completely with 6 mol of ROH, the resulting product is essentially an equimolar (1 mol of each) mixture of $(RO)_2P(=O)OH$ and $(RO)P(=O)(OH)_2$.

The combination of $(RO)_2P(=O)OH$ and $(RO)P(=O)(OH)_2$ is called mixed mono- and dialkyl phosphoric acids. The nature of R depends on the alcohol used in the reaction. When ethyl alcohol (C_2H_5OH) is used, the product is a mixture of the mono- and diethyl phosphoric acids [$C_2H_5OP(=O)(OH)_2$ and $(C_2H_5O)_2P(=O)OH$]. The ammonium salt of the mono- and diethyl phosphoric acid is very water-soluble. This salt has been promoted as a textile lubricant; thus, fabric can be fed through the machinery more easily. The ammonium salt is also a flameproofing agent. When the alcohol used in the reaction with P_4O_{10} is polyoxyethylenated nonylphenol

$$C_9H_{19}\text{–}\bigcirc\text{–}(OCH_2CH_2)_nOH$$

the reaction product is a mixture of

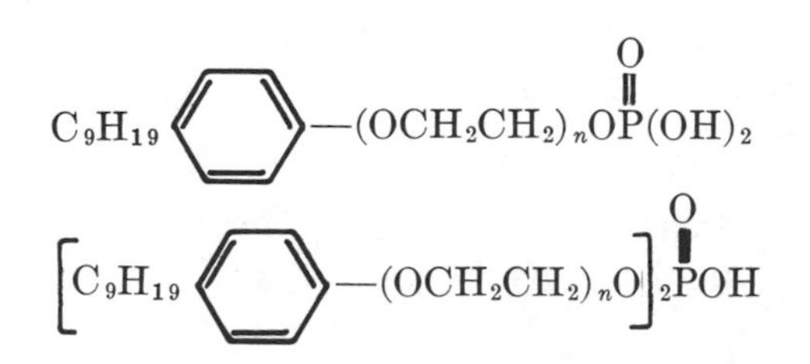

The sodium salts of this mixture are soluble in organic solvents and are good surfactants for dry cleaning; that is, the salts help to remove dirt during dry cleaning just as soap and detergents do during laundering (*13*, *14*).

The $(RO)_2P(=O)OH$ and $ROP(=O)(OH)_2$ mixture is acidic. Because it contains organic groups, it is soluble in organic systems. Many of these acids are used as catalysts in such polymerization reactions as the hardening of urea–formaldehyde resin. When mono- and dibutyl acid phosphates are used as the catalyst (about 3% concentration), they accelerate the hardening of the resin even at low (room) temperatures. Similar catalyst systems are also effective for curing or hardening melamine–formaldehyde resins.

The amine salts of the mono- and diisooctyl phosphoric acid mixture are effective corrosion inhibitors in gasoline pipes in refineries. The amine salts of a mixture of mono- and ditridecyl phosphoric acid, when added to gasoline, act as antistalling agents for automobiles by preventing carburetor icing in cold weather. Most of the other specific applications for mono- and dialkyl phosphoric acids and their salts depend on the nature of the alkyl group as well as the nature of the metal salt.

One interesting compound is prepared by neutralizing medium-length chains (with 8–12 carbons) of mixed mono- and dialkyl phosphoric acid with ethylene oxide. The product is a good emulsifier and wetting agent that foams only slightly—an important property in dye baths where too much foam causes overflow. The mixture is also an effective emulsifier for polymerizing vinyl acetate and for copolymerizing vinyl chloride with vinyl acetate. Because the mixture is also quite stable in acids, it is used as a wetting agent in acid metal cleaning systems.

When 1 mol of P_4O_{10} reacts with less than 6 mol of alcohol, the product contains unbroken POP bonds. In this respect, the product can be regarded as an organic polyphosphate. The sodium salts of the reaction products of 3 mol of P_4O_{10} with 10 mol of octyl alcohol ($C_8H_{17}OH$) are good wetting agents and effective stabilizers against the degradation of vinyl plastics by heat. Because this type of compound also sequesters heavy metals, it makes a good stabilizer in peracetic acid manufacture. Sequestering action ties up heavy metal impurities; thus, their catalytic decomposition of the peracetic acid is prevented.

The reaction product of 4 mol of 2-ethylhexyl alcohol and 1 mol of P_4O_{10} has been sold as an extractant of tetravalent uranium from solutions of wet-process phosphoric acid. Although named dioctyl pyrophosphoric acid, the reaction product is actually a mixture of phosphoric acid esters. The reaction product from capryl alcohol and P_4O_{10} was also sold for this application (*15*).

In recent years, two other extraction systems have come into favor. One uses the mixture of mono- and dioctylphenyl phosphoric acids (*16*). This mixture is prepared from the reaction product of 1 mol of P_4O_{10} and 6 mol of octylphenol. The second process uses trioctylphosphine oxide. These processes will be mentioned more fully at the end of this chapter.

Thiophosphoryl Chloride

Thiophosphoryl chloride, $PSCl_3$, is a liquid that boils at 125 °C. It is prepared by the method described earlier, the addition of elemental sulfur to PCl_3 in the presence of a catalyst. Although this reactive compound has not yet gained commercial importance, many of the sulfur-containing organic phosphorus herbicides and insecticides that are described in Chapters 18 and 20 could be made from $PSCl_3$. As an example, intermediates such as the dialkyl phosphorochloridothioates can be made from $PSCl_3$ by the following process:

$$\underset{\text{thiophosphoryl chloride}}{PSCl_3} + 2ROH + 2NaOH \longrightarrow \underset{\text{dialkyl phosphorochloridothioate}}{(RO)_2\overset{\overset{S}{\|}}{P}Cl} + 2NaCl + H_2O$$

These reactive intermediates are now made in a two-step process starting with P_4S_{10}. This process is described later in this chapter. The reason for the choice of the route from P_4S_{10} is because the two-step reaction produces the intermediates in good yield and requires a relatively unsophisticated isolation technique, that is, simple distillation of the intermediate from an essentially undistillable residue.

The preparation of such intermediates from $PSCl_3$ affords equivalent yields. However, obtaining such yields requires the careful separation of the coproducts, trialkyl phosphorothioates [$(RO)_3PS$] and alkyl phosphorodichloridothioates [$ROP(S)Cl_2$]. These coproducts often boil at temperatures close to those of the desired intermediate. Therefore, the separation is difficult. However, with recycling of the $ROP(S)Cl_2$, the waste diposal problems from this route are significantly less than from the P_4S_{10} route. Therefore, should new plants be built, the economics of the $PSCl_3$ route are likely to be relatively more favorable.

Miscellaneous Uses of Organic Compounds from Phosphorus Sulfides

Classes of Phosphorus Sulfides. Phosphorus sulfides are the products of elemental phosphorus and sulfur, and most give off a rotten-egg odor. Although these materials are malodorous, they are essential intermediates for insecticides to keep our flower gardens beautiful and to help farmers save their cotton crops for our clothing rather than as food for boll weevils. Also, phosphorus sulfides are ingredients for additives to automobile engine lubricants. Phosphorus sulfides are also used by miners to extract minerals from sand and clay mixtures. The better-known phosphorus sulfides are shown in Chart 15.I.

These structures suggest that phosphorus sulfides are formed by inserting sulfur atoms into the phosphorus, P_4, tetrahedron. Phosphorus pentasulfide (actually P_4S_{10}) is the result of the breaking of all the phosphorus to phosphorus (–P–P–) bonds in the P_4 tetrahedron along with the oxidation of all of the phosphorus atoms from a valence of 0 to a valence of 5. The other three phosphorus sulfides can be thought of as the result of the controlled but incomplete reaction of sulfur with the (P_4) phosphorus tetrahedron. In fact, the reaction conditions for manufacturing the various phosphorus sulfides are more or less the same. Except for tetraphosphorus pentasulfide, which is prepared by the reaction of $P_4S_3I_2$ with $[(CH_3)_3Sn]_2S$ (*18*), the phosphorus sulfides are prepared by heating phosphorus with sulfur at 300–400 °C. The controlling factor in the reaction is the ratio of sulfur to phosphorus.

In addition to the four phosphorus sulfides shown in Chart 15.I, at least four others are described in the literature, including P_4S_2, P_4S_9, P_4S_4, and P_2S (*17, 18*). These sulfides not only are of scientific interest but also are of commercial interest, because they can be present as contaminants in the commercial phosphorus sulfides. When present, they can affect the course and rate of reaction of the commercial phosphorus sulfides.

In general, stoichiometric quantities of the two elements as represented by the specific formulas are heated together to obtain the desired compound. Depending on the particular phosphorus sulfide obtained, it can be purified by recrystallization from solvents such as carbon disulfide or by distillation. The commercially important phosphorus sulfides are phosphorus sesquisulfide (P_4S_3) (used in safety matches) and phosphorus pentasulfide (P_4S_{10}).

Organic Compounds from Phosphorus Pentasulfide and Their Uses. Phosphorus pentasulfide (P_4S_{10}) is a light yellow crystalline

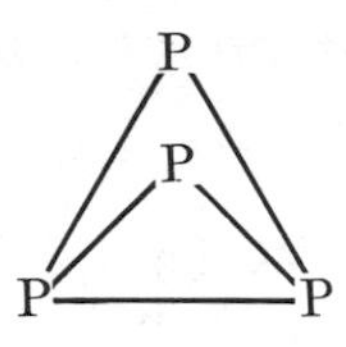

P_4
phosphorus

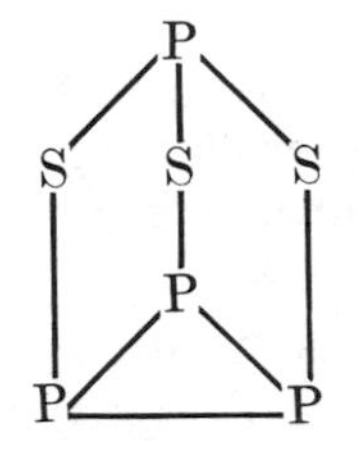

P_4S_3
phosphorus
sesquisulfide

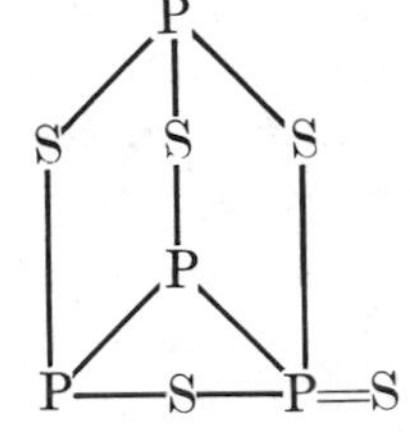

P_4S_5
tetraphosphorus
pentasulfide

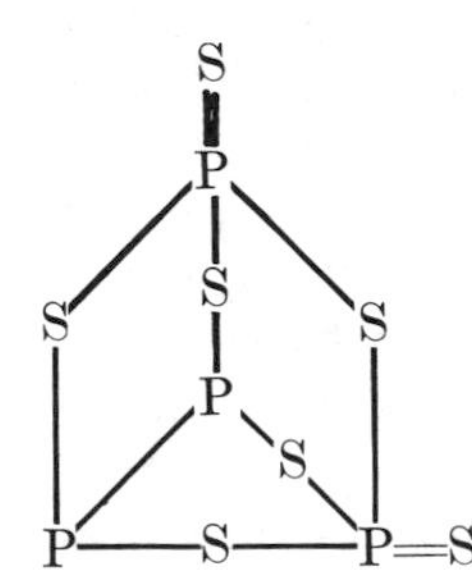

P_4S_7
tetraphosphorus
heptasulfide

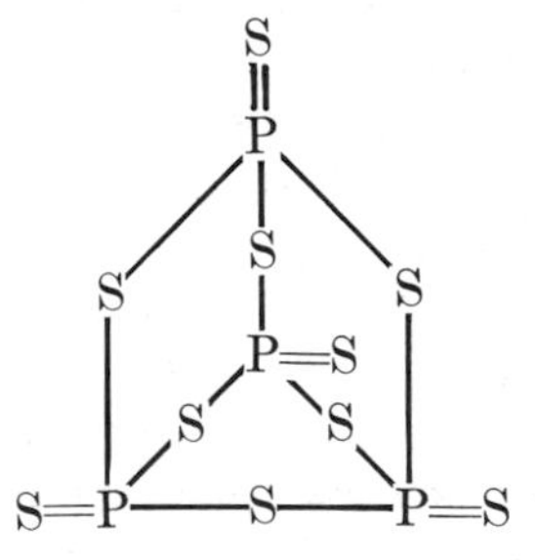

P_4S_{10}
phosphorus
pentasulfide

Chart 15.1

solid. It reacts rapidly with atmospheric moisture to liberate hydrogen sulfide gas. Its characteristic odor is thus that of rotten eggs, similar to that of hydrogen sulfide. This unsavory odor carries over to its derivatives. I have worked with many of these derivatives, and for most of them, if you say that they smell like a skunk, you are insulting the skunk.

Despite their bad odor, these derivatives have properties that make them valuable in many applications. For example, the insecticide malathion is one of the derivatives of phosphorus pentasulfide.

Phosphorus pentasulfide is a widely used intermediate for synthesizing insecticides, lubricating oil additives, and making flotation agents. In all these applications, phosphorus pentasulfide reacts with an alcohol to form the dialkyl phosphorodithioic acid, $(RO)_2P(=S)SH$:

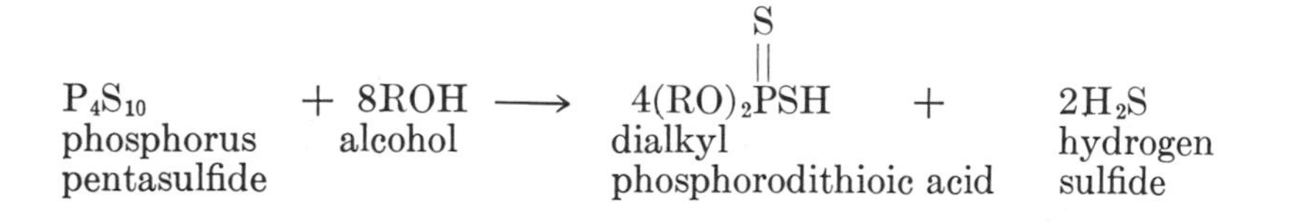

The alkyl alcohol (ROH) can be methyl (CH_3OH), ethyl (C_2H_5OH), propyl (C_3H_7OH), or a longer chain alcohol such as octyl [$CH_3(CH_2)_6CH_2OH$]. The alcohol used determines the final application of the product. The low molecular weight alkyl alcohols (methyl, ethyl, and propyl) are generally used to prepare insecticides.

INSECTICIDES. Insecticides such as malathion, Thimet, and Trithion are synthesized as shown below and on page 242, starting from phosphorus pentasulfide. Here, phosphorus pentasulfide reacts with methyl alcohol to form dimethyl phosphorodithioate [$(CH_3O)_2P(=S)SH$], which is added to ethyl maleate to produce malathion. This insecticide is toxic to insects, but it has little toxicity to mammals. Therefore, it is widely used to control garden and agricultural pests. In home gardening, therefore, the difference between having roses as a thing of beauty to enjoy or as food for insects is the use of an insecticidal dust containing malathion. Further discussion of the toxicity is given in Chapter 18.

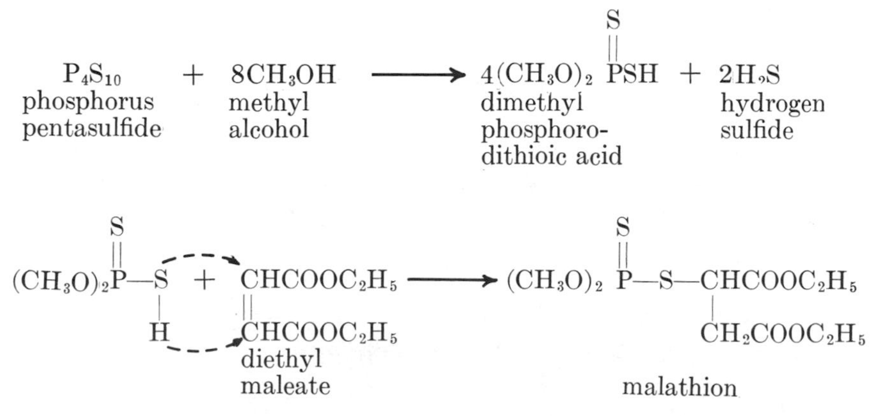

For Thimet, the diethyl phosphorodithioate is prepared by the reaction of phosphorus pentasulfide with ethyl alcohol. Diethyl phosphorodithioate then reacts with formaldehyde and ethyl mercaptan to eliminate 1 mol of water and give Thimet:

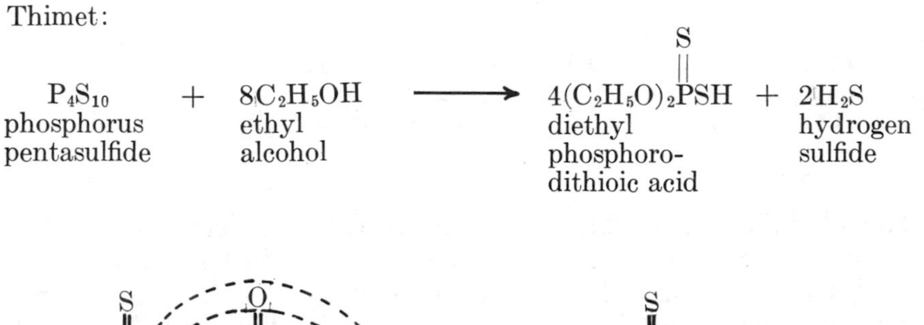

$$(C_2H_5O)_2P(=S)SH + HC(=O)H + HSC_2H_5 \longrightarrow (C_2H_5O)_2P(=S)SCH_2SC_2H_5 + H_2O$$

formaldehyde ethyl mercaptan Thimet

Trithion is also prepared from *O,O*-diethyl phosphorodithioate [$(C_2H_5O)_2P(=S)SH$]. Many other important insecticides are prepared from the diethyl or dimethylphosphorodithioate, which in turn are made by using phosphorus pentasulfide as the intermediate. Malathion, Thimet, and Trithion are examples.

Trithion:

$$(C_2H_5O)_2P(=S)SH + ClCH_2S\text{-}C_6H_4\text{-}Cl \longrightarrow (C_2H_5O)_2P(=S)\text{—}SCH_2S\text{-}C_6H_4\text{-}Cl + HCl$$

Dialkyl phosphorodithioic acids can be chlorinated to form dialkyl phosphorochloridothionate [$(RO)_2P(=S)Cl$]:

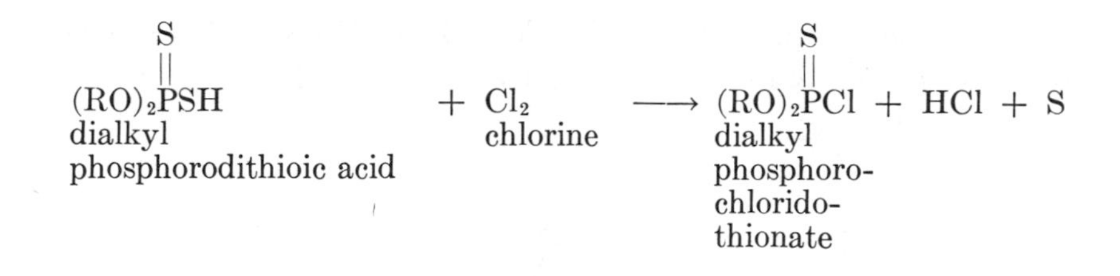

These dialkyl phosphorochloridothionates are also intermediates in the synthesis of some well-known insecticides. For example, dimethyl phosphorochloridothioate [$(CH_3O)_2P(=S)Cl$] is the intermediate for methyl parathion:

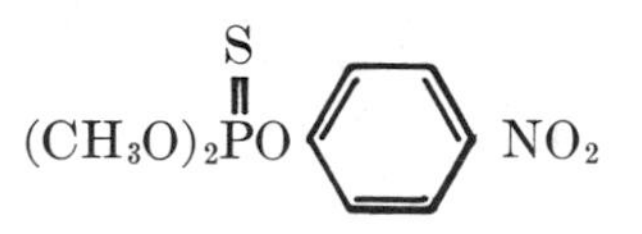

Dimethyl phosphorochloridothioate forms methyl parathion by reaction with sodium *p*-nitrophenolate

NaO–C_6H_4–NO_2

as shown:

$(CH_3O)_2P(=S)Cl$ + NaO–C_6H_4–NO_2

dimethyl phosphorochloridothionate + sodium *p*-nitrophenolate

⟶ $(CH_3O)_2P(=S)O$–C_6H_4–NO_2 + NaCl

methyl parathion

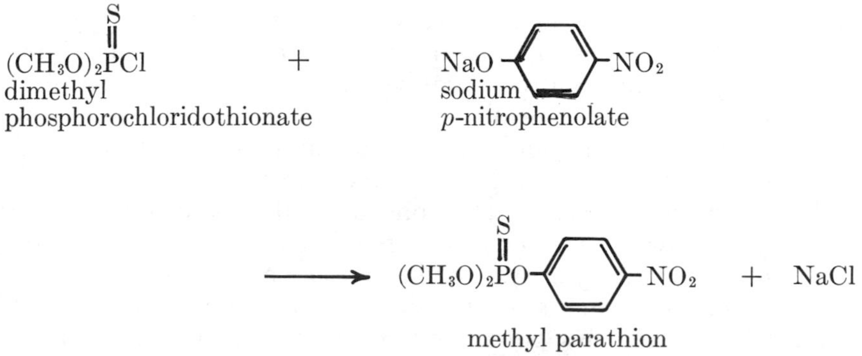

Similarly, diethyl phosphorochloridothioate [$(C_2H_5O)_2P(=S)Cl$] is used as the intermediate to make parathion:

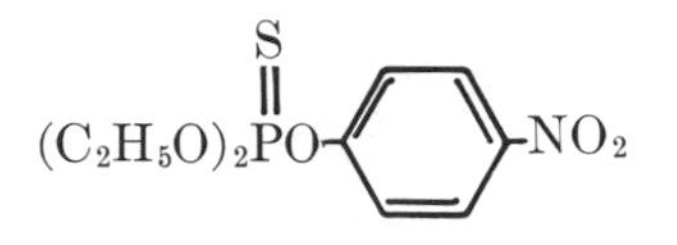

The parathions are used extensively to control cotton pests.

LUBRICANT ADDITIVES. Although other salts have been investigated, the zinc salts of dialkyl phosphorodithioates are particularly effective as a lubricant additive (*19*). The sodium salts are ineffective; the stannous (tin) salts are very good but are too expensive for this use (tin is much more expensive than zinc). The zinc salts are prepared by the reaction of zinc oxide (ZnO) with the dialkyl phosphorodithioate [$(RO)_2P(=S)SH$]:

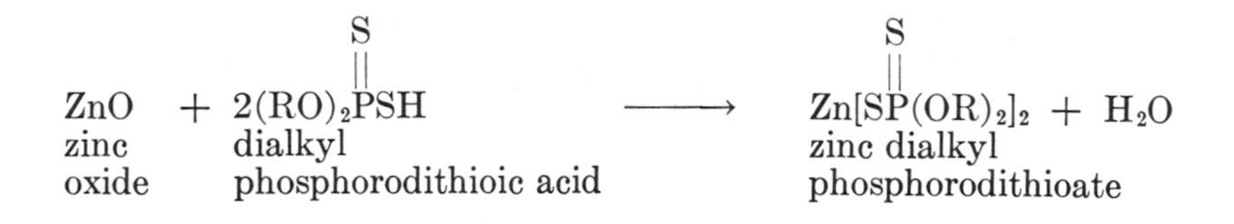

The nature of the alkyl groups depends largely on the final application of the compound. Isoamyl, hexyl, or cyclohexyl derivatives have all been used for almost 40 years.

The use of zinc dialkyl phosphorodithioate as an antiwear additive in lubricating oil is related to the process of coating the surface of the metal by phosphatization, as discussed in Chapter 9. In a new engine, the metal surface is coated with a phosphatic surface that reduces wear on the metal surfaces in the engine parts during break-in. When the engine is running, this phosphatic coating is gradually replaced by another coating from the reaction of zinc dialkyl phosphorodithioate with the metal surface; this new coating does not chip off. In bearings, gears, and valve trains, a smooth surface greatly reduces frictional wear and prolongs their useful life.

The function of the zinc dialkyl phosphorodithioate as an extreme pressure additive in gear lubricants is related to its antiwear properties. In an engine's gear system, the pressure at the point of contact between the two gears is so high that it can actually extrude all of the lubricant from this point. This results in direct metal-to-metal contact, which results in high wear. The phosphate additive coats the surface of the metal by chemical reaction; the organic portion of the phosphate additive holds the oily lubricant through chemisorption.

The protective coating formed by zinc dialkyl phosphorodithioate also retards corrosion of the bearing surface by the oxidation products of the oil. Further, the dithioate additives prevent or reduce (through chemical reactions) the formation of the corrosive decomposition products in the lubricating oil. This latter reaction is called the "antioxidant" reaction and helps to maintain the quality of oil during long usage. Through the use of additives such as these, crankcase oil changes are reduced. The formulation of lubricants for use in crankcases and gear boxes is a highly developed science. The use of dialkyl phosphorodithioates, prepared from phosphorus pentasulfide, represents just a part of this development. Non-phosphorus derivatives are also important in lubricant formulations.

ORE FLOTATION AGENTS. Flotation agents are compounds used to concentrate desired minerals in crude ores. When copper is mined, for example, the copper sulfide ore is contaminated with large amounts of sand, clay, and other minerals. Separation of the ore from these contaminants efficiently and economically is necessary.

The ore flotation process is based on the property of a properly treated surface having a high affinity for air bubbles and a low affinity for water. In copper sulfide flotation, for example, finely ground ore is treated in a water slurry with such compounds as the sodium or ammonium salts of dialkyl phosphorodithioate (called collectors). These salts selectively wet the surface of the copper sulfide particles but do not wet the surface of the contaminants. Air is then bubbled into the water slurry. The air bubbles attach themselves to the treated surface of the copper sulfide and not to the contaminants. The air bubbles with the copper sulfide attached rise as a froth to the top and are then separated from the contaminants that have been wetted by the water and have settled to the bottom. The concentration of dialkyl phosphorodithioate collector used for flotation is only 0.01–0.10 lb/ton of ore.

Flotation is used to recover such other valuable minerals as molybdenum, lead, zinc, silver, and gold, as well as silicate-based minerals like mica and feldspar. The particular flotation collector used depends on the mineral to be collected. Generally, most of the dialkyl phosphorodithioates used for this purpose are prepared from ethyl, isopropyl, and *sec*-butyl alcohols. The important factor is high selectivity for the preferential coating of the specific mineral for air flotation.

Other agents such as the dithiocarbonates and xanthates are also used as collectors. However, phosphorodithioates prepared from the reaction of phosphorus pentasulfide and alcohols are also very important (*20–23*).

Phosphines and Phosphine Oxides

Until recently, the organophosphines have been primarily laboratory curiosities. However, a number have attained commercially significant uses in recent years.

Triphenylphosphine and Analogues. Triphenylphosphine is a white solid melting at 79 °C. It has found use as a ligand in transition metal catalysts and as an intermediate for the preparation of selected olefins.

Triphenylphosphine can be prepared by the reaction of PCl_3 and chlorobenzene in the presence of elemental sodium

$$3\,C_6H_5Cl + 3Na + PCl_3 \longrightarrow (C_6H_5)_3P + 3NaCl$$

or from the reaction of PCl_3 with phenyl magnesium bromide:

$$3C_6H_5MgBr + PCl_3 \longrightarrow (C_6H_5)_3P + 3MgClBr$$

Because several of the intermediate phosphine halides are commercially available, that is, $C_6H_5PCl_2$ and $(C_6H_5)_2PCl$, these reagents may be used in place of the PCl_3.

An important event in the history of homogeneous catalysis was the observation that phosphines accelerated the catalytic activity of certain transition metal catalysts, particularly those of group 7. A major landmark was the discovery that a rhodium complex with triphenyl phosphine, $RhCl[P(C_6H_5)_3]_3$, known as Wilkinson's catalyst (*24*), was capable of catalyzing the hydrogenation of olefins at 25 °C under 1 atm of hydrogen pressure. This finding directed attention to other phosphine complexes, including those of other transition metals, and stimulated an exploration into the effect of such catalysts on other processes. The investigation of the effect of the phosphine ligands on structures showed that substantially greater catalytic activity is obtained by replacing the monodentate phosphine ligands, such as triphenylphosphine, with the chelating bidentate-type diphosphines, such as 1,2-bis(diphenylphosphino)ethane, $(C_6H_5)_2PCH_2CH_2P(C_6H_5)_2$ (*25*). The enhancement effect of phosphine ligands appears to apply to all of the group 8 metal ions.

Studies showed that these metal phosphine complexes enhance not only the hydrogenation of olefins but also many other processes.

These processes include the hydrogenation of acetylenes, dienes, aldehydes, and ketones; the hydroformylation of olefins, known as the "oxo" process; and the insertion of CO into carbon–chlorine bonds (*26*).

An interesting result of the study of these metal phosphine ligands was the finding that an asymmetric (asymmetric refers to four different groups attached to a central atom) phosphine ligand is able to induce asymmetric hydrogenation (*27, 28*). This finding is of particular importance for the preparation of biologically active material such as amino acids.

Two commercially important amino acids have been prepared as a result of this finding. L-Dopa is used in the treatment of Parkinson's disease. This disease occurs in the middle-aged and elderly. It is characterized by slowness, diminished ability to move purposefully, muscular rigidity, and tremors. Although no cure for this disease exists at present, L-dopa, the L isomer of 3,4-dihydroxyphenylalanine, greatly reduces the symptoms.

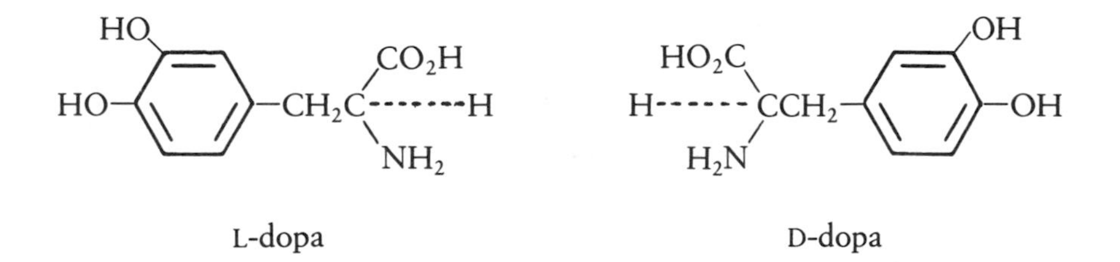

D-Dopa, shown on the right, is a mirror image of L-dopa. Although D-dopa contains the same groups about the center carbon atom, these two stereoisomers cannot be superimposed on each other. In the typical organic sythesis of dopa, both isomers are formed in equal amounts. The result is called a racemic mixture. In the treatment of Parkinson's disease, only the L-dopa has a positive effect; the D isomer provides no benefits. Also, treatment usually requires 3–5 g/day of L-dopa (*29*). If we treat the patient with the racemic mixture of D and L isomers, the patient would have to have to take double this amount to get the needed amount of L-dopa. This dosage could place an undue hazard on an already infirm person. Therefore, importantly, the patient must be supplied with the L form only.

In their study of the use of asymmetric phosphine ligands to induce asymmetric synthesis, Knowles and co-workers (*30–32*), Kagan and Dang (*33*), and Fryzuk and Bosnick (*34*) developed processes for producing L-amino acid derivatives by the asymmetric hydrogenation of (acylamino)acrylic acids to amino acids with stereospecificity in excess of 95%. In 1977, Knowles et al. (*35*) received a U.S. patent for the first commercial asymmetric synthesis of L-dopa. The process starts with vanillin. The process scheme is shown in Scheme 15.I.

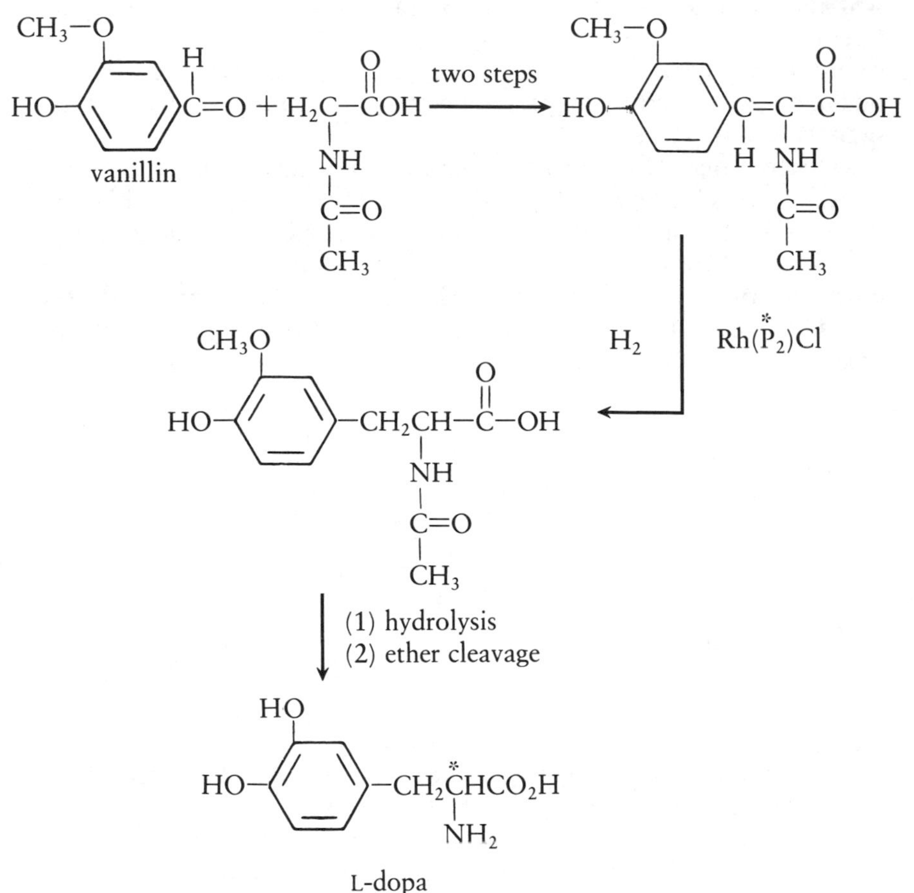

Scheme 15.I. The asterisk refers to an asymmetric center; the term $Rh(P_2{}^*)^+$ *refers to rhodium complexed with two asymmetric phosphine ligands.*

In this first patent, Knowles described the effectiveness of a monophosphine, such as **I,** where the asymmetry was on the single phosphorus atom:

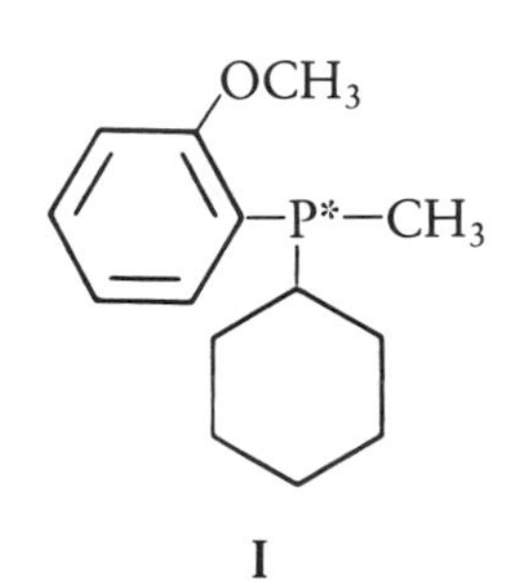

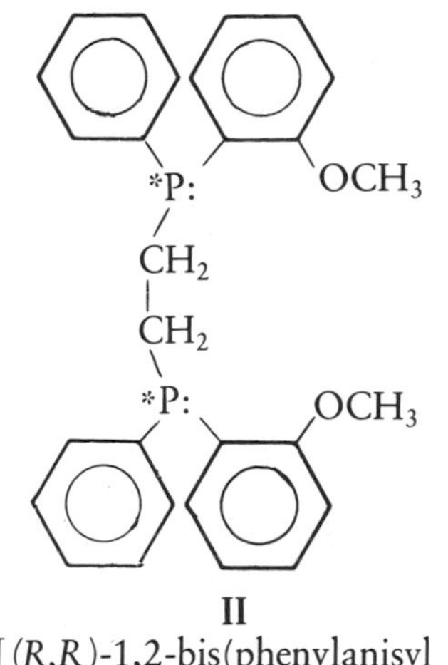

I
(cyclohexyl(anisyl)methylphosphine)

II
[(*R*,*R*)-1,2-bis(phenylanisyl phosphine)ethane (DIPAMP)]

In a later patent (*36*), Knowles et al. showed that diphosphines, such as (*R*,*R*)-1,2-bis(phenylanisylphosphine)ethane (DIPAMP, **II**), which have asymmetry on both phosphorus atoms, were even more effective. Kagan and Dang (*33*) showed that the diphosphine, DIOP, a bidentate phosphine ligand in which the asymmetry was on the non-phosphorus portion of the ligand, was equally effective.

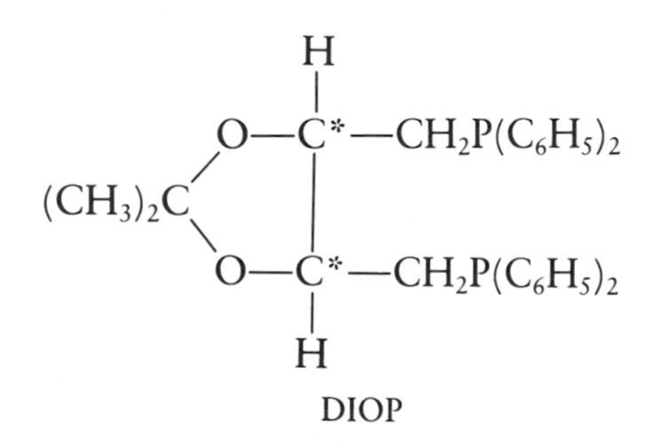

DIOP

This asymmetric synthesis process has also been applied to the preparation of L-phenylalanine. L-Phenylalanine is a component of the artificial sweetener, Aspartame, shown on page 250. However, because processes such as fermentation are proving to be more eco-

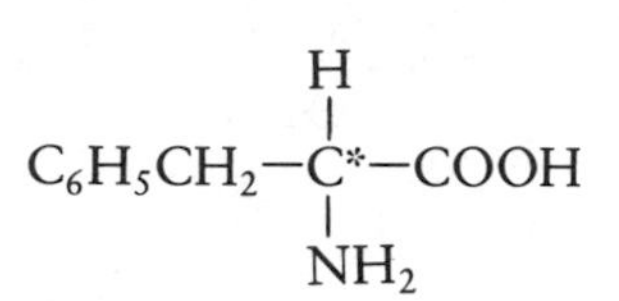

L-phenylalanine

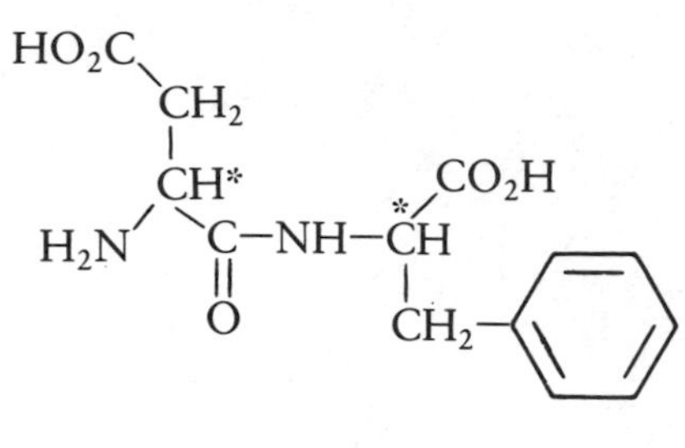

aspartame

nomical, L-phenylalanine is not made commercially today by asymmetric induction.

A second important use of triphenylphosphine is in a reaction discovered by Wittig and Geissler (*37*). The reaction has become one of the most important methods for the synthesis of compounds containing a C═C bond. The reaction involves a phosphorus ylid (**IV**) prepared from triphenylphosphine and an organic halide, X, as follows:

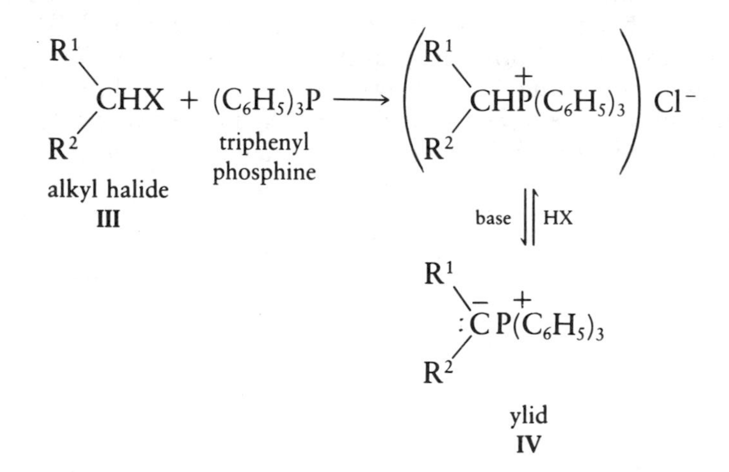

Such ylids react with aldehydes and ketones to yield unsaturated compounds.

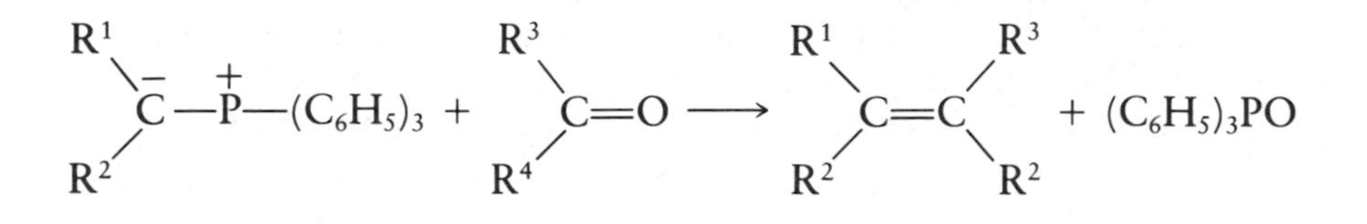

The reaction has allowed the successful synthesis of many natural products. An example is vitamin A, a vitamin that is essential for proper retinal function and plays a strong role in maintaining the integrity of epithelial cells. An industrial-scale synthesis of vitamin A is based on the reaction (*38*)

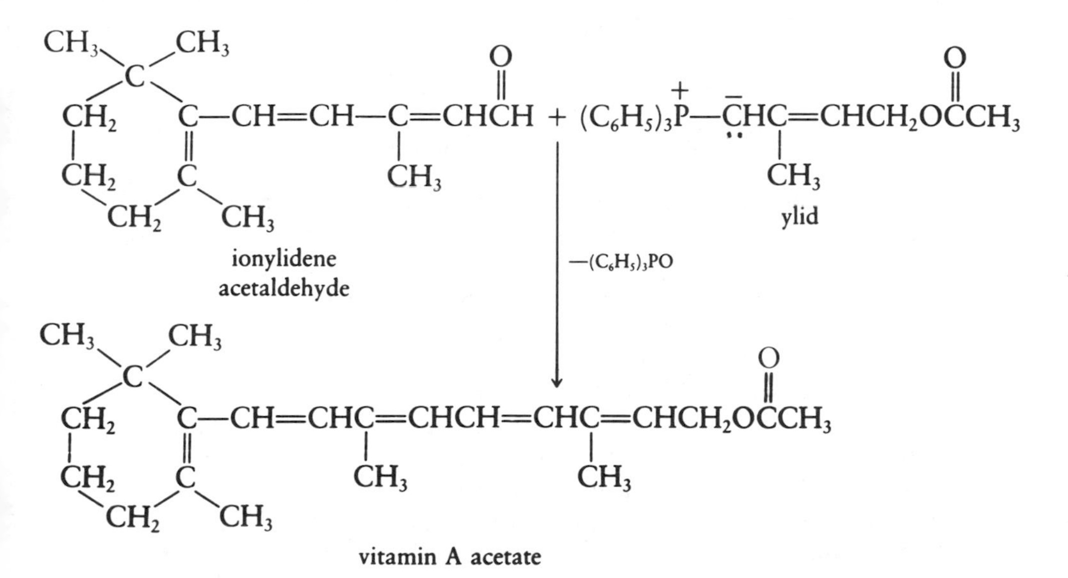

This olefin synthesis process is also used to prepare pheromones (sex attractants) for insects. Many of these pheromones are derivatives of unsaturated fatty alcohols, such as $RCH{=}CH(CH_2)_8$-$CH_2CH_2OC({=}O)CH_3$, a class of pheromone active on Lepidoptera (moths and butterflies).

A third use of an organic phosphine was a gasoline additive. In the 1960s, tributylphosphine was added to gasoline as an octane demand reducer and ignition improver. The compound is made from the catalytic addition of phosphine to butylene.

$$3C_2H_5CH{=}CH_2 + PH_3 \xrightarrow[\text{catalyst}]{\text{peroxide}} (C_4H_9)_3P$$

This proved to be too expensive when compared to the less expensive phosphate esters, and this use ended after a few years.

Trioctylphosphine Oxide. This compound can be prepared from the reaction of $POCl_3$ with octylmagnesium bromide (*39*)

$$3C_8H_{17}MgBr + POCl_3 \longrightarrow (C_8H_{17})_3PO + 3MgBrCl$$

or from the radical-initiated addition of 1-octene to phosphine (*40*), followed by an oxidation step:

$$3C_6H_{13}CH{=}CH_2 + PH_3 \longrightarrow (C_8H_{17})_3P$$

$$(C_8H_{17})_3P + H_2O_2 \longrightarrow (C_8H_{17})_3PO + H_2O$$

This product is currently being used as a metal extractant. It is particularly effective in extracting uranium from wet-process phosphoric acid solutions. Trioctylphosphine oxide (TOPO) is often used in conjunction with bis(2-ethylhexyl) phosphate (DEPA). This reaction is referred to as the DEPA–TOPO process. This process competes with the mono- and dioctylphenyl phosphate (OPAP) process. OPAP extracts U^{4+}, whereas DEPA–TOPO extracts U^{6+}.

The "black" wet-process phosphoric acid is first given a pretreatment step (an oxidation step) to remove the organic matter. The uranium valence is then adjusted to the desired state, either by reduction to U^{4+} or by oxidation to U^{6+}. A multistage extraction is then conducted by using a kerosene solution of the phosphorus-containing extractant. So that the uranium can be stripped from the kerosene solution, the valence of the uranium is reversed, that is, U^{6+} extracted by the DEPA–TOPO process is reduced to U^{4+} and stripped from the DEPA–TOPO–kerosene solution.

Literature Cited

1. Papay, A. U.S. Patent 3 446 739, May 27, 1969 (to Mobil Oil).
2. Papay, A.; Hollinghurst, R. U.S. Patent 3 652 410, March 28, 1972 (to Mobil Oil).
3. Johnson, M. K. *C.R.C. Crit. Rev. Toxicol.* **1975,** *3*, 289.
4. Sullivan, J. D. U.S. Patent 3 773 864, Nov. 28, 1973 (to Monsanto).
5. Dounchis, H. U.S. Patent 3 931 023, Jan. 6, 1979 (to FMC).
6. Codding, J. W.; Haas, W. V., Jr.; Heumann, F. K. *Ind. Eng. Chem.* **1958,** *50* (2), 145.
7. Lanhan, W. G.; Runion, T. C. ORNL-479, U.S. Atomic Energy Commission Report, Oak Ridge National Laboratory, Oak Ridge, TN, 1949.
8. Stoltz, E. M., Jr. *Proc. 2nd Int. Conf. PUAE* **1958,** *3*, 234.
9. Armstrong, H. I.; Spink, P. W.; Waychoff, W. F. *Mod. Plastics* **1956,** *34*, 144.
10. Gramrath, H. R.; Carver, J. K. U.S. Patent 2 596 141, 1952 (to Monsanto).
11. Trademan, L. U.S. Patent 3 089 907, May 14, 1963 (to Chemagro).
12. Brun, W. A; Cruzado, H. J.; Muzik, T. J. *Trop. Agric.* (*London*) **1961,** *38*, 68–81.

13. Nunn, L. G., Jr.; Hesse, S. H. U.S. Patent 3 004 056, Oct. 10, 1961 (to General Aniline and Film).
14. Nunn, L. G., Jr. U.S. Patent 3 004 057, Oct 10, 1961 (to General Aniline and Film).
15. Cronan, C. S. *Chem. Eng.* **1959,** *66* (9), 108.
16. Hurst, F. J.; Crouse, D. J. *Ind. Eng. Chem., Proc. Des. Dev.* **1974,** *13*, 286.
17. Vincent, H. *Bull. Soc. Chim. Fr.* **1972,** *No. 2*, 4517.
18. Griffin, A. M.; Marshall, P. C.; Sheldrick, G. M. *J. Chem. Soc., Chem. Commun.* **1976,** 809.
19. Martin, T. W.; Norman, G. R.; Weilmuenster, E. A. *J. Am. Chem. Soc.* **1945,** *67*, 1662.
20. Whitworth, F. T. U.S. Patent 2 038 400, April 21, 1936.
21. Derby, I. H.; Cunningham, O. D. U.S. Patent 2 134 706, Nov. 1, 1969 (to P. C. Reilly).
22. Taggart, A. F. *Handbook of Mineral Dressing*, 8th ed.; Wiley: New York, 1964; pp 12-08 to 12-10.
23. Aplan, F. F. *Encyclopedia of Chemical Technology*, 3rd ed.; Wiley: New York, 1980; Vol. 10.
24. Young, J. F.; Osborn, J. A.; Jardine, F. J.; Wilkinson, G. *J. Chem. Soc., Chem. Commun.* **1965,** 131–132.
25. Halpern, J. *Phosphorus Sulfur* **1983,** *18*, 307–309.
26. Pignolet, L. J. *Homogeneous Catalysis with Metal Phosphine Complexes*; Plenum: New York, 1983.
27. Knowles, W. S.; Sabacky, M. J. *J. Chem. Soc., Chem. Commun.* **1968,** 1445–1446.
28. Horner, L.; Siegel, H.; Buthe, H. *Angew. Chem., Int. Ed. Engl.* **1968,** *7*, 942.
29. *The Merck Manual of Diagnosis and Therapy*, 13th ed.; Merck: Rahway, NJ, 1977; pp 1448–1449.
30. Knowles, W. S.; Sabaxky, M. J.; Vineyard, B. D. *J. Chem. Soc., Chem. Commun.* **1972,** 10.
31. Knowles, W. S.; Sabacky, W. S.; Vineyard, B. D. *Adv. Chem. Ser.* **1972,** *132*, 274–282.
32. Vineyard, B. D.; Knowles, W. S.; Sabacky, M. J.; Bachman, G. C.; Weinkauff, D. J. *J. Am. Chem. Soc.* **1977,** *99*, 5946.
33. Kagan, H. B.; Dang, T.-P. *J. Am. Chem. Soc.* **1972,** *99*, 6429.
34. Fryzuk, M. D.; Bosnick, B. *J. Am. Chem. Soc.* **1977,** *99*, 6262.
35. Knowles, W. S.; Sabacky, M. J.; Vineyard, B. D. U.S. Patent 4 005 127, Jan. 25, 1977 (to Monsanto).
36. Knowles, W. S.; Sabacky, M. J.; Vineyard, B. D. U.S. Patent 4 008 281, Feb. 15, 1977 (to Monsanto).
37. Wittig, G.; Geissler, G. *Liebigs Ann. Chem.* **1953,** *580*, 44.
38. Pommer, H. *Angew. Chem.* **1960,** *72*, 811.
39. White, J. C. ORNL-2161, 1956, U.S. Atomic Energy Commission.
40. Rauhut, M. M.; Currier, H. A.; Semsel, A. M.; Wystrach, V. P. *J. Org. Chem.* **1961,** *26*, 5138.

Chapter Sixteen

The Code of Life: DNA and RNA

ONE OF OUR MORE ENDURING NURSERY RHYMES proposes that little girls are made of sugar and spice and everything nice and that little boys are made of snakes and snails and puppy dogs' tails. The unromantic chemist, however, sees us as various molecular arrangements of the following elements: 65% oxygen, 18% carbon, 10% hydrogen, 3% nitrogen, 1.5% calcium, 1% phosphorus, and 1.5% other elements. To shatter our illusions further, the net value of the chemical ingredients in the human body is calculated as about $9.00. This figure is for a 150-pound person and is based on 1984 prices. In 1936, during the depression, the human body was worth only $0.98; in 1974, about $5.60.

Oxygen, carbon, and hydrogen constitute the greatest percentage of the body. Most of the hydrogen and oxygen, of course, is present as water. Phosphorus represents only 1% of our body weight, but, as we shall see, a very important 1%.

About 23% of the human skeleton is mineral matter. The phosphorus content of this mineral matter, calculated as tricalcium phosphate, $Ca_3(PO_4)_2$, represents 87% of the total. Similarly, our teeth are basically calcium phosphates.

The proper biological functioning of our bodies depends on the action of the many phosphorus-containing compounds, or biophosphates, in our systems. The most important are DNA (deoxyribonucleic acid) and RNA (ribonucleic acid). These compounds are located in the cells of living organisms and carry the genetic code that determines what we are. Another important biochemical is ATP, or adenosine triphosphate, the energy carrier. The chemistry and some of the functions of DNA and RNA are discussed in this chapter.

1002–0/87/0255$06.00/1

Phosphorus energy transfer agents such as ATP are covered in the next chapter.

Phosphorus enters our bodies mainly through our food. Every bite of meat or mouthful of vegetables has in it some phosphorus-containing compounds because phosphorus is also essential to plants and animals. Plants take up phosphorus from soil nutrients (fertilizers, decayed organic matter, etc.); animals eat plants, and we eat animals and plants. We also consume phosphorus compounds such as calcium phosphates (discussed in earlier chapters), which are incorporated as supplemental nutritive additives in food. Some phosphate is excreted and ends up in streams, rivers, and lakes, where plant life such as algae utilize it as one of their 16–17 required nutrients. The excess growth of algae leads to eutrophication, a subject that was covered in Chapter 8.

What role does phosphorus-containing DNA play in our biological system? When an egg cell is fertilized by a sperm cell, a new life begins. The fertilized egg absorbs various nutrients from its surroundings and begins to grow in an orderly manner. Different eggs grow into different animals—for example, some eggs hatch into chickens, others grow into elephants or monkeys, and some become human beings. Even humans show differences. Some are male; some are female. Some become pretty blondes; some handsome brunettes. The nutrient molecules or raw materials available to fertilized eggs are chemically not very different. What then determines how these raw materials are arranged into a unique final shape?

The answer came after years of accumulated research by many scientists. According to present findings, small units in the fertilized egg called genes transmit hereditary characteristics. The active components in the gene that serve as templates for growth and differentiation are the DNA molecules.

DNA is a very large molecule. Its presently accepted basic structure was established in 1953 by James B. Watson, Maurice H. F. Wilkins, and Francis H. C. Crick. (Their work so advanced the science of molecular biology that they were awarded a Nobel prize in 1962. The exciting story of the unraveling of the true DNA structure is told in *The Double Helix*, by James Watson.) As proposed, a DNA molecule consists of two very long thin chains, twisted around each other as a regular double helix. This spiral-staircase arrangement is shown in Figure 16.1. These chains, called nucleic acids, are formed by the combination of smaller building blocks called nucleotides. A nucleotide in turn is a combination of a nitrogen-containing base, a five-carbon sugar (for DNA, deoxyribose), and phosphoric acid. Four different nitrogen-containing bases are present in a DNA molecule: the two purines, adenine (A) and guanine (G); and the two pyrimidines, thymine (T) and cytosine (C). A nucleotide with adenine as the nitrogen-containing base is shown in **I.**

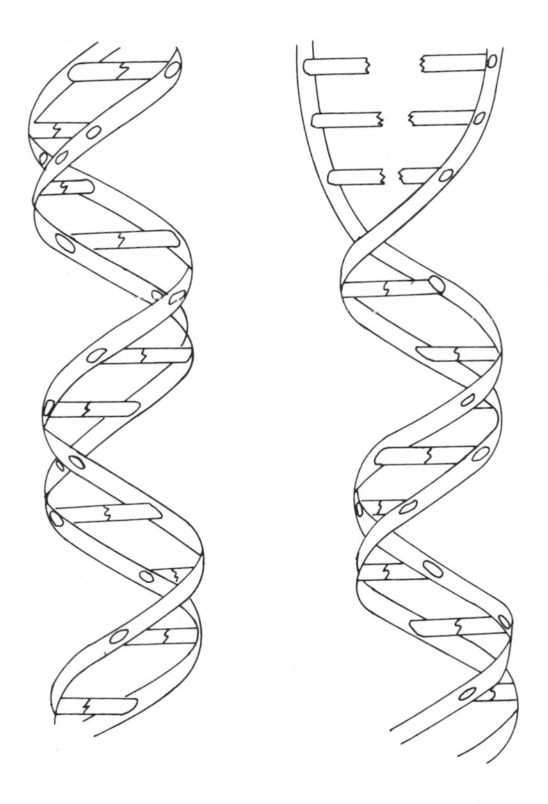

Figure 16.1. Pictorial representation of the DNA double helix.

That portion of the nucleotide that contains only the nitrogen base and the deoxyribose sugar is called the nucleoside. From the standpoint of phosphorus chemistry, we can visualize phosphoric acid as having the important function of joining the thousands of nucleosides together through phosphodiester bonds, which act as bridges to connect adjacent deoxyribose units. This combination forms the long chain of nucleic acid. The phosphate is esterified to the 5′-hydroxy group of one nucleotide and to the 3′-hydroxyl group on the adjacent deoxyribose unit. The numbering is shown in **I.** By convention, the prime numbers are used to number the deoxyribose ring; the unprimed numbers are reserved for numbering the nitrogen-containing bases, the purine or pyrimidine rings. A small section of a nucleic

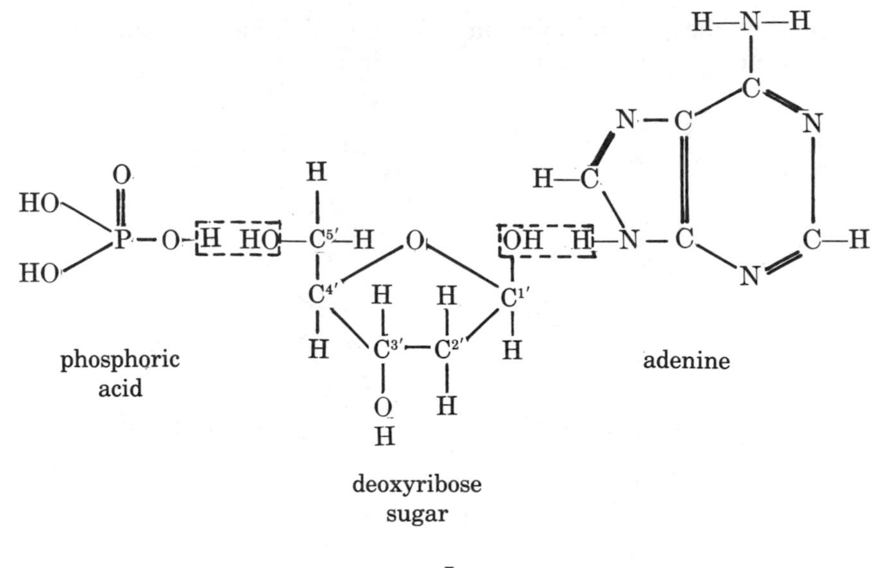

I

acid chain is shown in **II**. Although the four different nitrogen-containing bases are shown, in an actual nucleic acid chain, the arrangement of these bases is specific for a specific nucleic acid. The particular arrangement of these bases at various positions in the nucleic acid chain provides the basic genetic code of life.

In Figure 16.1, both sides, or banisters, of the staircase are of identical composition because each is a long chain of alternate phosphate and deoxyribose molecules. Each step, indicated by the horizontal bars in Figure 16.1, consists of two nitrogen bases linked together. One base is connected to deoxyribose through a ring nitrogen on the left banister, and the other is similarly connected to the deoxyribose on the right banister. These two bases are then connected to each other by hydrogen bonds. A hydrogen bond is formed by the attractive force between the positive hydrogen of one base and the negative oxygen or nitrogen of the complementary base. Because of the physical size of the four nitrogen bases and the location of those groups that can partake in the hydrogen bonds on those bases, the base thymine in one chain always bonds with the base adenine in the other chain, and the base cytosine always bonds with the base guanine. In other words, thymine is complementary to adenine, and cytosine is complementary to guanine. Because the bases are complementary to each other, the two chains of the double-helical DNA are idential either in base composition or in sequence. Also, the chains are antiparallel, that is, their internucleotide phospho-

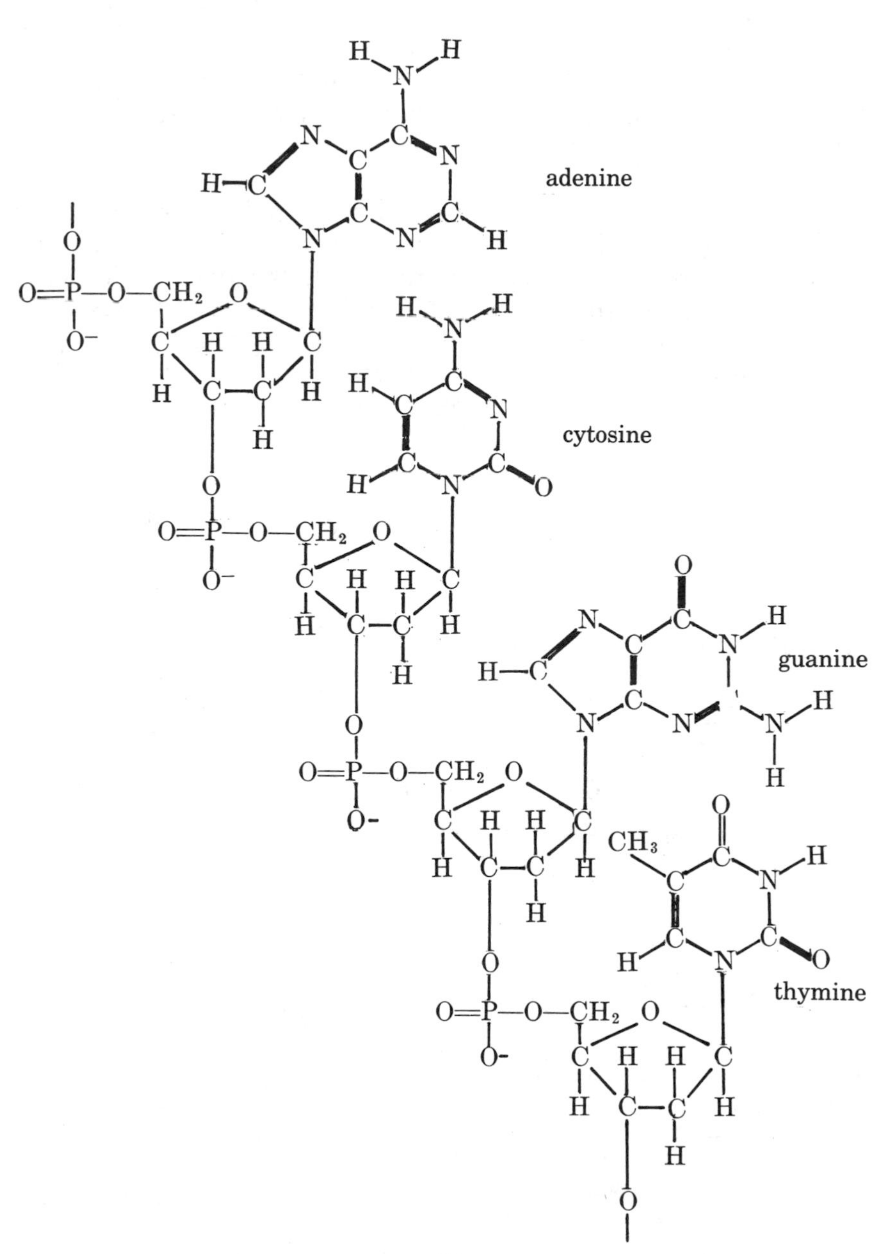

II

diester bridges run in opposite directions. Because the nucleic acid chain is so long, many different combinations and arrangements of these bases are possible. The difference between human and fish, frog and yeast, and male and female depends on the percentage of each base and its specific position in the DNA molecule, the length of the DNA molecule, and the number of different DNA molecules in each cell. Hereditary information is transmitted via these arrangements of bases in DNA.

The present view is that the bases are bonded together by fairly weak hydrogen bonds, as shown in Chart 16.I. When cell division begins, these hydrogen bonds rupture, and the two DNA chains separate (Figure 16.1b). Each chain uncoils to form two independent single DNA strands. Each strand is now a template for arranging and organizing the smaller nucleotide units present in the cell into a complementary chain. The left chain is the template through which

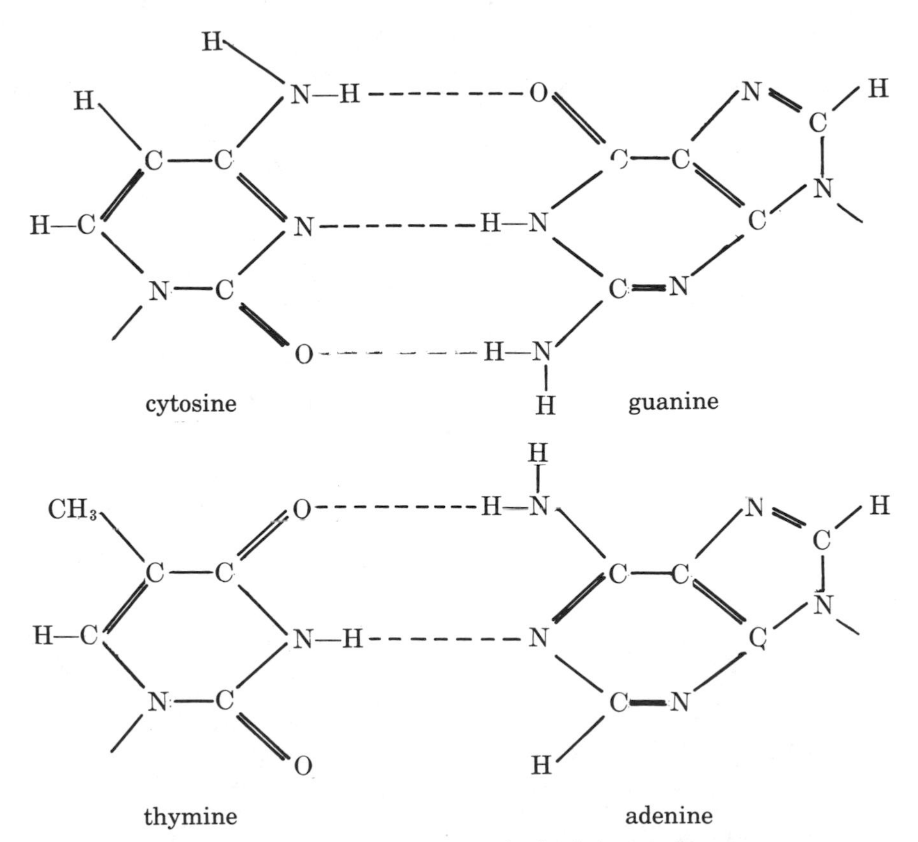

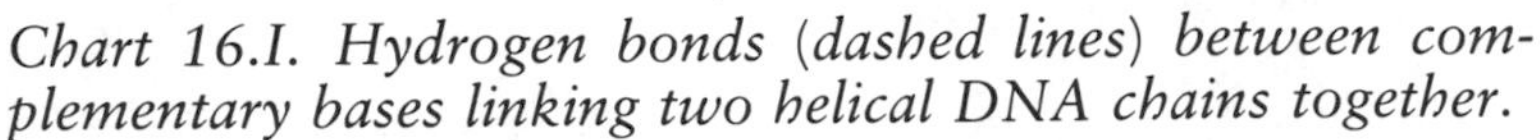
Chart 16.I. Hydrogen bonds (dashed lines) between complementary bases linking two helical DNA chains together.

the cell synthesizes a new complementary right chain, and the right chain does the same for the new left chain. Thus, a new identical double helix of nucleic acid chains is synthesized from the original.

Because a DNA chain contains thousands of nucleotide units and the synthesis of DNA goes on in billions of living cells each second, does nature ever make a mistake, so that the daughter DNA double helix produced is not exactly like the mother DNA? According to Jacques Monod, Nobel prizewinner for medicine and physiology in 1965, some errors do occur, and when they do, the new DNA molecule with the slight difference will produce a cell with the new characteristic. Because DNA molecules generally produce daughter molecules identical with themselves, once an error occurs, subsequent DNA generations and their respective cells will contain the differences caused by the original error. This invariance to change is retained until the next error occurs. One source of mutations is the result of such errors. Other sources include the effects of radiation and chemicals on the DNA. A succession of many such errors and coincidences were required for the living cell to evolve into humans as we know them today. This theory is compatible with Darwin's theory of evolution. According to him, the evolution of a living creature is determined by natural selection—that is, survival of the fittest. Many past errors of DNA reproduction fizzled out because the mutation produced could not survive its hostile environment.

In addition to DNA, the other important phosphorus-containing molecule that is essential to the biological process is RNA, ribonucleic acid. RNA's structure is similar to DNA's structure, with the backbone of the helix (the banister of the spiral staircase) constructed of ribose sugar molecules (instead of deoxyribose sugar) joined by phosphoric acid. Ribose differs from deoxyribose in having only one oxygen atom at carbon 2:

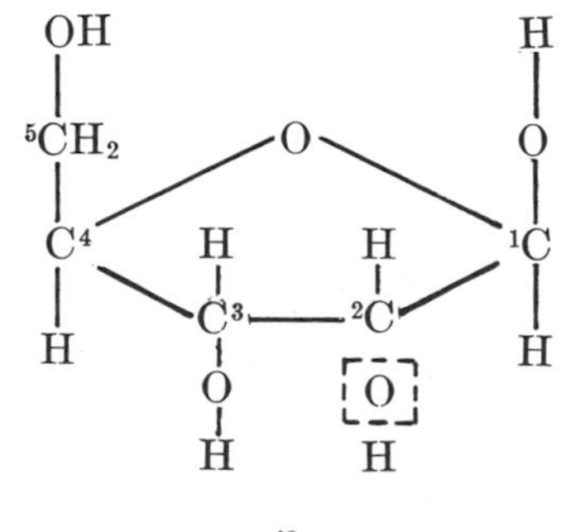

ribose

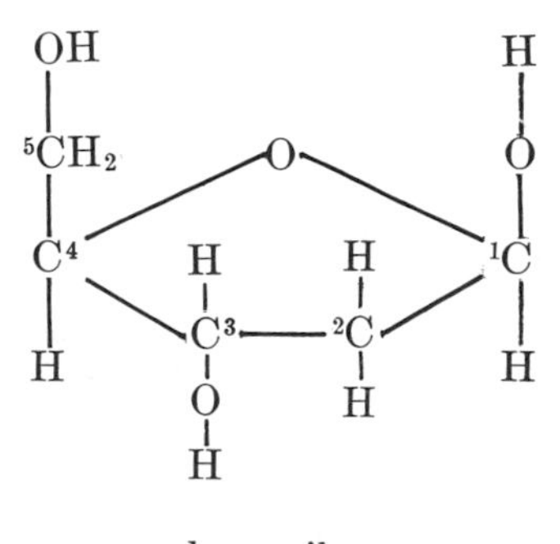

deoxyribose

Also, in RNA, one of its four nitrogen bases is uracil rather than the thymine base present in DNA.

RNA, of which three major forms are known, exists mostly as a single-stranded polynucleotide. However, when a small section exists in a double helix structure, the steps in the very short spiral staircase section are also formed by complementary base pairs. As in DNA, cytosine is bonded to guanine, but adenine, instead of bonding to thymine, is bonded to uracil (U):

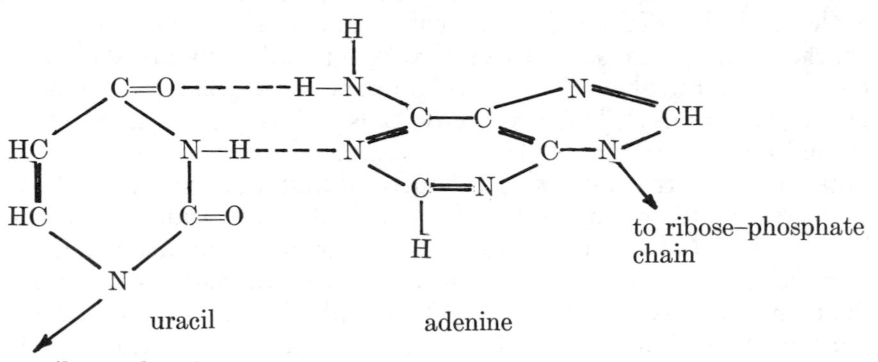

Many of the functions of RNA in a biological system are still undiscovered. However, one of its major functions is to effect the synthesis of protein molecules from amino acids. An amino acid is an organic compound containing an amino group, $-NH_2$, and a carboxylic acid group, $-COOH$. For example, the amino acid glycine has the structure H_2N-CH_2-COOH, and the amino acid alanine has the structure $H_2N-C(CH_3)HCOOH$. When two amino acids join "head to tail" and a water molecule is eliminated between them, an amide linkage forms:

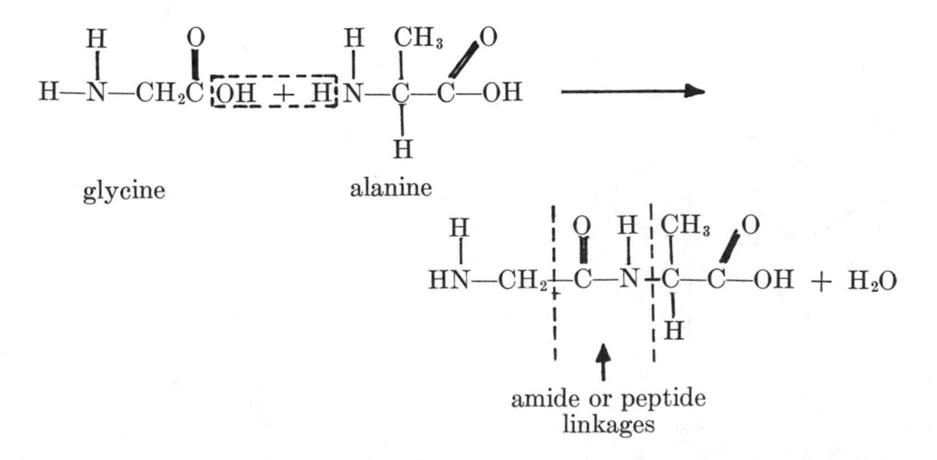

This amide linkage is also called a peptide bond. About 20 different amino acids are commonly used to build a protein molecule. When hundreds of these amino acids are joined, they form a polypeptide or a protein molecule. In a protein molecule, the amino acids are connected in a preordered sequence that determines the function of the protein, that is, structural, hormonal, or catalytic. For example, liver cells make proteins suitable for use in the liver. The liver proteins have chain lengths, amino acid sequences, and geometric structures different from those proteins made for the pancreas. Many enzymes, the catalysts that accelerate biochemical reactions, are produced in the liver and pancreas and also are protein molecules. They differ from each other with respect to size, amino acid sequence, and three-dimensional geometric structure.

How do cells make protein molecules with these differences? The gross mechanism for protein synthesis has been elucidated by molecular biologists through some very ingenious laboratory experiments. Protein synthesis involves three distinct types of RNA molecules:

(1) Ribosomal RNA, or rRNA. When combined with a protein, it is called a ribosome. Ribosomes are located in the cytoplasm of the cell, and they are the factories where proteins are made. The function of the rRNA in ribosomes is not fully understood.

(2) Messenger RNA, or mRNA. As the name indicates, mRNA carries the message from the DNA, that is, the genetic information code. mRNA functions as the template by which ribosomes translate the genetic information of the DNA into the amino acid sequence of proteins. The nucleotide sequence of mRNA is complementary to the genetic message in a segment of the template strand of DNA. The code for the sequence of amino acids in the protein is a series of trinucleotides in the mRNA chain, called a codon.

(3) Transfer RNA, or tRNA. Molecules of tRNA are relatively small single strands of ribonucleotides. These molecules consist of about 75–100 phosphorus-containing nucleotides. Each of the 20 amino acids found in proteins has at least one tRNA to bind it and transfer it to the ribosome. Each tRNA contains one trinucleotide sequence, called an anticodon, that is a complement to the trinucleotide sequence of the mRNA that codes for one specific amino acid. The number of different types of bases in tRNA is much greater than in mRNA. Not only does tRNA have the four major RNA bases but also it has a large number of rare bases; these rare bases can constitute up to 10% of the total base count. The structure of a yeast RNA is shown in Figure 16.2.

When the body needs a certain protein, a signal is sent to the cell. Inside the cell, specific enzymes come into play, and the DNA double helix, with its coded messages, uncoils. Portions of the two strands

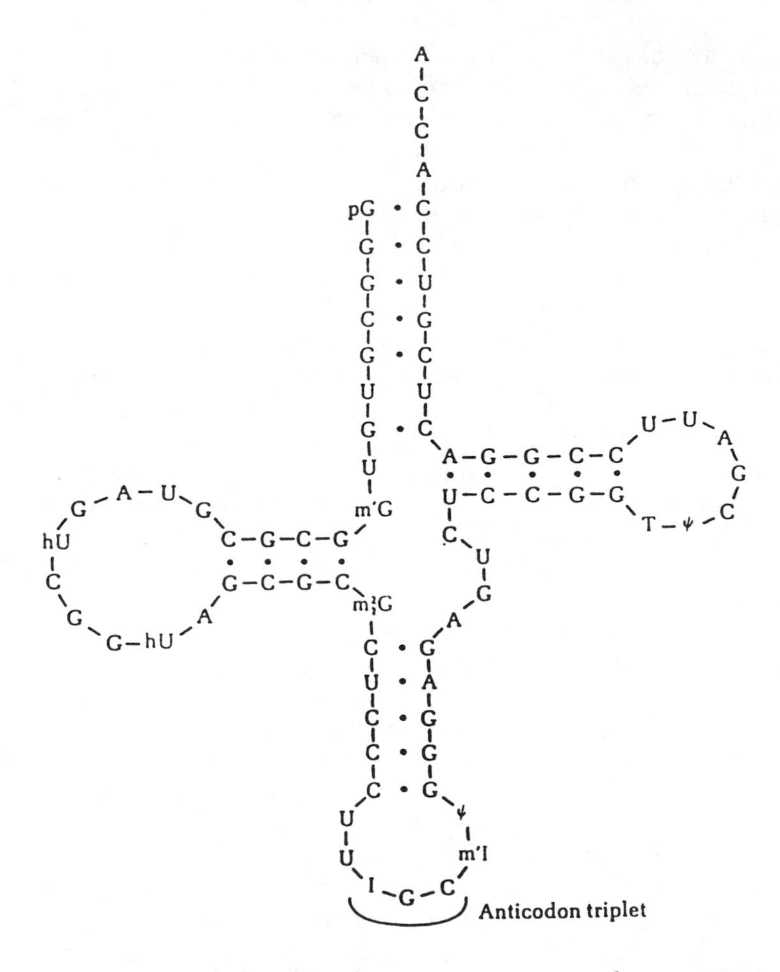

Figure 16.2. The nucleoside sequence in yeast alanine tRNA1. In addition to adenine (A), guanine (G), uracil (U), and cytosine (C), the following symbols are used for the nucleosides: ψ is pseudouridine, I is inosine, T is ribothymidine, hU is 5,6-dihydrouridine, m^1I is 1-methylinosine, m^1G is 1-methylguanosine, and m_2^2G is 2,2-dimethylguanosine. The dots between the parallel sections show the points where hydrogen bonding occurs between complementary bases. All tRNAs have a similar cloverleaf structure. The specific anticodon for alanine shown is the nucleotide triplet capable of recognizing a codon for alanine in the mRNA molecule.

separate. The synthesis of mRNA begins. The process for the synthesis of mRNA is called transcription. The single nucleotides present in the cytoplasm of the cell attach themselves quickly to one of the DNA strands via complementary nitrogen bases and, when condensed by the enzyme DNA-directed RNA polymerase, form a single strand of mRNA. Thus, the sequence of bases in one strand of the chromosomal DNA is enzymatically transcribed in the form of a single strand of mRNA. Because DNA contains two strands, different mRNA strands can be transcribed from each of the two DNA strands. The mechanism by which only one of the two strands of DNA is chosen for transcription is unknown.

The message carried by the sequential arrangement of nitrogen bases in the original DNA is passed on to the new mRNA strand. Every three consecutive bases represent a code word (codon) for a specific amino acid. After the mRNA molecule is formed, it leaves the nucleus and attaches itself to a ribosome. Meanwhile, the separated portion of DNA double helix in the cell rewinds and is available for later use. After it is picked up by the ribosome, the mRNA strand begins to call for the proper amino acids according to the coded message it contains.

The tRNA, which carries the called-for amino acid and also contains the three-nucleotide anticodon that is specific for the amino acid, answers the call. This amino acid bearing tRNA is called aminoacyl-tRNA ester, because the tRNA is esterified to the carboxylic acid of the amino acid through a hydroxy group of the terminal nucleotide, which is always adenosine. The ester lodges its anticodon onto the three-nucleotide codon of the mRNA strand. The mRNA then calls for the next tRNA with its amino acid. When it arrives at the site adjacent to the first amnio acid, the two amino acids join to form a peptide bond. The tRNA for the first amino acid leaves, and the second tRNA now has two amino acids attached to it. After each peptide bond is formed, the ribosome moves to the next codon of the mRNA to bring this codon in alignment for the next aminoacyl-tRNA. Repeating this process brings the amino acids to the ribosome in the sequence dictated by the coded message from the mRNA, and the amino acids rapidly build into a protein chain. When completed, the protein chain is released into the cytoplasm, where it arranges into its proper geometric structure for use. The freed mRNA, when attached to another ribosome, can be used to assemble another protein molecule.

Although this description of protein synthesis is simplified, it shows how important DNA and RNA are in the biological system. Once the nature and function of these phosphorus compounds are understood, scientists should be able to unlock the secret of life itself.

More detailed and sophisticated discussion of the molecular biology of DNA and RNA is available in the original publications of such distinguished scientists as E. F. Hoppe-Seyler and F. Miescher, who first isolated nucleic acid in 1869; Oswald Avery and his co-workers, who in 1943 resolved the role of DNA as a carrier of genetic information; J. B. Watson, M. H. F. Wilkins, and F. H. C. Crick, who established the presently accepted double helix structure for DNA; F. H. C. Crick and his co-workers, who in 1961 provided experimental evidence to support the concept of triplet code; S. Ochoa, who in 1955 discovered a catalyst to manufacture synthetic RNA; A. Kornberg, who discovered a catalyst to produce synthetic DNA; M. W. Nirenberg and J. H. Matthaei, who in 1961 used Ochoa's synthetic RNA process to crack the genetic code for making a polypeptide; W. Arber, H. Smith, and D. Nathans, who discovered the DNA restriction endonucleases and their usefulness in the dissection of genes; F. Sanger, who in 1977 reported the first base sequence of DNA molecule; Sanger and his colleagues and A. Maxam and W. Gilbert, who developed methods for gene sequencing; and R. Okazaki and J. Cairns, who both contributed to the understanding of the mechanism of DNA replication in bacteria. The publications of these scientists will lead you to additional important contributions by other molecular biologists.

A good summary of the functions of DNA and RNA and some of the research done by various scientists to arrive at the understanding of these functions are described in references *1–5*.

We have seen how the DNA carries the genetic code that in large part ultimately controls our inherited features plus the foundation of growth, development, and reproduction. The gene, composed of DNA, is the self-reproducing particle that constitutes the unit of heredity. The genes are located in the cell chromosomes. These chromosomes are fine, long filaments that consist of hundreds or thousands of diverse genes, connected in a single file and in a given order. The genes may be directly attached to one another or may be joined by a fine fiber. These genes are activated or suppressed as needed.

Because the genes are composed of DNA and because the DNA is a condensation polymer of nucleotides, genes should be able to be produced in the laboratory. H. Gobind Khorana and his colleagues did exactly that. In July of 1970, he and his colleagues announced the total synthesis of the gene for an alanine transferase RNA (*6*), and in 1972, they reported the total synthesis of a tyrosine suppressor transfer RNA (*7–19*). This latter gene was incorporated into a mutated bacteriophage by using the procedures of genetic engineering. When so incorporated, it restored the ability of the mutant phage to reproduce; thus, the synthetic gene had the same biological capability as the natural gene.

The synthesis of these genes entailed making segments of the gene by chemical synthesis and uniting these segments by use of an enzyme, DNA ligase. The chemical synthesis provided over 90% of the internucleotide bonds present in both strands of the DNA. For those interested in more details, references 6 and 7 give details of the synthesis principles used.

In recent years, substantial developments have been made in genetic engineering. Foreign genes or foreign segments of DNA can now be incorporated (ligated) either with stable replicatory plasmids or with bacteriophage DNA through the selective breaking and re-forming of the desired phosphate ester bonds. These segments are then introduced into living cells. Once incorporated within the cell, these foreign visitors can function in a way to cause the cell to produce proteins of type and concentration quite different from those produced by the untreated cells. For example, such cells can be used in fermentation processes to produce biologically functional proteins, such as insulin, growth hormones, interferons, blood clotting factors, antibodies, and other pharmaceutically important compounds.

Currently, a major source of the foreign genes and DNA segments is from other living species. However, the ability to synthesize genes affords much greater breadth to such work. Now automated DNA synthesizers, called "gene machines" (*20*), exist that automate many of the techniques used by Khorana to synthesize DNA. Today, these gene machines cannot actually synthesize genes, but they are able to synthesize DNA segments. These segments can be of help in expediting the development and the fulfillment of the process of genetic engineering.

Literature Cited

1. Levine, L. *Biology of the Gene*; C. V. Mosby: St. Louis, MO, 1969.
2. Srb, A. M.; Owen, R. D.; Edgar, R. S. *General Genetics*, 2nd ed.; W. H. Freeman: San Francisco, 1965.
3. Weissbach, H.; Pestka, S. *Molecular Mechanism of Protein Biosynthesis*; Academic: New York, 1977.
4. Kornberg, A. *DNA Replication*; W. H. Freeman: San Francisco, 1980.
5. Lehninger, A. L. *Principles of Biochemistry*; Worth: New York, 1982.
6. Agarwal, K. L., et al. *Nature (London)* **1970,** *227*, 27.
7. Khorana, H. G. *Science (Washington, D.C.)* **1979,** *203*, 614.
8. Khorana, H. G., et al. *J. Biol. Chem.* **1976,** *251* (3), 568–570.
9. van de Sande, J. H.; Caruthers, M. H.; Kumar, A.; Khorana, H. G. *J. Biol. Chem.* **1976,** *251* (3), 571–586.

10. Minamoto, K.; Caruthers, M. H.; Ramamoorthy, B.; van de Sande, J. H.; Sidorova, N.; Khorana, H. G. *J. Biol. Chem.* **1976,** *251* (3), 587–598.
11. Agarwal, K. L.; Caruthers, M. H.; Fridkin, M.; Kumar, A.; van de Sande, J. H.; Khorana, H. G. *J. Biol. Chem.* **1976,** *251* (3), 599–608.
12. Jay, E.; Cashion, P. J.; Fridkin, M.; Ramamoorthy, B.; Agarwal, K. L.; Caruthers, M. H.; Khorana, H. G. *J. Biol. Chem.* **1976,** *251* (3), 609–623.
13. Agarwal, K. L.; Caruthers, M. H.; Büchi, H.; van de Sande, J. H.; Khorana, H. G. *J. Biol. Chem.* **1976,** *251* (3), 624–633.
14. Sekiya, T.; Hesmer, P.; Takeya, T.; Khorana, H. G. *J. Biol. Chem.* **1976,** *251* (3), 634–641.
15. Loewen, P. C.; Miller, R. C.; Panet, A.; Sekiya, T.; Khorana, H. G. *J. Biol. Chem.* **1976,** *251* (3), 642–650.
16. Panet, A.; Kleppe, R.; Kleppe, K.; Khorana, H. G. *J. Biol. Chem.* **1976,** *251* (3), 651–657.
17. Caruthers, M. H.; Kleppe, R.; Kleppe, K.; Khorana, H. G. *J. Biol. Chem.* **1976,** *251* (3), 658–666.
18. Kleppe, R., et al. *J. Biol. Chem.* **1976,** *251* (3), 667–675.
19. Ramamoorthy, B.; Lees, R. G.; Kleid, D. G.; Khorana, H. G. *J. Biol. Chem.* **1976,** *251* (3), 676–684.
20. Cook, R. M. *Res. Dev.* **1984,** *May*, 106.

Chapter Seventeen

ATP: The Energy Carrier

THE SYNTHESIS OF DNA (deoxyribonucleic acid), RNA (ribonucleic acid), and protein molecules requires work or energy. For that matter, all human activity—breathing, eating, running, playing tennis or golf—requires energy. Energy is also required when a plant synthesizes sugar or starch molecules or when a firefly emits light. A major source of energy for all these activities is the biotriphosphates. Energy is given off when one triphosphate bond is hydrolyzed to form a monophosphate and a diphosphate.

For most biological processes, the energy carrier or energy-transfer agent in the cell is a biotriphosphate called adenosine triphosphate, or ATP (**I**). ATP is composed of three kinds of building blocks: the base adenine, the five-carbon sugar D-ribose (these two compounds, when combined, form one of the nucleosides for RNA), and a chain of three phosphate groups. The phosphate groups are connected by P–O–P bonds known as anhydride linkages. (Adenosine diphosphate, or ADP, also an important biological phosphorus compound, has only two phosphate groups.) ATP resides in cells in concentrations between 0.5 and 2.5 mg/cm^3. ATP serves as a major link between energy-yielding and energy-requiring chemical reactions in these cells. When one phosphate group in ATP is hydrolyzed at neutral pH at 25 °C to ADP and inorganic orthophosphate (P_i), 7,300 cal/mol of energy is liberated. This free energy of hydrolysis is used to perform biological work.

ATP is not the only phosphorus energy carrier in the biological system. Other carriers are shown in Table 17.I. Some compounds have a higher free energy of hydrolysis than ATP and some lower. The minus sign in Table 17.I indicates that the system loses or gives off energy on hydrolysis. AMP is adenosine monophosphate, another important biophosphate.

A phosphate group from a compound with a larger negative free energy of hydrolysis can be transferred to form phosphate compounds

1002–0/87/0269$06.00/1

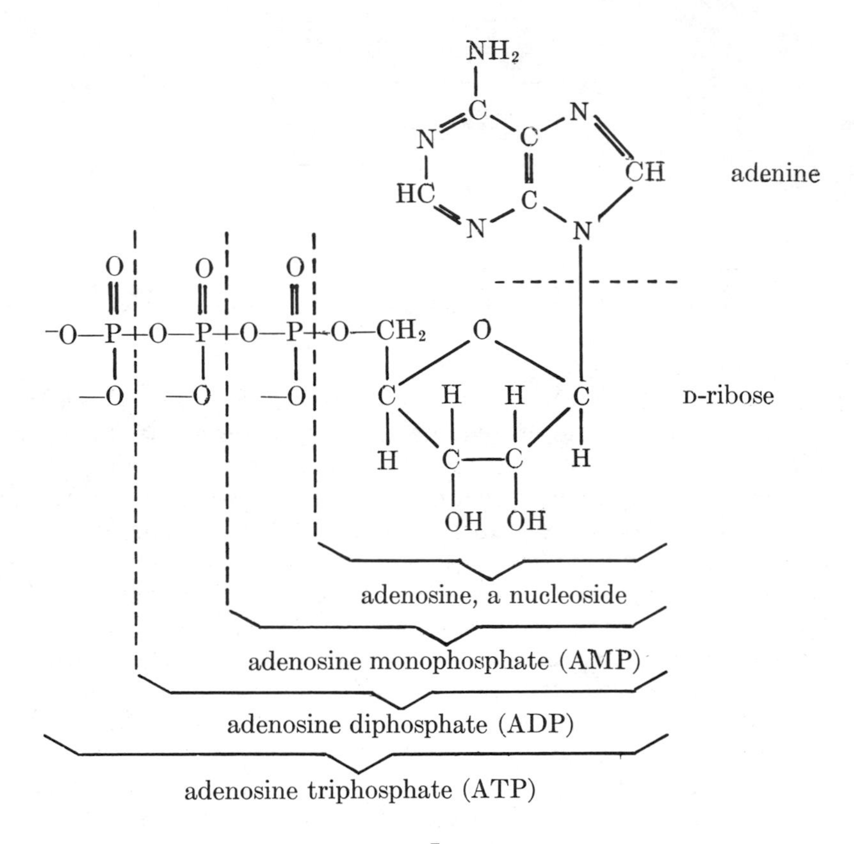

I

Table 17.I. Free Energy of Hydrolysis of Biophosphate Compounds

Compound	*Free Energy (cal/mol)*
Phosphoenolpyruvate	−14,800
3-Phosphoglyceroyl phosphate	−11,800
Phosphocreatine	−10,300
Acetyl phosphate	−10,100
ATP → ADP + P_i	−7,300
ADP → AMP + P_i	−7,300
Glucose 1-phosphate	−5,000
Fructose 6-phosphate	−3,800
AMP → adenosine + P_i	−3,400
Glucose 6-phosphate	−3,300
3-Phosphoglycerate	−3,100
Glycerol 1-phosphate	−2,200

with smaller negative free energies of hydrolysis. With each phosphate transfer, energy is also transferred.

How do organic phosphates with a high negative free energy of hydrolysis obtain their energy? In particular, how does ATP obtain its bond energy? The answer lies in the food we eat. In general, the breakdown of food molecules in the cell is always coupled with the eventual formation of ATP from ADP and inorganic orthophosphate. Two examples show how energy released by this breakdown is partially conserved by the formation of ATP. Both examples involve the breakdown of a glucose molecule (a six-carbon sugar). One example is by an anaerobic (absence of oxygen) cell, the other by an aerobic cell.

Anaerobic Oxidation of Glucose

One mole of glucose in a living cell can break down under anaerobic conditions to form 2 mol of lactate, a three-carbon compound, by the process called glycolysis. This breakdown involves the net formation of 2 mol of ATP from 2 mol of ADP and 2 mol of orthophosphate, as shown in the following overall reaction:

$$\text{glucose} + 2\text{HPO}_4^{2-} + 2\text{ADP}^{3-} \longrightarrow 2(\text{lactate}^-) + 2\text{ATP}^{4-} + 2\text{H}_2\text{O}$$

Simple breakdown of glucose to lactate is accompanied by a liberation or change in free energy of $-47{,}000$ cal/mol. Because 2 mol of ATP is formed from ADP and orthophosphate, and each mole of ATP requires at least 7,300 cal, the summation is $14{,}600/47{,}000 \times 100 = 31\%$. In other words, about 31% of the energy released when a molecule of glucose breaks down to lactate is conserved as ATP. The lactate then leaves the cell as waste.

The anaerobic breakdown of glucose into two lactates is known as the Embden–Meyerhof cycle, named after two German biochemists who predicted and experimentally proved the overall pattern of glucose breakdown in the 1930s. The mechanism is not as simple as shown. It requires 11 enzyme-catalyzed steps; each step involves specific enzyme catalysts. Also, each step involves a phosphorylated species—that is, all of the intermediates in the first 10 steps, the glycolysis reaction sequence, require the participation of a phosphate group in the form of an ester of phosphoric acid. The 11 steps are shown in equations 1–11. The enzyme that catalyzes each reaction is shown above the arrow.

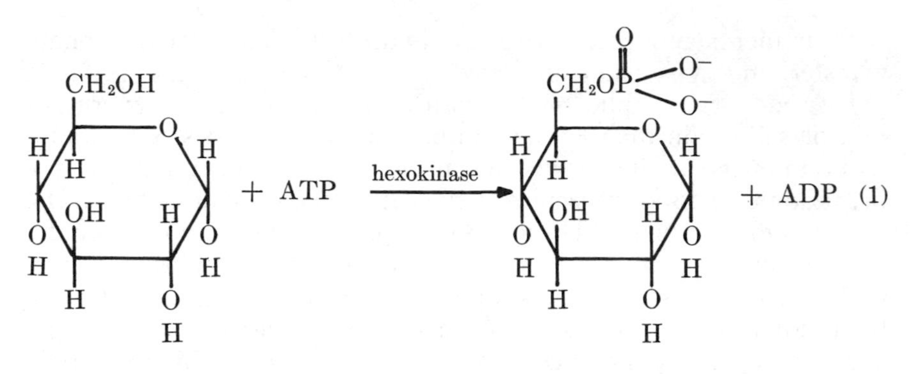

glucose

glucose 6-phosphate

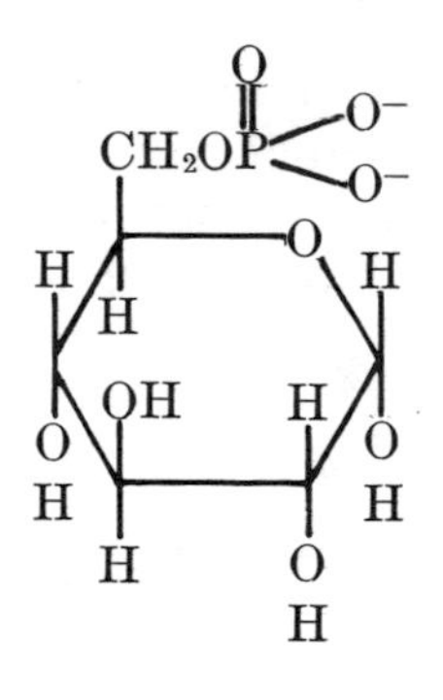

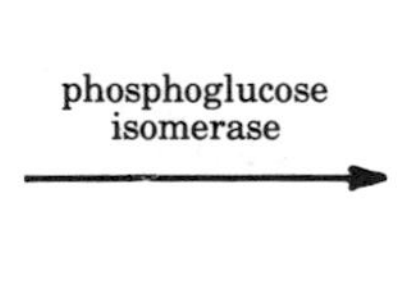

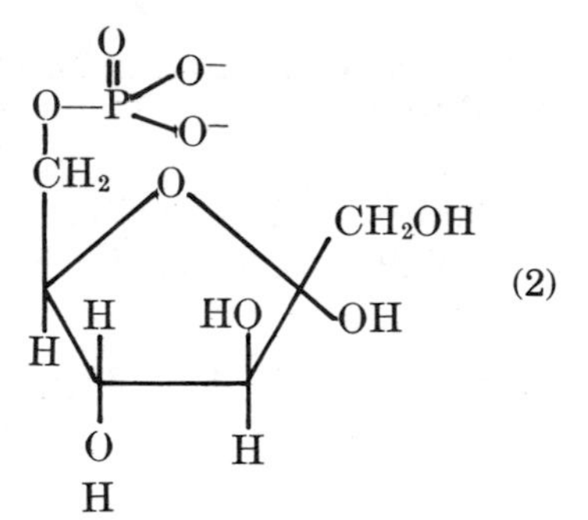

glucose 6-phosphate

fructose 6-phosphate

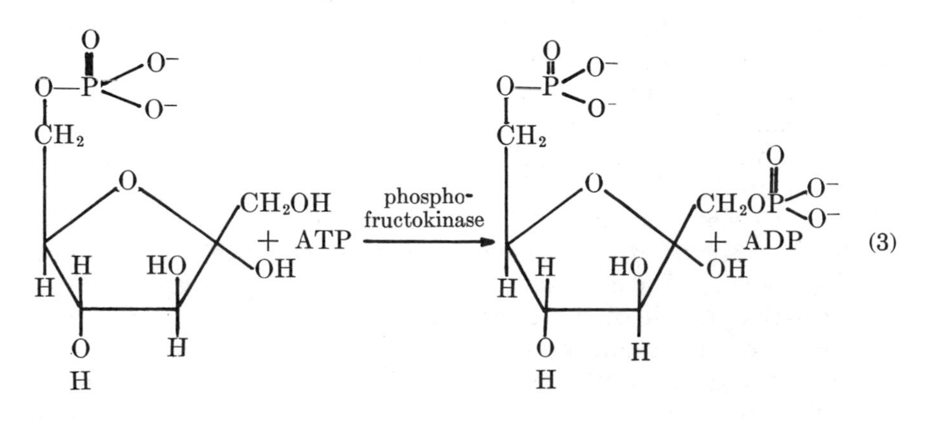

fructose 6-phosphate

fructose 1,6-diphosphate

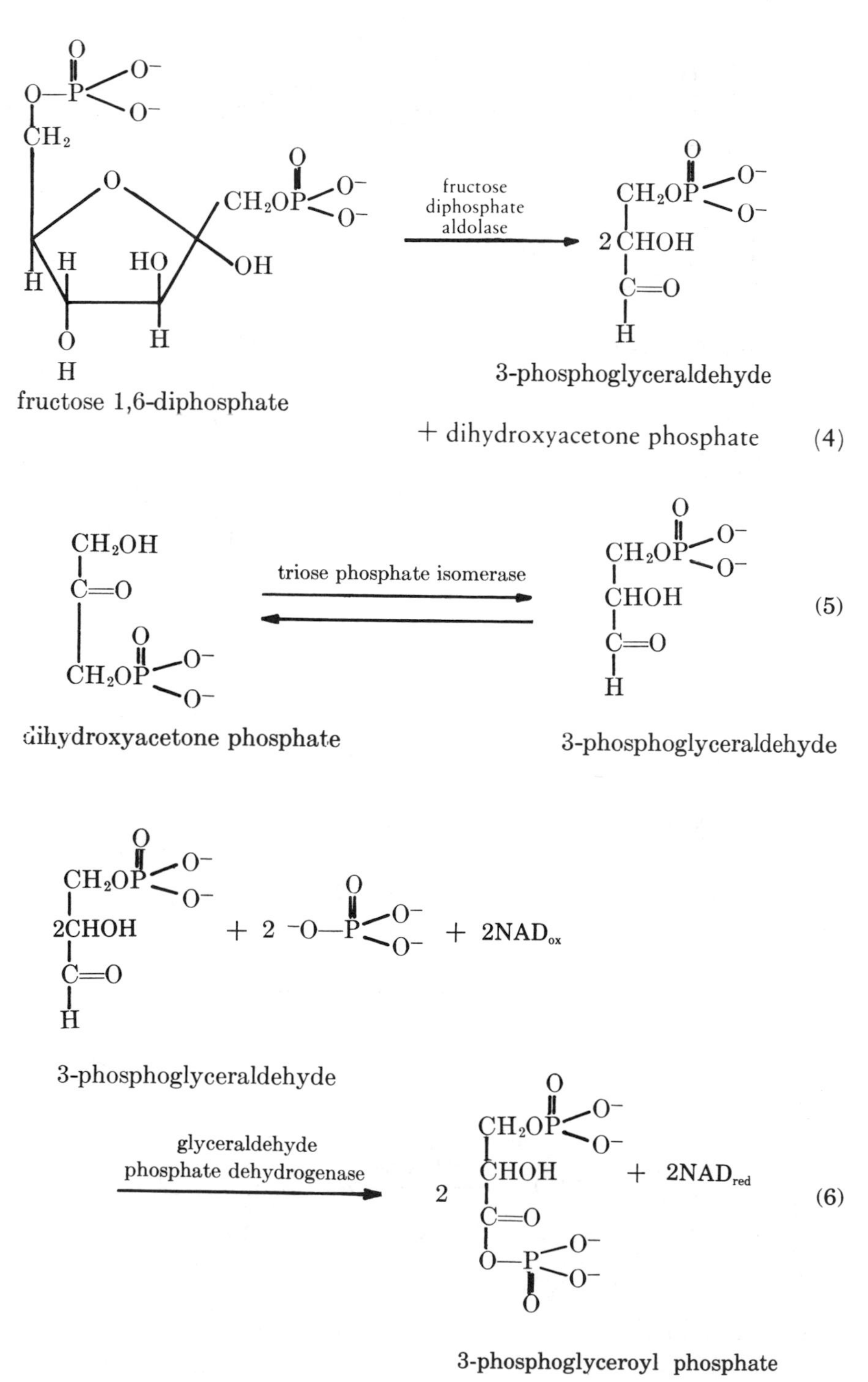

fructose 1,6-diphosphate
fructose diphosphate aldolase
3-phosphoglyceraldehyde
+ dihydroxyacetone phosphate (4)
triose phosphate isomerase
dihydroxyacetone phosphate
3-phosphoglyceraldehyde
(5)
3-phosphoglyceraldehyde
+ 2NAD$_{ox}$
glyceraldehyde phosphate dehydrogenase
+ 2NAD$_{red}$
(6)
3-phosphoglyceroyl phosphate

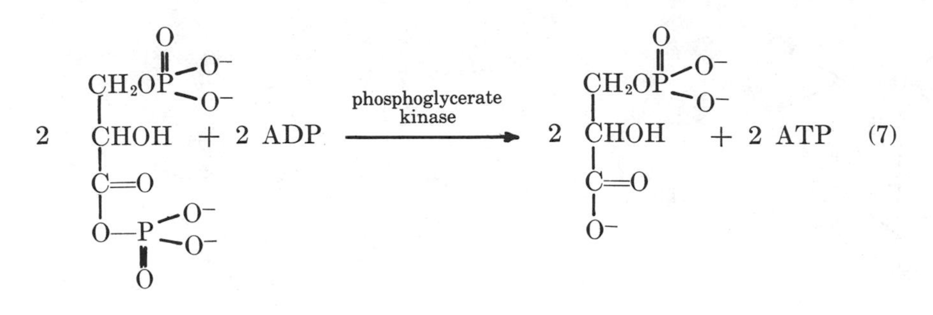

3-phosphoglyceroyl phosphate 3-phosphoglycerate

$$2\ \mathrm{CH_2OPO_3^{2-}{-}CHOH{-}COO^-} \xrightarrow{\text{phosphoglyceratemutase}} 2\ \mathrm{CH_2OH{-}CHOPO_3^{2-}{-}COO^-} \quad (8)$$

3-phosphoglycerate 2-phosphoglycerate

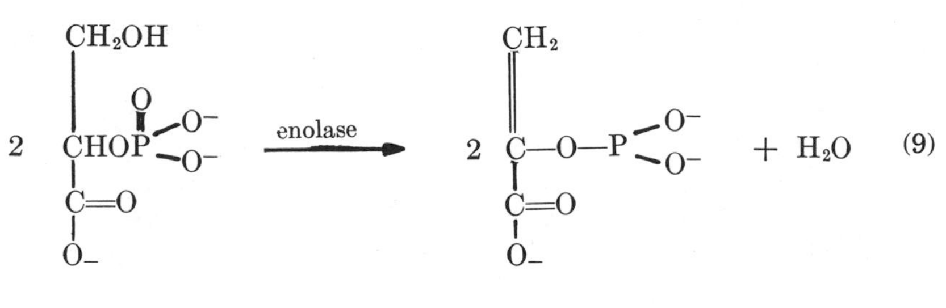

2-phosphoglycerate phosphoenolpyruvate

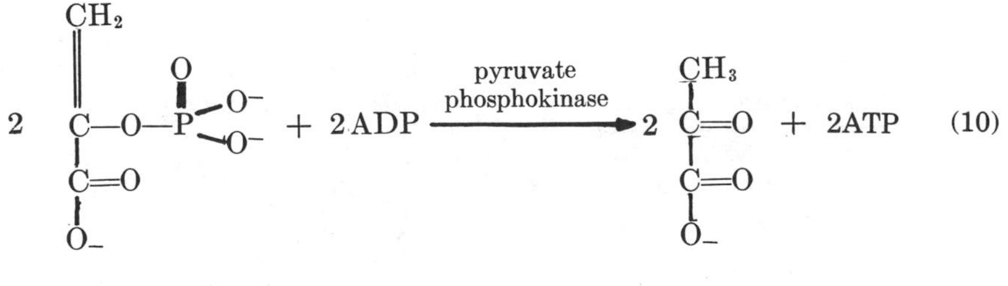

phosphoenolpyruvate pyruvate

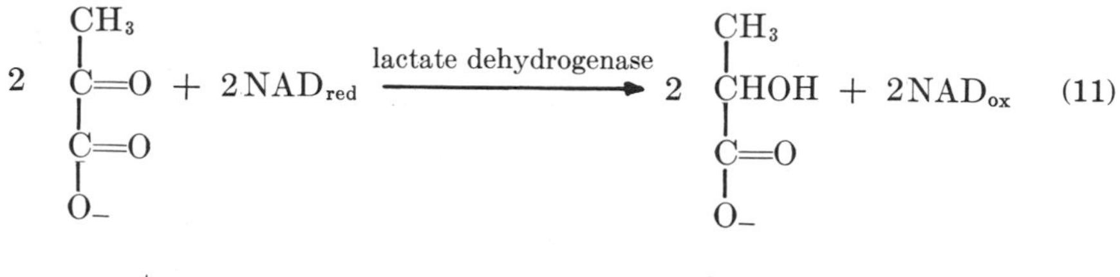

These sequences show that even the anaerobic breakdown of glucose requires energy to offset several of its reaction steps, and this energy is supplied by ATP. Equations 1 and 3 each show that energy from 1 mol of ATP is needed. Thus, energy from 2 mol of ATP is needed early in the process. However, when the mechanism reaches equations 7 and 10, 4 mol of ATP is produced. The net gain in the 11 steps is 2 mol of ATP from the energy supplied when 1 mol of glucose is converted to 2 mol of lactate.

Another phosphorus-containing compound that enters this sequence of reactions, via equations 6 and 11, is nicotinamide adenine dinucleotide, NAD (**II**). NAD is an electron carrier—that is, the nicotinamide moiety of the NAD is the electron carrier. In equation 6, the aldehyde group –C(=O)H is oxidized to the carboxylic acid –C(=O)O– with the loss of two electrons per molecule. The oxygen in this oxidation is supplied by a water molecule, and the electrons are removed from the two hydrogen atoms of the same water molecule. For this reason, biochemists also call such an oxidation reaction a dehydrogenation reaction.

Each oxidized molecule of NAD, or NAD_{ox}, can accept two electrons to become reduced NAD, or NAD_{red}. In equation 11, pyruvate is reduced to lactate, and each molecule reduced requires two electrons. The needed electrons are brought from equation 6 by NAD_{red}. After NAD_{red} donates the two electrons to equation 11, it becomes NAD_{ox} and is ready to carry electrons again from equation 6.

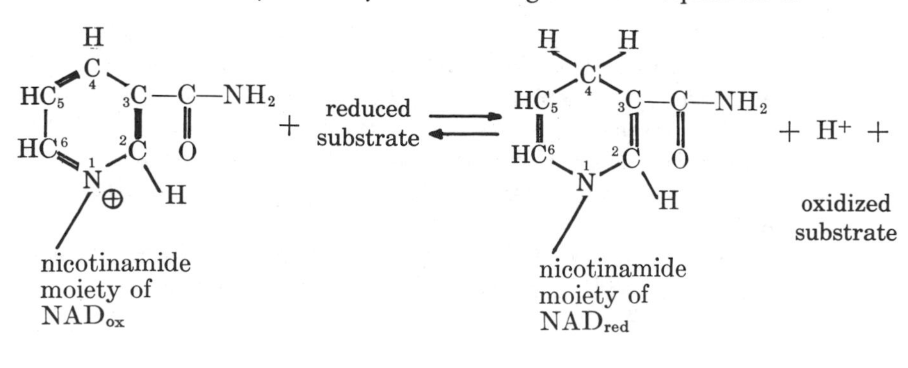

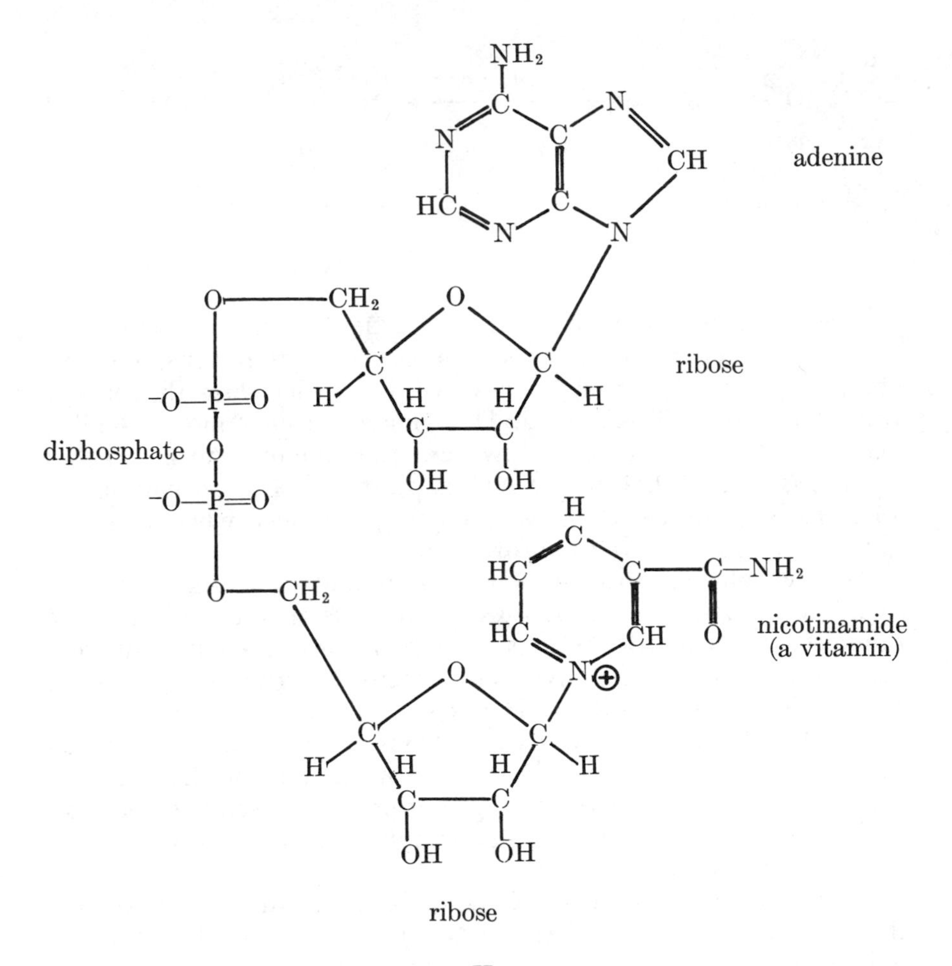

II

Note that the energy resulting from the oxidation of 3-phosphoglyceraldehyde in equation 6 is first conserved by the formation of 3-phosphoglyceroyl phosphate. As indicated in Table 17.I, the latter compound has a free energy of hydrolysis of −11,800 cal/mol. Some of this energy is transferred to the phosphate bond energy in ATP when it donates a phosphate to ADP to form ATP as shown in equation 7.

One possible reason why 3-phosphoglyceroyl phosphate has such a high free energy of hydrolysis is its structural makeup:

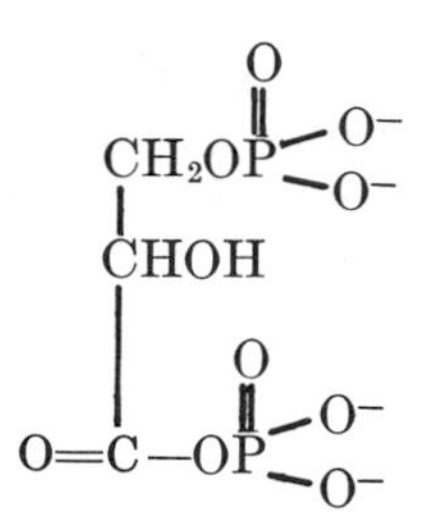

The two phosphate groups close together in the molecule, each with two negative charges, repel each other. Also, the –C(=O)–O–P(=O)< linkage has a high density of electrons. These negative charges repel each other and thus are ready for hydrolysis. After hydrolysis, the negative charges impede the phosphate [$–OP(=O)(O^-)_2$] and carboxyl groups [$C(=O)O^-$] from recombining. In addition, the electrons in these two groups arrange into a more stable configuration, which makes recombination more difficult.

Aerobic Oxidation of Glucose

The anaerobic breakdown of 1 mol of glucose results in the formation of 2 mol of lactate. However, for the degradation of glucose by aerobic cells—that is, cells requiring oxygen—glucose is not converted to lactate but to pyruvate, the product of equation 10. Pyruvate is then oxidized to CO_2 and H_2O; the total energy released is 686,000 cal. Again, much of this energy is conserved in ATP. Aerobic cell oxidation also depends on enzyme catalysts at each step. These enzymes are fixed in a specific geometric arrangement in the mitochondria or the power plants of the cell.

When we eat our daily meals, we are often quite particular about whether the food is properly prepared. We place great emphasis on the flavor and texture. We complain if the meat is tough, the salad not crisp, or the bread stale. As far as the body cells are concerned, however, food is just a source of raw materials and energy. Body cells oxidize (burn) carbohydrates (from starches and sugars), fatty acids (from fats), and amino acids (from meat and proteins) to yield the energy needed for various functions. The mechanism of this oxidative process is the Krebs tricarboxylic acid cycle (also called the citric acid cycle) (Figure 17.1). This cycle was deduced and proved in the 1930s by Hans A. Krebs, who was awarded a Nobel prize for this work. According to the Krebs mechanism, food is not converted directly to energy. Preliminary enzymatic reactions break food molecules into two-carbon segments. These two-carbon segments are available as acetyl groups [$CH_3C(=O)$–]. The Krebs cycle can accept only the acetyl group in the form of a derivative of coenzyme A, or CoA, another biophosphate:

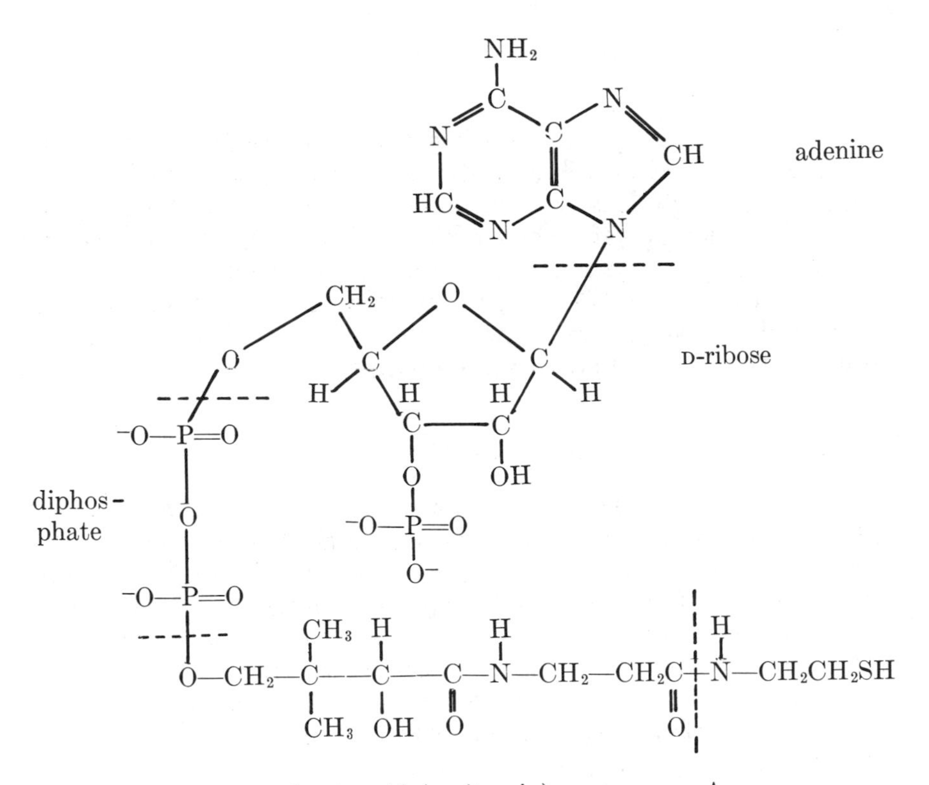

An illustration of coenzyme A as a carrier of an acetyl group for the Krebs cycle is shown in the case of acetic acid from the enzymatic oxidation of pyruvic acid of equation 10 in the glycolysis reactions:

$$CH_3C(=O)\text{—}COOH + CoA\text{—}SH + NAD_{ox} \xrightarrow[\text{and other enzymes}]{\text{pyruvate dehydrogenase}}$$

$$CoA\text{—}SHC(=O)CH_3 + CO_2 + NAD_{red}$$

Electrons removed from pyruvic acid are accepted by NAD_{ox} and carried in NAD_{red}. The acetyl group, carried by coenzyme A, can be

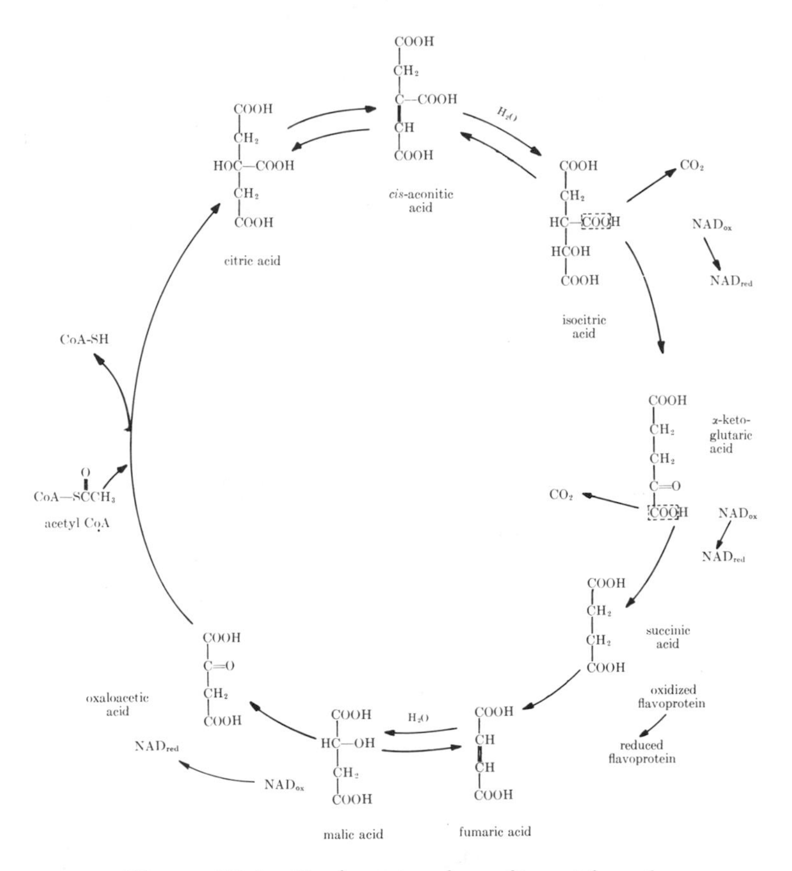

Figure 17.1. Krebs tricarboxylic acid cycle.

transferred to any acetyl acceptor during cell metabolism. An example of coenzyme A as an acetyl acceptor is shown in the Krebs tricarboxylic acid cycle.

As its main functions, the Krebs cycle (Figure 17.1) oxidizes the acetyl group into CO_2 and water and releases energy, most of which is conserved as ATP.

The Krebs cycle begins when acetyl-CoA donates the acetyl group to oxaloacetic acid, a four-carbon atom dicarboxylic acid, to form citric acid, a six-carbon tricarboxylic acid; this action releases coenzyme A to bring in another acetyl group for the next cycle. The individual steps in the cycle are shown in equations 12–18, and each step is catalyzed by a specific enzyme. Analysis of all the equations is not necessary for the following discussion. The important point is that when a compound is oxidized, it gives up electrons (electron donor), and when a compound is reduced, it accepts electrons (elec-

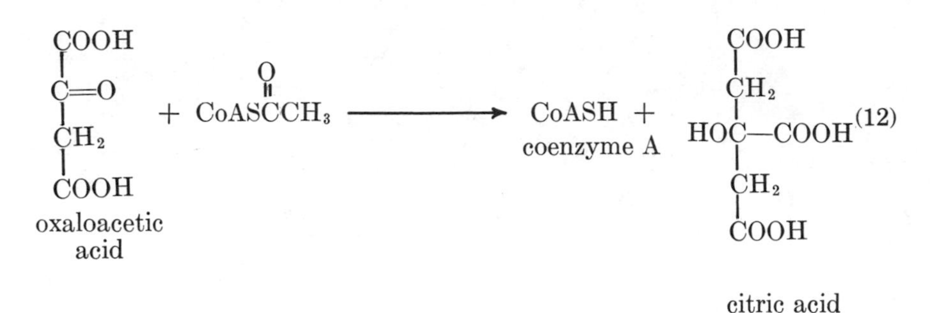

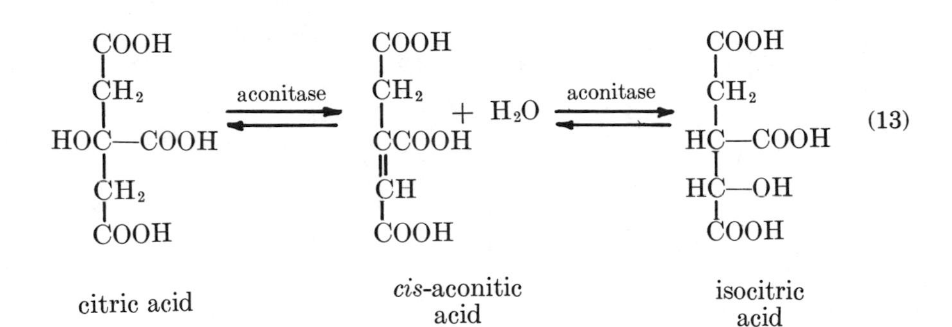

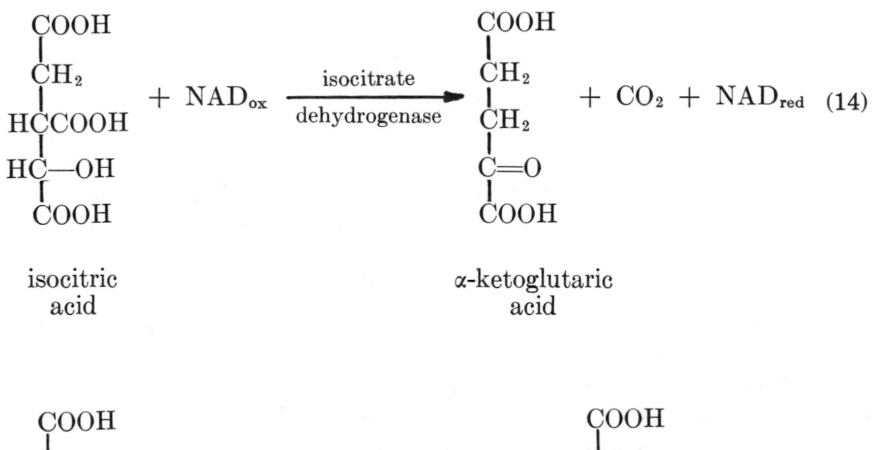

$$\begin{array}{c}\mathrm{COOH}\\|\\\mathrm{CH_2}\\|\\\mathrm{CH_2}\\|\\\mathrm{C{=}O}\\|\\\boxed{\mathrm{COO}}\mathrm{H}\end{array} + \mathrm{NAD_{ox}} \xrightarrow[\text{dehydrogenase}]{\alpha\text{-ketoglutarate}} \mathrm{CO_2} + \begin{array}{c}\mathrm{COOH}\\|\\\mathrm{CH_2}\\|\\\mathrm{CH_2}\\|\\\mathrm{COOH}\end{array} + \mathrm{NAD_{red}} \quad (15)$$

α-ketoglutaric acid succinic acid

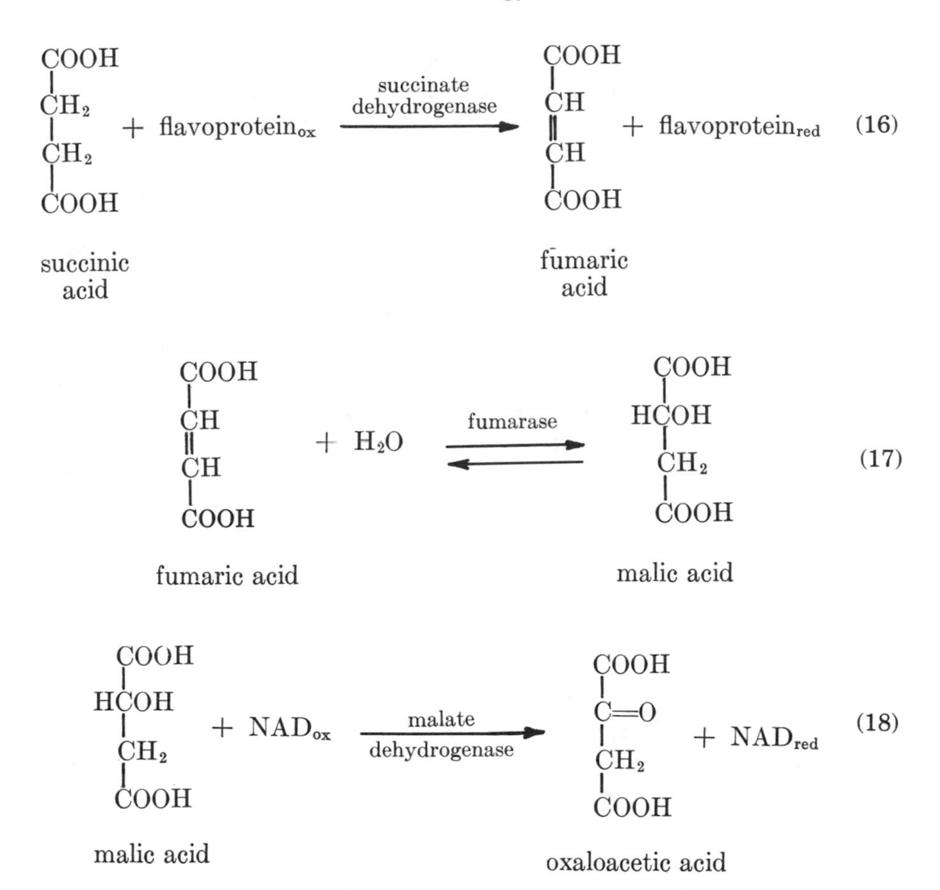

tron acceptor). An oxidation–reduction reaction, therefore, involves a transfer of electrons from one element or one compound to another. Each electron donor tends to give up electrons, and each electron acceptor tends to accept electrons. In other words, some compounds are very hard to oxidize or reduce and others are very easy.

A characteristic affinity for electrons is measured in terms of electromotive force (emf), called standard oxidation–reduction potential; emf values can be arranged from the most negative to the most positive. The most negative (the easiest to lose electrons) compounds can lose electrons to those compounds lower or less negative in the series. Every time electrons are transferred down the series, energy is liberated. The amount depends on the difference in the standard oxidation–reduction potential (emf) between the two compounds. In this respect, this situation is analogous to that of the biophosphorus compounds listed earlier, in which the compound highest on the list with the greatest free energy of hydrolysis can transfer a phosphate group to form a compound lower in the series. As shown previously, energy is also liberated with each phosphate transfer.

When an acetyl group in the Krebs cycle is oxidized to CO_2 and H_2O, four oxidation or dehydrogenation steps occur in each cycle (equations 14–16 and 18). In these oxidation steps, the lost electrons do not go directly from the hydrogen atom to the oxygen atom to form water. Instead, in equations 14, 15, and 18, the electrons first go to the biophosphate electron carrier, NAD_{ox}, to convert it to NAD_{red}; in equation 16, the electrons are passed on to an oxidized flavoprotein to be carried as reduced flavoprotein. In the standard oxidation–reduction series, NAD_{red} is the most negative and loses its electrons most easily. Next in the series is the electron carrier flavoprotein, also a phosphorus-containing compound. Oxygen is the least negative in the series, so much so that it goes to the positive side in the oxidation–reduction potential scale. In between flavoprotein and oxygen are electron carriers called cytochromes; each cytochrome has an active group containing an iron atom. In its oxidized form, iron has a valence of 3+ (Fe^{3+}); Fe^{3+} can accept an electron and be reduced to Fe^{2+}.

When a compound in the Krebs cycle is oxidized, it loses two electrons. These electrons do not go directly to the oxygen atom. They go first to NAD, then to flavoproteins, and down the series through a couple of cytochromes before they reach oxygen and reduce it to form water. Many details in this sequence of electron transfers are still not clearly understood, but the principle of the total transfer is known. As a sidelight, cyanide is a deadly poison because it blocks the enzymatic transfer of electrons from cytochrome to oxygen. In other words, cyanide inhibits the last step in the respiratory sequence.

The importance of the oxidation reactions in the Krebs cycle to the biological system, of course, lies in the release of energy. When 2 mol of electrons travels from NAD_{red} to oxygen, 47,000 cal of energy is released; a large portion of this amount is conserved by the formation of 3 mol of ATP from ADP and orthophosphate. As you will recall, each mole of ATP formation conserves 7,300 cal of energy. What would happen if a poison were introduced that permits the transport of electrons but that interferes with the formation of ATP from ADP and orthophosphate? When such a poison, 2,4-dinitro-

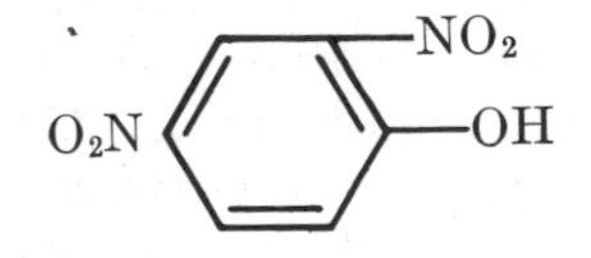

phenol, for example, is introduced into a rat, the respiration rate of the rat is increased and is accompanied by a rapid increase in body temperature. Because the oxidation energy is not conserved as ATP, the energy is lost as heat.

Energy is released when glucose is broken down by anaerobic cells. Energy is also released from the aerobic cell oxidation via the Krebs cycle mechanism of the acetyl group from the foods we consume. From the standpoint of phosphorus chemistry, a good portion of the released energy is conserved by the formation of ATP from ADP and orthophosphate. When the body needs energy to perform a function, such as flexing a muscle, the energy is supplied by the breakdown of ATP to ADP and orthophosphate.

The roles of other biophosphates, such as NAD, which serves as an electron carrier, and coenzyme A, which serves as an acetyl carrier, have also been described. Many additional phosphorus compounds, indispensable to the functioning of the body, are known, and no doubt more compounds remain to be discovered. The 1971 Nobel prize in physiology and medicine was awarded to Earl W. Sutherland of Vanderbilt University at Nashville, Tennessee, for his discovery of cyclic adenosine monophosphate (cAMP) and adenyl cyclase, the enzyme that synthesizes cAMP from ATP:

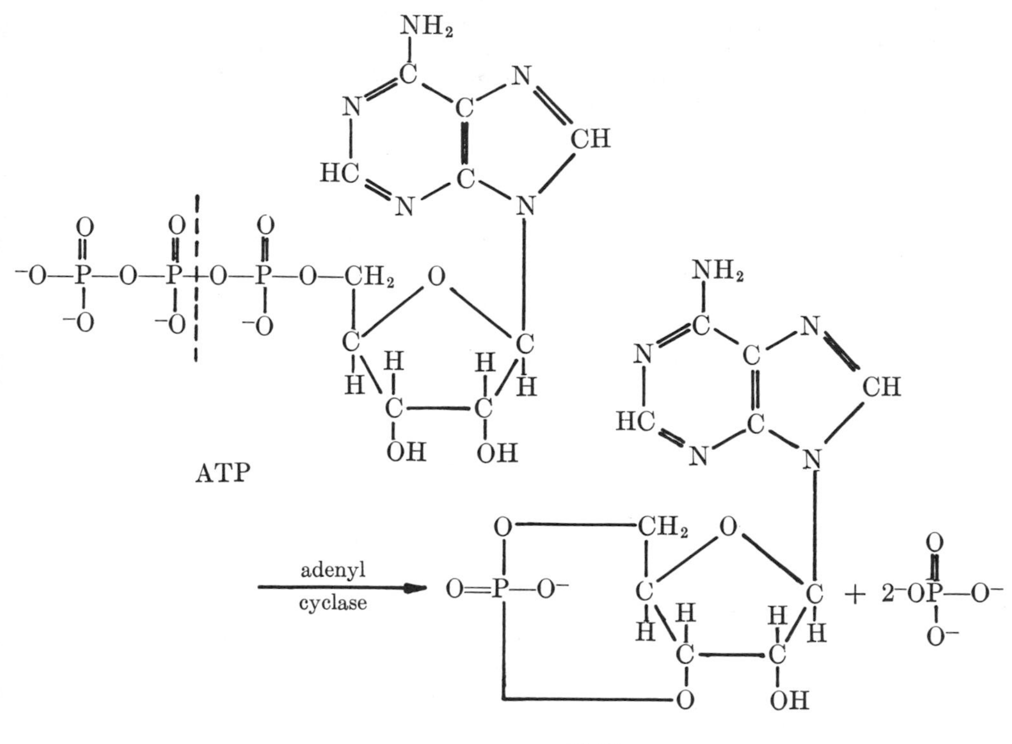

This unique phosphorus-containing molecule mediates the action of many hormones and regulates important activities in almost every animal cell. The elucidation of the cAMP structure is an interesting example of cooperative efforts among scientists and their alertness in coordinating seemingly unrelated facts. Because AMP is present in cells in extremely small amounts, difficulties arose in accumulating enough of it for analysis and characterization. Finally, Professor Sutherland had enough evidence that cAMP was a nucleotide and sent his results to Leon Heppel, then at the National Institutes of Health. Dr. Heppel had previously received a letter from David Lipkin of Washington University, St. Louis, MO, describing a nucleotide that he had prepared chemically by treating ATP with barium hydroxide, a base. One day while clearing his desk, Dr. Heppel ran across both letters and recognized that both scientists were probably working on the same compound. Heppel got the two scientists to communicate with each other, and from their cooperative effort the chemical structure of cAMP was established.

General References

1. Lehninger, A. L. *Bioenergetics*; W. A. Benjamin: New York, 1965.
2. Lehninger, A. L. *Principles of Biochemistry*; Worth: New York, 1982.
3. Feder, J. In *Topics in Phosphorus Chemistry*; Grayson, M.; Griffith, E. J., Eds.; Wiley: New York, 1983; Vol. II.

Chapter Eighteen

Organic Phosphorus Nerve Gases and Insecticides

YOU HAVE HEARD OF NERVE GASES so potent that a few drops will kill a horse and so dangerous that no country dares to use it first in actual warfare. You have benefited indirectly from organic phosphorus insecticides because they increase the yield of agricultural products and reduce the suffering and destruction caused by insect pests. You may be surprised to learn, however, that the reseach that led to the development of organic phosphorus insecticides was carried out in parallel with the basic research that led to the development of nerve gases intended for killing people.

The story of this research illustrates some basic scientific methods. The first thing research chemists must consider is whether or not someone else is thinking and working along a similar or even the exact same line. Even though the original premise formulated for the problem by the two groups may or may not be the same, the chances are good that similar, perhaps identical, results will be obtained.

The second consideration is that most successful compounds are a result of a systematic search sparked by some early clue. This search is accompanied by deductive reasoning based on known theories as well as hypotheses developed as the problem progresses. The active lead from the early clue may be discovered accidentally in one's own laboratory, or the lead may be inspired by literature on related subjects.

Organic phosphorus poison research programs were carried out during World War II both in England and in Germany. In England, the program was led by Saunders at Cambridge University (*1*). Gerhard Schrader of the I. G. Farben Works at Elberfeld directed the German program. Because the two countries were at war with each other, neither group supposedly knew what the other group was doing.

1002–0/87/0285$06.00/1

The Cambridge program aimed to develop a potent chemical warfare agent. In a lecture on the subject, Dr. Saunders stated facetiously that his original reason for picking phosphorus–fluorine compounds was that phosphorus is known to be very poisonous. Thus, a combination of phosphorus and fluorine, another known dangerous chemical, should yield some very poisonous compounds. Of course, such a premise is not always borne out by facts. For example, chlorine is a known poisonous gas, and sodium metal burns explosively in water, yet their combination, sodium chloride, is harmless table salt.

Nevertheless, some premise must be formulated at the beginning of all chemical research, and such a premise is usually based on the best available knowledge. If the original premise is proved wrong after a few experiments, it can be discarded; thus, knowledge is gained even from negative results. New premises are then formulated and evaluated until a clue is discovered.

Dr. Saunders did indeed find that some organic phosphorus–fluorine compounds were quite toxic, especially the dialkyl phosphorofluoridates with the general formula $(RO)_2P(=O)F$. Here was the lead he was looking for. A literature search showed that in 1932 in Germany, Lange and his assistant, von Kruger, had reported the synthesis and toxic properties of diethyl phosphorofluoridate, $(C_2H_5O)_2P(=O)F$ (2). Because Lange and von Kruger had other research goals in mind, the toxicity angle was not followed, but now Saunders' research group pursued this lead assiduously. They synthesized and evaluated many dialkyl phosphorofluoridate derivatives. They prepared substances in which R groups in the $(RO)_2P(=O)F$ formula were different. They made compounds in which the fluorine atom was attached directly to the phosphorus atom. They made other compounds in which the fluorine atoms were located on various carbons in the R group. Some of these compounds were toxic, and some compounds were not. Toxicity was first evaluated through animal tests.

For some compounds, Dr. Saunders even used himself as the guinea pig. Several times he almost died after entering a chamber that he thought contained a sublethal dose of a toxic organic phosphorus compound. At one stage of his research, the British government questioned their continued support of his project. Saunders invited the project officer to a session in the test chamber. When the officer discovered firsthand how powerful the agents were, Dr. Saunders had no further difficulty with continued support.

After synthesizing and evaluating many organic phosphorus fluorine compounds of various structures, the researchers chose diisopropyl phosphorofluoridate, $(i\text{-}C_3H_7O)_2P(=O)F$, commonly called DFP, as the compound with the correct balance of toxicity, chemical and physical properties, low cost, and ease of manufacturing for a chemical warfare agent. Doubtless some of Dr. Saunders' other compounds were not made public because of military secrecy. Some compounds, however, were promoted as agricultural insecticides. In fact, one compound, diisopropyl phosphorodiamidic fluoride, $[(i\text{-}C_3H_7)_2N]_2P(=O)F$, was found to be an effective soil insecticide and nematocide. This compound was used for a time in cocoa plantations in South Africa. However, the compound was later found to cause nonreversible damage to some nerves in humans. This damage resulted in total paralysis with no known cure. Diisopropyl phosphordiamidic fluoride therefore was withdrawn from use as an agricultural pest control chemical.

After a compound is determined to be effective and selected for further development, a practical process must be developed for its large-scale, economical manufacture. Several processes are usuallly possible. The process that Saunders' group developed was based on the use of phosphorus trichloride (PCl_3) as the intermediate. One mole of phosphorus trichloride reacts with 3 mol of isopropyl alcohol ($i\text{-}C_3H_7OH$) to form diisopropyl phosphonate:

$$PCl_3 + 3i\text{-}C_3H_7OH \longrightarrow \underset{\text{diisopropyl phosphonate}}{(i\text{-}C_3H_7O)_2P(=O)H} + i\text{-}C_3H_7Cl + 2HCl$$

Diisopropyl phosphonate then reacts with elemental chlorine to form diisopropyl phosphorochloridate:

$$(i\text{-}C_3H_7O)_2P(=O)H + Cl_2 \longrightarrow \underset{\text{diisopropyl phosphorochloridate}}{(i\text{-}C_3H_7O)_2P(=O)Cl} + HCl$$

Diisopropyl phosphorochloridate can be converted to DFP by treatment with sodium fluoride (NaF), in a solvent such as benzene:

$$(i\text{-}C_3H_7O)_2P(=O)Cl + NaF \xrightarrow[\text{solvent}]{\text{benzene}} (i\text{-}C_3H_7O)_2P(=O)F + NaCl$$

DFP is extremely toxic. Animals die violently after an infinitesimal dose of this compound is injected. To be exact, the lethal dose needed to kill 50% (LD_{50}) of the rats exposed for 10 min is 0.36 ppm (slightly more than one-third of 1 ppm) in air.

In Germany, organophosphorus compounds were studied by Gerhard Schrader to develop an effective insecticide. Consequently, many extremely toxic compounds were discovered and diverted for use as chemical warfare agents. Here, the borderline between research on insecticides and research on chemical warfare agents became obscured indeed.

Schrader (*3*) gives an interesting account of how he happened to be working on organic phosphorus compounds as insecticides. Unlike Saunders, he did not begin by speculating that phosphorus–fluorine compounds might be toxic. He arrived at the toxic organic phosphorus compounds through years of systematic research. His research program was initiated in 1934 when the director of research of the German I. G. Farben factory in Leverkusen instructed him to seek a new type of insecticide because of undesirable properties in insecticides then in use.

At that time, inorganic researchers at the I. G. Farben laboratory were working on cheaper methods for preparing fluorine and its compounds. Schrader decided, therefore, to undertake the systematic synthesis of flourine-containing compounds. This program was done in collaboration with biologists at Leverkusen who tested the compounds on various insects. Before long, these researchers found a fluorine–sulfur compound—methanesulfonyl fluoride (CH_3SO_2F)—that was an effective fumigant against insect pests in stored grains. Unfortunately, in large-scale evaluation, methanesulfonyl fluoride was found to be absorbed by the grain, which thus became too toxic for human consumption. Schrader did not give up. He reasoned that because the grain with absorbed methanesulfonyl fluoride is poisonous, why not evaluate them as poisoned bait for rats and mice? However, upon further evaluation, the absorbed methanesulfonyl fluoride was found to evaporate gradually from the grain, and toxicity thus decreased with time and finally disappeared. So the idea of using the compound absorbed on grains as a poisoned bait was abandoned.

The systematic study of organic fluorine–sulfur compounds to find a viable insecticide was continued. Organic fluorine compounds containing a sulfur–nitrogen bond, sulfur–carbon bond, and sulfur–oxygen bond were synthesized. Toxicity was discovered in many of these compounds, and some were almost good enough for commercial insecticide exploitation. Schrader stated that he finally exhausted the possibilities of preparing fluorine compounds with sulfur as the central atom. In the periodic table of chemical elements, phosphorus is next to sulfur. Because these neighbors have related properties, Schrader thought that organic fluorine compounds containing phosphorus as the central atom might also be biologically toxic. The first compound he synthesized in this new series that contained a phosphorus–fluorine bond showed some insecticidal activity. This compound was bis(*N*,*N*-dimethyl)phosphorodiamidic fluoride, $[(CH_3)_2N]_2P(=O)F$.

The fact that this compound is effective at the 0.05% level to give 100% kill of aphids is not so impressive by present-day standards. We now have insecticides that are effective at the level of a few hundred thousandths of 1%. But in the 1930s, the effective concentration of 0.05% was a breakthrough.

Schrader and co-workers theorized that one reason for the effectiveness of $[(CH_3)_2N]_2P(=O)F$ is that this compound is a mixed anhydride of *N*,*N*-dimethylphosphorodiamidic acid and hydrofluoric acid:

$$[(CH_3)_2N]_2P(=O)\boxed{OH + H}F \longrightarrow [(CH_3)_2N]_2P(=O)F + H_2O$$

Even though the method was not used for the synthesis of this compound, the method did lead to the concept that an anhydride structure was necessary for the phosphorus compound to be an active insecticide. The concept of an anhydride structure was known as the "acyl" rule. Subsequent research showed that the acyl groups could be extremely varied. These groups included fluoride, cyanide, phosphate, substituted phenoxy, substituted pyridyl, mercaptans, heterocyclic, enol, and so on.

Following the principle of the acyl rule, Schrader's group synthesized and evaluated many mixed anhydrides based on phosphorus as the central atom. Each new structural modification was based on the results of the biological evaluation of the previous series of compounds. This systematic research proved extremely fruitful. The research led eventually to the development of organic phosphorus compounds suitable as commercial insecticides that contained no fluorine atoms at all. Some of the very potent compounds were diverted for potential military use.

Among the more important insecticides discovered in Schrader's laboratory and introduced commercially immediately after World War II were the following:

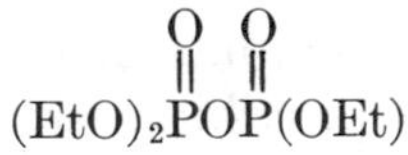

$$(EtO)_2\overset{O}{\overset{\|}{P}}O\overset{O}{\overset{\|}{P}}(OEt)_2 \qquad (C_2H_5O)_2\overset{S}{\overset{\|}{P}}O\text{-}C_6H_4\text{-}NO_2 \qquad (CH_3O)_2\overset{S}{\overset{\|}{P}}O\text{-}C_6H_4\text{-}NO_2$$

tetraethyl pyrophosphate 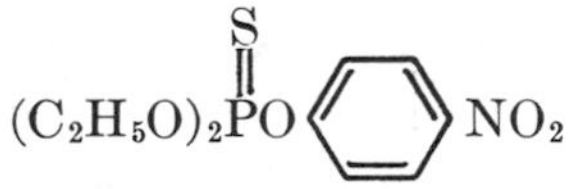 — ethyl parathion 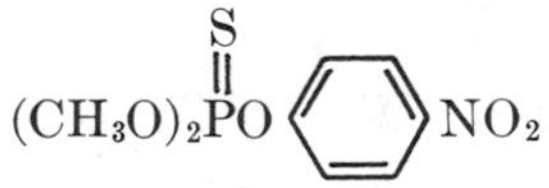 — methyl parathion

At the end of World War II, the news of Schrader's work on organic phosphorus insecticides reached the United States. This news created considerable excitement. The American farmers with their extensive agricultural developments do have a great need for effective insecticides for various crops. Also, a sufficient chemical industry interest in fulfilling those needs existed. The program on phosphorus insecticides pursued by American chemical industries followed three general routes: (1) they modified their existing chemical plants or built new plants to manufacture some of the effective insecticides already discovered by the Germans; (2) they began to evaluate organic phosphorus compounds they had on hand that were made for different applications, that is, those compounds that had structures that fitted into hypotheses, such as the acyl rule proposed by the Germans; and (3) they began a program of synthesis of new compounds, at first following the hypotheses proposed by the Germans as guides for the new structures.

In the industrial exploitation of those active insecticides already discovered by Schrader, the American companies used the Schrader technology or introduced their own process modifications. The first compound produced was the so-called "hexaethyl tetraphosphate". Schrader's method involved heating 3 mol of triethyl phosphate $[(C_2H_5O)_3P{=}O]$ with 1 mol of phosphorus oxychloride. Three moles of ethyl chloride (C_2H_5Cl) was evolved as a byproduct. However, even though the mixture had an empirical composition of hexaethyl tetraphosphate, no such compound formed. In the reaction for making the compound, various reorganizations of the reactants occurred; these reorganizations led to a mixture of compounds. The active component was tetraethyl pyrophosphate (TEPP). As produced by Schrader's method, the mixture contained 20–25% TEPP.

Because the company at which we work is a manufacturer of phosphorus compounds, our chemists tried various synthetic technologies. They found that the reaction of 4 mol of triethyl phosphate with 0.5 mol of P_4O_{10} resulted in a mixture containing about 35–40% TEPP. This method became a preferred manufacturing process because it had the advantage of producing a larger amount of the active component as well as not producing ethyl chloride as a byproduct. Subsequently, because of its high toxicity to humans and rapid hydrolysis in water, TEPP lost favor as an important commercial insecticide.

The other Schrader compounds that were produced commercially after World War II were ethyl parathion and methyl parathion (the synthesis is described in Chapter 15). These two compounds, especially methyl parathion, were used extensively as agricultural insecticides. In 1966, 19.4 million pounds of ethyl parathion and 35.8

million pounds of methyl parathion were produced. These two compounds were highly toxic to warmblooded animals, including humans. Because early manufacturers were unfamiliar with compounds of such high toxicity, several workers died from accidental poisoning. In addition, many users of these insecticides on farms and in fruit orchards did not quite believe how toxic these compounds were and did not follow the labeled precautions. Unfortunately, several additional deaths resulted.

As a consequence of their relatively high toxicity and the introduction of new, more effective and less toxic insecticides, the usage of these compounds continued to decrease. In 1983, the markets in the United States for ethyl and methyl parathion were 7.2 and 17.0 million pounds, respectively. The largest use for methyl parathion was in cotton and for ethyl parathion was in wheat (*4*).

In the evaluation of organophosphorus compounds already on the shelves, at first compounds with structures that fit into Schrader's acyl rule hypothesis were picked. Later, as large screening and evaluating facilities were established, all organophosphorus compounds on hand were screened. In many cases, even nonphosphorus compounds were included. Many of these compounds had been synthesized originally for other purposes, such as pharmaceuticals, oil additives, flotation agents for separating metal ores from sand and clay, plasticizers, and so on. Most compounds were found to be insufficiently active for their intended use. However, they were still on the shelf and available for testing as pesticides. In the evaluations, after a compound having some activity was found, chemists started modifying the structure to see whether they could find analogues or related compounds that were more effective. The effectiveness, of course, is measured in terms of more toxic to pests, less toxic to humans, and more economical to manufacture.

Two outstanding organophosphorus insecticides were discovered by this approach. One insecticide is the well-known malathion:

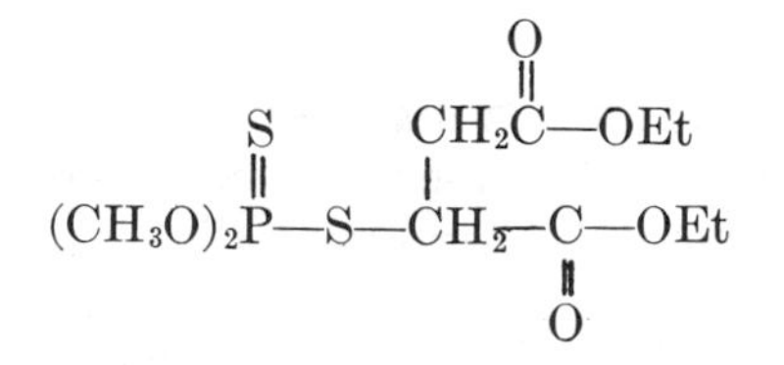

This compound is related to a class of compounds originally synthesized for testing as flotation agents. Malathion is a remarkable compound because it is toxic to a wide range of insects and yet is essentially harmless to mammals. It is, therefore, ideal for use in home gardens. The rotten-egg odor encountered in most insect dusts used in rose gardens is usually the result of the presence of malathion in the formulation.

In recent years, a great deal of publicity on malathion has occurred because of the fruit fly infestation in citrus crops in California and Florida. The Environmental Protection Agency (EPA), an agency of the U.S. government, has shown that malathion is the insecticide least harmful to the environment and least dangerous to humans for the control of fruit flies. However, vocal environmentalists still objected to aerial spraying of malathion. In California, their original objections delayed the application and caused the spread of fruit flies to a wider area. That infestation, of course, resulted in an economic loss of over $100 million to the California fruit growers. Fortunately, in Florida, the spread of fruit fly infestation was stopped by applying malathion without such unnecessary delays. The consumption of malathion for the agricultural market in 1982 was 13.5 million pounds.

Another compound that is a relative of a compound originally synthesized as a lubricating oil additive has the formula $(C_2H_5O)_2$-$P(=S)SCH_2SC_2H_5$. This compound is known as Thimet and is an extremely effective systemic insecticide but is quite toxic to warm-blooded animals.

Many effective pesticides appear to have been discovered accidentally. However, if chemists and biologists had depended entirely on luck, the odds of discovering a successful compound probably would have been quite low. Thus, the American chemists obtained their original leads on what made an effective organic phosphorus insecticide from the Germans. Following such leads, the American chemists found compounds they had on hand to have biological activity. They then formulated their own hypotheses for such activity. Next, model compounds were designed to check these hypotheses. This process led to the third phase of research on organic phosphorus pesticides, that of synthesizing new types of compounds. As a guide for the new structural route, researchers asked themselves two major questions: (1) What structural type of organic phosphorus compound is effective, that is, which functional group or groups in the compound are essential to activity? (2) What is the mechanism for the toxic action, that is, how do the compounds kill insects or warm-blooded animals?

An important breakthrough was the development of the hypothesis that the toxicity of the organic phosphorus insecticide resulted from the inhibition of the enzyme cholinesterase (*see* Chapter 20). In the decades after 1940, chemists and biologists learned much about the structure–activity relationship of the organic phosphorus insecticides.

These researchers found that for the compound to be active through the cholinesterase inhibition route, it must have the proper balance of lipid and water solubility so that it can pass through the various membranes of the insect to reach the site of the enzyme at the nerve endings. The compound should have the proper physical size and proper electrical charge. It should have the proper molecular geometry so that it can fit like a lock and key onto the enzyme. It should have a leaving group that is sufficiently stable so that it will not leave before the compound reaches the proper site in the enzyme. Yet, after the compound is locked in place in the enzyme, the leaving group should then leave to allow the rest of the molecule to phosphorylate the –OH group of the serine amino acid in the enzyme molecule.

Chemists and biologists use these as guides in synthesizing new candidates for insecticides.

In the early days and up to the mid-1950s period of research for an economically viable organic phosphorus insecticide, a commercially useful compound could be found for every 3000–4000 compounds synthesized and evaluated. Today, with the existence of so many good compounds already on the market, any new compounds must have improved properties or economics or both to be able to displace or supplement existing ones. Researchers are continually looking for insecticides that act rapidly, are easy to apply, have sufficiently long shelf life, and can be shipped easily. Preferably, the new compound also does not have unforeseen side effects, that is, the compound should not be toxic to nontarget organisms or induce resistance in targeted pests. Thus far, no single discovery of a compound approaching this ideal has been made. As time goes by, government regulations on new compounds have also become more strict. With these more stringent criteria, for each new pesticide marketed, more than 10,000–12,000 compounds have to be synthesized, screened, and evaluated. The effort in time and money involved to develop and register such a compound is estimated to be in the order of 7–8 years and as much as $20–30 million. However, the potential for profit for a commercially successful compound is also very high. As a consequence, the research continues toward the discovery of the ideal compounds, compounds that are selective, effective, economical, and safe for humans.

Literature Cited

1. Saunders, B. *Some Aspects of the Chemistry and Toxic Action of Organic Compounds Containing Phosphorus and Fluorine*; Cambridge University: Cambridge, England, 1957.

2. Lange, W.; von Kruger, G. *Ber. Dtsch. Chem. Ges.* **1932,** *65*, 1598.
3. Schrader, G. *Die Entwirklung neuer insektizider Phosphorsaurester*; Verlag Chemie GmbH: Bergstrasse, Weinheim, West Germany, 1963.
4. *Chemical Economics Handbook*; SRI International: Menlo Park, CA, 1984; 573.3002 0.

Insecticides mean the difference between profit and loss. Untreated cotton, left, yielded one-tenth bale per acre; treated, a bale per acre. (Courtesy of United States Department of Agriculture.)

Chapter Nineteen

Some Interesting Toxic Organic Phosphorus Compounds

I HAVE HAD SOME EXPERIENCE in research on organic phosphorus compounds that were possible insecticides. When I first heard that tetraethyl pyrophosphate, TEPP [$(C_2H_5O)_2P(=O)OP(=O)(OC_2H_5)_2$], was effective as an insecticide, I looked through the literature for known methods of preparation. If a practical way to make TEPP existed in the pure form, I wanted to adapt the method to large-scale industrial production. I found none, but laboratory research resulted in a practical and economical process. Unfortunately, the use of pure TEPP as an insecticide did not develop into a very large business, and no large plant was ever needed. However, I used the new methods to synthesize analogues of TEPP such as tetrapropyl pyrophosphate [$(C_3H_7O)_2P(=O)OP(=O)(OC_3H_7)_2$] and tetrabutylpyrophosphate[$(C_4H_9O)_2P(=O)OP(=O)(OC_4H_9)_2$].These compounds were less toxic than TEPP to mammals, but unfortunately they were also less toxic to insects.

Using the same chemical principle as used in that process, I developed a process for synthesizing a sulfur analogue of TEPP, tetraethyl dithionopyrophosphate, thio-TEPP [$(C_2H_5O)_2P(=S)OP(=S)(OC_2H_5)_2$]. This compound was also quite toxic to mammals. Schrader's work (*1*) showed that he too had made the sulfur analogue of TEPP but by a different method. Because I had a pretty good process for thio-TEPP, I decided to synthesize some tetrapropyl dithionopyrophosphate [$(C_3H_7O)_2P(=S)OP(=S)(OC_3H_7)_2$]. Because of the results obtained with the oxygen analogues, this compound was expected to be less toxic than thio-TEPP to warmblooded animals and also less toxic to insects. In other words, I did not expect

1002–0/87/0295$07.00/1

much. To my surprise, I found that for practical purposes this compound is not toxic to warmblooded animals at all but is quite toxic to several species of insects. In fact, tetrapropyl dithionopyrophosphate is extremely effective against the chinch bug, a pest that destroys lawns in the southern states along the eastern seaboard. This compound is now known as Aspon.

I have also discovered a compound with the formula

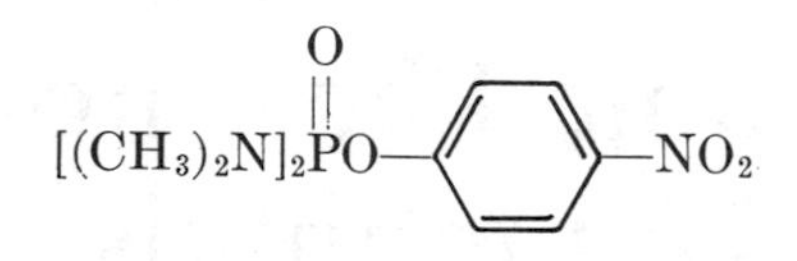

This compound does not harm most insects, but it is quite toxic to warmblooded animals. I can find no theoretical explanation for this finding, and the compound seems to have no practical use.

In Table 19.I are listed some of the organic phosphorus nerve gases. All of these compounds are liquids at room temperature.

The term nerve gas has been used for this class of compound. The substances are cholinesterase inhibitors (*see* Chapter 20 on the mechanism of toxic action). They cause death by blocking nerve function activity. Death comes either very quickly or agonizingly slowly depending on the degree of exposure. The symptoms are impaired vision, convulsion, dizziness, and confusion. Death is usually the result of respiratory failure. As a class, nerve gases are the most deadly synthetic poisons known. Of the compounds in Table 19.I, agent VX is the most lethal. It is considerably more toxic than DFP or sarin.

The relative toxicity (2), as measured by injection to kill 50% of the species (LD_{50}) and expressed in milligram per kilogram of body weight is as follows: for DFP, 0.450; for sarin, 0.017; and, for agent VX, 0.009. Interestingly, a close relative of agent VX, a phosphate rather than a phosphonate

$$\begin{array}{l} C_2H_5O \diagdown \overset{O}{\overset{\|}{}} \\ \qquad\qquad PSCH_2CH_2N(C_3H_7\text{-}i)_2 \\ C_2H_5O \diagup \end{array}$$

has an LD_{50} of 0.08 mg/kg. The lower toxicity of this compound is still many times that of DFP, the original organic phosphorus nerve gas developed by the British.

Table 19.I. Organophosphorus Nerve Gases

Common Names	*Formula*	*Chemical Names*
Tabun	$(C_2H_5O)((CH_3)_2N)P(=O)-CN$	*O*-ethyl *N*,*N*-dimethylphosphoramidocyanidate
Soman	$(CH_3)_3C\text{-}CH(CH_3)\text{-}O\text{-}P(=O)(CH_3)\text{-}F$	1-methyl-2,2-dimethylpropyl methylphosphonofluoridate
Sarin	$(CH_3)_2CH\text{-}O\text{-}P(=O)(CH_3)\text{-}F$	*O*-isopropyl methylphosphonofluoridate
DFP	$[(CH_3)_2CH\text{-}O]_2P(=O)\text{-}F$	*O*,*O*-diisopropyl phosphorofluoridate
Agent VX	$(CH_3)(C_2H_5O)P(=O)\text{-}SCH_2CH_2N(C_3H_7\text{-}i)_2$	*O*-ethyl *S*-2-(diisopropylamino)ethyl methylphosphonothiolate

The high toxicity of agent VX may be desirable as a chemical warfare agent; however, this toxicity also creates many problems. In the manufacture of this compound and in the handling and loading of it into weapons such as rockets, mortars, artillery shells, bombs, spray tanks, and mines, the high toxicity represents an extreme hazard to the workers. One solution to this problem is the development of the binary system. Rather than produce and store the toxic nerve gas, two components of relatively low toxicity are produced and stored. These two components, where mixed, spontaneously react and the reaction product then rearranges to form the toxic nerve gas.

This system requires the carrying out of a chemical reaction in the weapon during the short period the weapon is on its way toward the target. The reactions involved are as follows for agent VX:

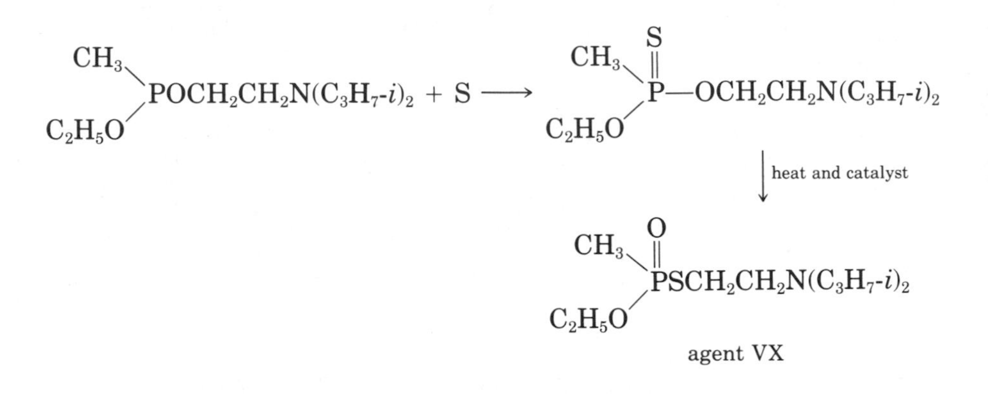

One approach for making such a binary system is to have the intermediate, *O*-ethyl *O*-[2-(diisopropylamino)ethyl] methylphosphonite, a relatively nontoxic compound, stored in one compartment of the weapon. In a separate compartment, sulfur powder along with the catalyst is stored. After the missile is fired, a mechanism is activated to permit mixing of the two reactants. The reaction is exothermic and therefore provides sufficient heat, along with the catalyst, to cause the reaction product formed from the two reactants to rearrange into the final agent VX. The chemistry involved is rather simple when carried out in the laboratory. To have the reaction occur in the missile over a short period of a few seconds is indeed a challenge to the weapon's designers.

Contact Insecticides

Contact insecticides, as the name implies, kill when the compounds actually touch an insect. Many of the organic phosphorus insecticides listed in Table 19.II, besides being toxic to a wide range of insect pests such as cotton boll weevils, aphids, mites, and caterpillars, are also quite lethal to warmblooded animals. Nevertheless, these compounds are still widely used because they are effective against a large number of insect species. [The term insect species is used loosely here. For example, mites and ticks are not classified as true insects by entomologists. They belong to the order Acarina. The poisons used

Table 19.II. Organophosphorus Contact Insecticides, Acaricides, and Nematicides

Common and Trade Names	*Formulas*	*Chemical Names*	*U.S. Consumption, 1982 (Million lb)*	*Acute Dermal LD_{50} (mg/kg)*
methyl parathion	$(CH_3O)_2P(=S)O-C_6H_4-NO_2$	*O,O*-dimethyl *O*-*p*-nitrophenyl phosphorothioate	17.0	300–400[a]
parathion	$(C_2H_5O)_2P(=S)O-C_6H_4-NO_2$	*O,O*-diethyl *O*-*p*-nitrophenyl phosphorothioate	7.2	55[b]
Sumithion	$(CH_3O)_2P(=S)-O-C_6H_3(CH_3)-NO_2$	*O,O*-dimethyl *O*-(4-nitro-*m*-tolyl) phosphorothioate	negligible	1300[b]
EPN	$(C_2H_5O)(C_6H_5)P(=S)O-C_6H_4-NO_2$	*O*-ethyl *O*-*p*-nitrophenyl phenylphosphonothioate	1.5	420[a]

Table 19.II (***Continued***)

Common and Trade Names	*Formulas*	*Chemical Names*	*U.S. Consumption, 1982 (Million lb)*	*Acute Dermal LD_{50} (mg/kg)*
chlorpyrifos, Dursban	C_2H_5O, C_2H_5O, S, P—O—, Cl, Cl, Cl, N	*O,O*-diethyl *O*-(3,5,6-trichloro-2-pyridyl) phosphorothioate	7.2	2000[a]
Diazinon	C_2H_5O, C_2H_5O, S, P—O—, N, N, H, C—$(CH_3)_2$, CH_3	*O,O*-diethyl *O*-(2-isopropyl-6-methyl-4-pyrimidinyl) phosphorothioate	5.8	600[a]
fensulfothion, Dasonit	C_2H_5O, C_2H_5O, S, P—O—, O, —S—CH_3	*O,O*-diethyl *O*-[4-(methylsulfinyl)phenyl] phosphorothioate	0.5	3–30[b]

fenthion, Baytex	$(CH_3O)_2P(=S)-O-C_6H_3(CH_3)-SCH_3$	*O,O*-dimethyl *O*-[3-methyl-4-(methylthio)phenyl] phosphorothioate	0.3	1680[c]
trichlorofon, Dipterex	$(CH_3O)_2P(=O)-CH(OH)-CCl_3$	*O,O*-dimethyl (2,2,2,-trichloro-1-hydroxy-ethyl)phosphonate	1.0	>500[b]
DDVP	$(CH_3O)_2P(=O)OCH=CCl_2$	2,2-dichlorovinyl dimethyl phosphate	0.1	107[a]
naled, Dibrom	$(CH_3O)_2P(=O)-OCH(Br)-C(Br)-Cl_2$	1,2-dibromo-2,2-dichloroethyl dimethyl phosphate	1.0	1100[a]

Table 19.II (*Continued*)

Common and Trade Names	*Formulas*	*Chemical Names*	*U.S. Consumption, 1982 (Million lb)*	*Acute Dermal LD_{50} (mg/kg)*
crotoxyphos, Ciodrin	$(CH_3O)_2P(=O)-O-C(CH_3)=CH-C(=O)-O-CH(CH_3)-C_6H_5$		0.7	~385
tetrachlorovinphos, Gardona	$(CH_3O)_2P(=O)-O-C(=CHCl)-C_6H_2Cl_3$ (2,4,5-Cl)	isomer of the 2-chloro-1-(2,4,5-trichlorophenyl)vinyl dimethyl phosphate	0.1	>2500[a]
profenophos, Curacron	$(C_2H_5O)(CH_3CH_2CH_2S)P(=O)-O-C_6H_3(2\text{-}Cl)(4\text{-}Br)$	*O*-(4-bromo-2-chlorophenyl) *O*-ethyl *S*-propyl phosphorothioate	0.1	422[a]

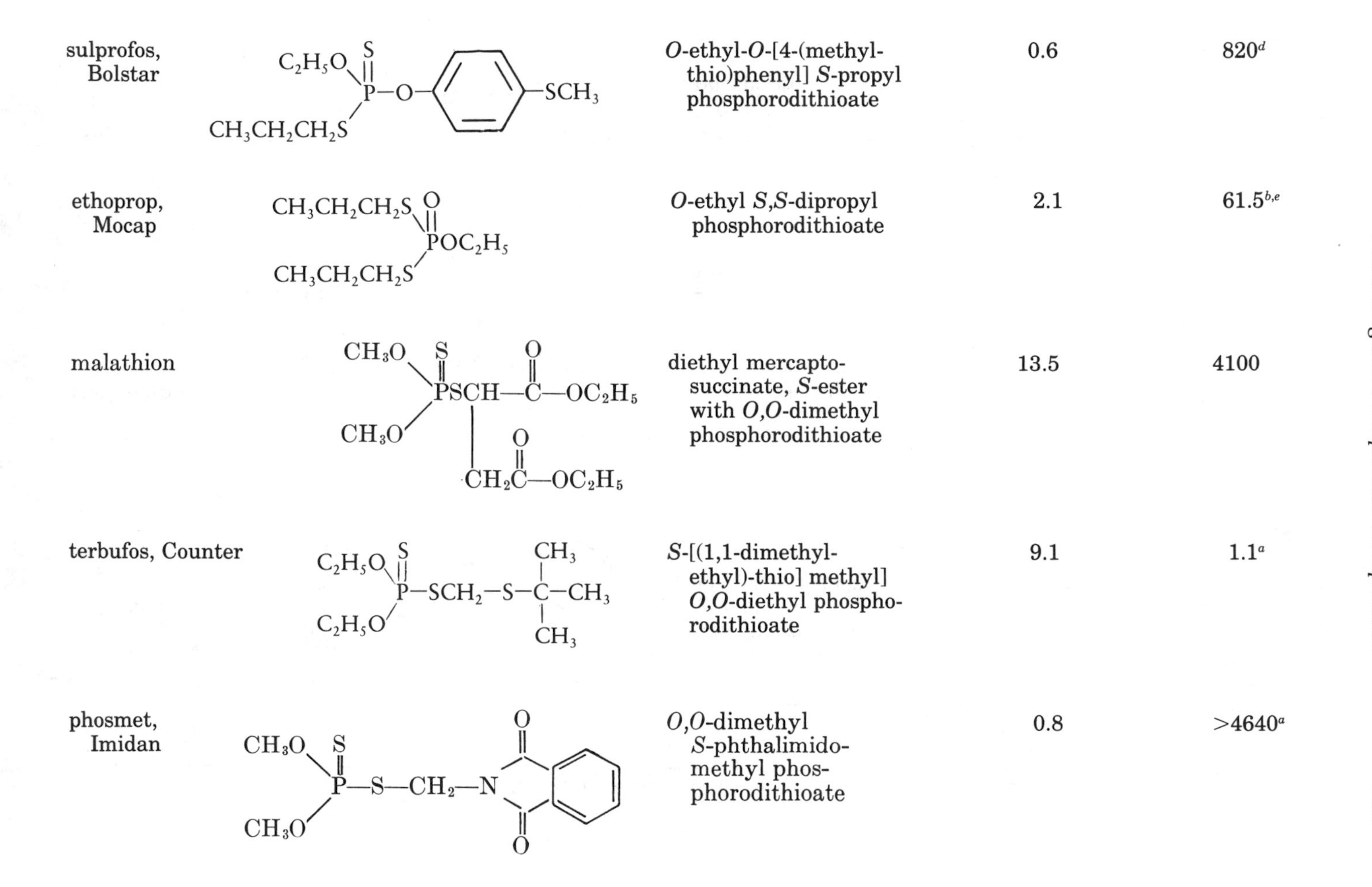

sulprofos, Bolstar	*O*-ethyl-*O*-[4-(methylthio)phenyl] *S*-propyl phosphorodithioate	0.6	820[d]
ethoprop, Mocap	*O*-ethyl *S*,*S*-dipropyl phosphorodithioate	2.1	61.5[b,e]
malathion	diethyl mercaptosuccinate, *S*-ester with *O*,*O*-dimethyl phosphorodithioate	13.5	4100
terbufos, Counter	*S*-[(1,1-dimethylethyl)-thio] methyl] *O*,*O*-diethyl phosphorodithioate	9.1	1.1[a]
phosmet, Imidan	*O*,*O*-dimethyl *S*-phthalimidomethyl phosphorodithioate	0.8	>4640[a]

Table 19.II (*Continued*)

Common and Trade Names	*Formulas*	*Chemical Names*	*U.S. Consumption, 1982 (Million lb)*	*Acute Dermal LD_{50} (mg/kg)*
azinphos-methyl, Guthion	$(CH_3O)_2P(=S)SCH_2N$ (ring: N–$C(=O)$–C_6H_4–$N{=}N$–)	*O,O*-dimethyl *S*-4-oxo-1,2,3-benzotriazin-3(4*H*)-ylmethyl phosphorodithioate	3.0	220[a]
methidathion, Supracide	$(CH_3O)_2P(=S)$–S–CH_2–N (ring: $O{=}C$–S–$C(OCH_3){=}N$–N)	O,O-dimethyl phosphorodithioate, *S*-ester with 4-(mercaptomethyl)-2-methoxy-Δ^2-1,3,4-thiadiazolin-5-one	0.7	200[a]
ethion	$(C_2H_5O)_2P(=S)$—S—CH_2—S—$P(=S)(OC_2H_5)_2$	*O,O,O′,O′*-tetraethyl *S,S′*-methylene bisphosphorodithioate	0.8	96[b,f]

phosalone, Zolone	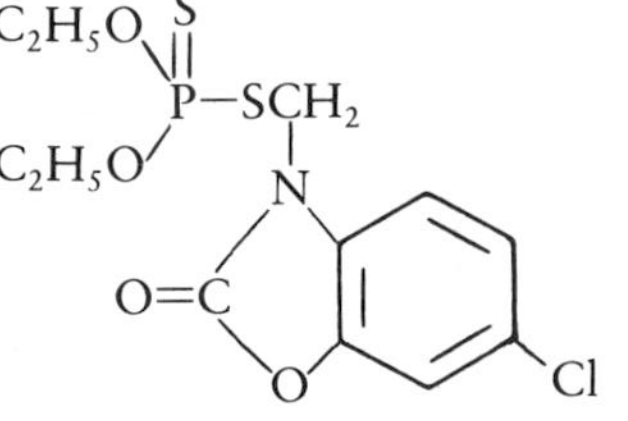	*S*-[(6-chloro-2-oxo-3-(2*H*)-benzoxazolyl)methyl] *O,O*-diethyl phosphorodithioate	0.6	120–170[c,e]
carbophenothion, Trithion	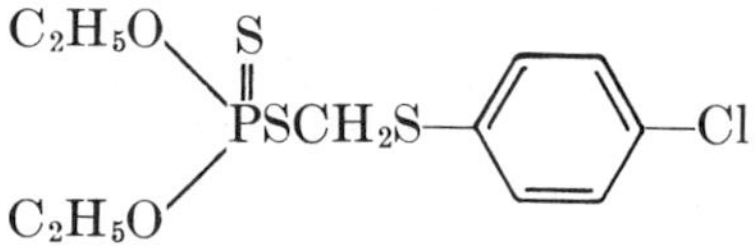 $(C_2H_5O)_2P(=S)SCH_2S-C_6H_4-Cl$	*S*-[(*p*-chlorophenylthio)methyl] *O,O*-diethyl phosphorodithioate	0.2	6.8–36.9[c,e]
dioxathion, Delnav	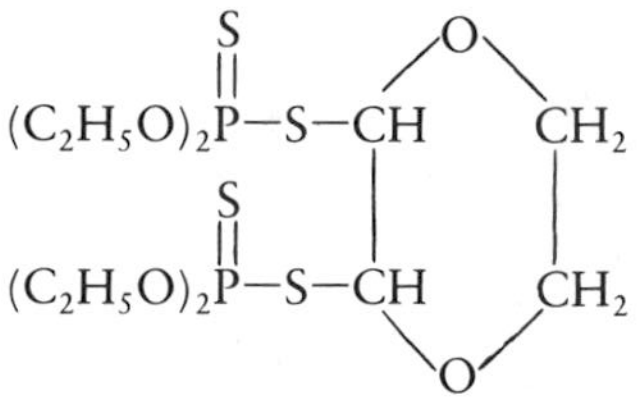 $(C_2H_5O)_2P(=S)-S-CH$, $(C_2H_5O)_2P(=S)-S-CH$ (ring: $CH-O-CH_2-CH_2-O-CH$)	1,4-*p*-dioxane-2,3-dithiol *S,S*-bis(*O,O*-diethyl phosphorodithioate)	0.1	235[c]
fonophos, Dyfonate	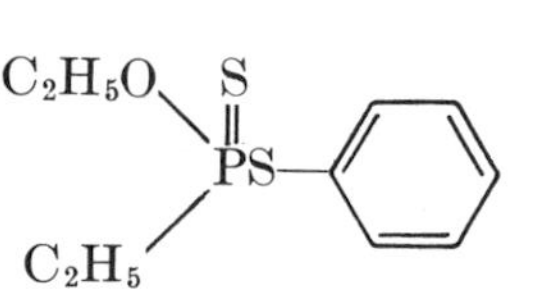	*O*-ethyl *S*-phenyl ethylphosphonodithioate	7.6	25[a]

Table 19.II. (*Continued*)

Common and Trade Names	*Formulas*	*Chemical Names*	*U.S. Consumption, 1982 (Million lb)*	*Acute Dermal LD_{50} (mg/kg)*
methamidophos, Monitor	$(CH_3S)(CH_3O)P(=O)-NH_2$	*O,S*-dimethyl phosphoramidothioate	1.3	118[a,g]
acephate, Orthene	$(CH_3S)(CH_3O)P(=O)-NH-C(=O)-CH_3$	*O,S*-dimethyl acetylphosphoramidothioate	2.2	>10, 250[a]
isofenphos, Amaze	$((CH_3)_2CH-NH)(C_2H_5O)P(=S)-O-C_6H_4-COOCH(CH_3)_2$	1-methylethyl 2-[[ethoxy [(1-methylethyl) amino]phosphinothioyl] oxy]benzoate	2.0	162–315[a]

[a]Rabbits.
[b]Rats.
[c]Male rats.
[d]Male rabbits.
[e]Acute oral LD_{50}.
[f]Acute oral technical LD_{50}.
[g]75% technical.

to kill them are called acaricides. Similarly, nematodes are not insects. They are minute worms that live in soil and infect plant roots. They are killed by nematocides.] Also, after application, organic phosphorus contact insecticides have only limited persistence in the environment as compared to the organic chlorine insecticides. Depending on the nature of the organic phosphorus insecticides, they decompose within a period ranging from hours to a few days. Also, insects have been slow in developing resistance to them. For example, the methyl and ethyl parathions have been in commercial use since the 1940s. They are still used on a large scale, though not at as large a volume as during the 1960–1970s. One reason for their continued use is that they are quite economical to produce. Also, the hazard of high toxicity to workers and users has been reduced substantially by the implementation of more sophisticated safe handling techniques from the point of manufacture to application in the field.

In cases where toxicity problems in the field are difficult to solve, other new and less toxic, though more expensive, organic phosphorus or non-phosphorus-containing insecticides are used.

An interesting case that I heard about was the use of methyl parathion to control insect pests in the cotton fields in Egypt. The miniature "baby" cotton bolls, with their low molecular weight cellulose molecules, have a sweet taste though not as sweet as candy. Children and farmers love to pick and eat them. After methyl parathion is sprayed on the cotton plants, it persists for a few days. People who picked and ate those "baby" cotton bolls too soon after spraying got very sick and in some cases died. As a consequence, methyl parathion was banned for this application. A less toxic organic phosphorus insecticide, Dursban, is used in its place. I understand that in some places even Dursban was subsequently displaced by still less toxic insecticides, such as the synthetic pyrethroids.

With the advent of newer and better insecticides, the U.S. usage of methyl and ethyl parathion has dropped respectively from 35,862,000 and 19,444,000 lb in 1966 to 17,000,000 and 7,200,000 lb in 1982.

O-Ethyl *O*-*p*-nitrophenyl phenylphosphonothioate, or EPN, a close structural relative of parathion, is toxic but not quite as dangerous as parathion. Parathion is used in Japan to control rice stem borers. However, because rice is grown in paddies, Japanese farmers also raise carp either in these water areas or in nearby ponds. Unfortunately, parathion also kills the carp. EPN has a wide enough margin of safety so that it kills only the borers. EPN is a phenylphosphonate. It has also been used for many years as an agricultural insecticide in the United States, particularly in the cotton fields. However, with the discovery that another phenyl phosphonate, Phosvel, delayed neurotoxicity, the usage of EPN has dropped precipitously. Phosvel is now banned.

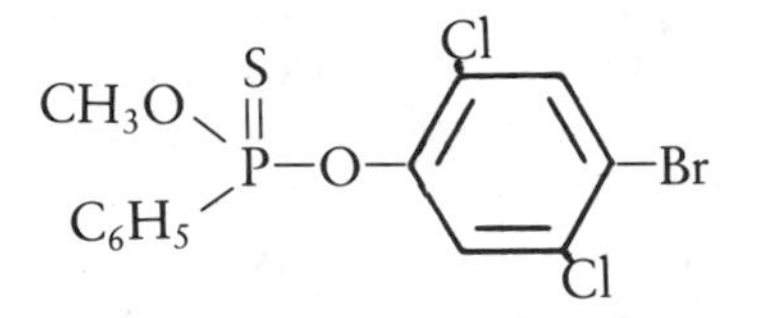

Sumithion has the same structure as methyl parathion except that it has a methyl (CH_3–) group substituted for a hydrogen in the meta position of the benzene ring in the *p*-nitrophenyl part of the molecule:

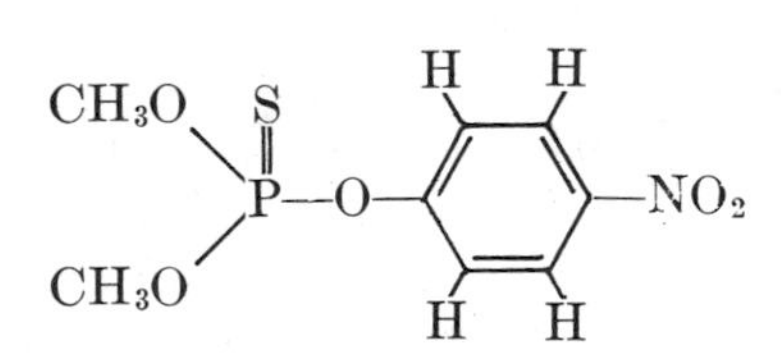

methyl parathion

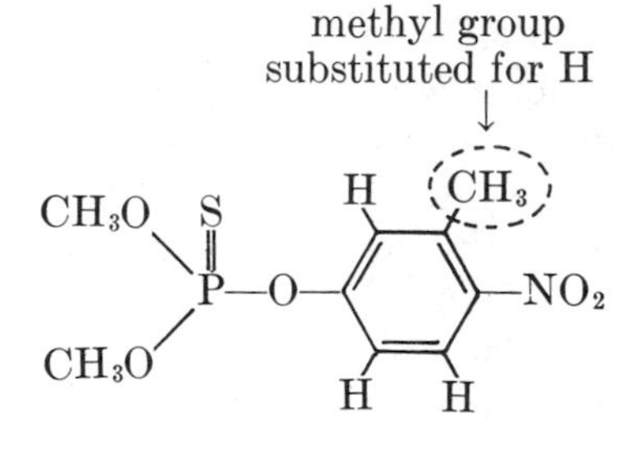

Sumithion

This substitution, surprisingly, makes Sumithion a compound that has low toxicity to warmblooded animals but is still able to kill a wide range of insects. In 1975, it was one of the insecticides used to control the spruce budworm disease. This disease is caused by the tiny larvae of the spruce budmoth, which strip the needles from the tree for food. Millions of acres of spruce and balsam fir that are used for making paper are threatened by this disease.

Sumithion, in spite of its having a much lower toxicity to warm-blooded animals than methyl parathion, has never become an important agricultural insecticide in the United States. The reason is largely economic. Sumithion is more expensive to make because the needed intermediate, *p*-nitrometacresol, is considerably more costly than the intermediate used for methyl parathion, *p*-nitrophenol.

Certain insecticides will render plants toxic to insects for many days. However, the U.S. Department of Agriculture forbids their use on fruits, for example, less than 30 days before harvest. At this point, other insecticides that kill pests and then are quickly hydrolyzed to harmless byproducts can be used. For example, Phosdrin can be applied to a field of lettuce to kill worms the night before harvest.

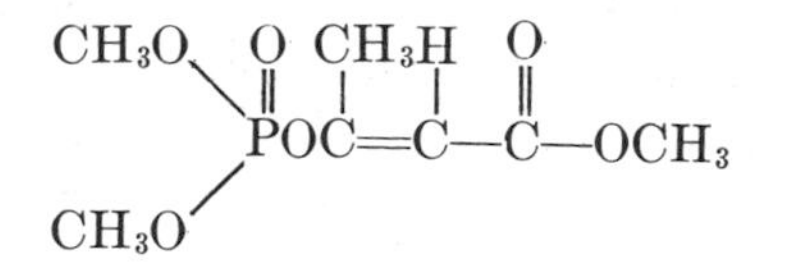

Malathion is used on a large scale to control garden insects. This compound is nontoxic to pets and children but because it has an unpleasant odor, it is not used indoors. Instead, compounds like Dipterex, Dursban, or Diazinon, which are relatively free of bad odor and are also quite nontoxic to warmblooded animals, are used.

An accidentally discovered decomposition product of dipterex has proved to be an interesting insecticide. In fact, many believe the toxicity of Dipterex is the result of its chemical decomposition to dimethyl dichlorovinyl phosphate (DDVP) via the reaction

$$(CH_3O)_2\overset{\overset{O}{\|}}{P}-\underset{H}{\overset{\overset{OH}{|}}{C}}CCl_3 \longrightarrow (CH_3O)_2\overset{\overset{O}{\|}}{P}OCH{=}CCl_2 + HCl$$

DDVP was also independently discovered and synthesized by the Perkow reaction, named after the German chemist who first published it in 1954. The Perkow reaction involves the action of trimethyl phosphite on chloral and is illustrated by

$$\underset{\text{trimethyl phosphite}}{(CH_3O)_3P} + \underset{\text{chloral}}{Cl_3CCHO} \longrightarrow \underset{\text{DDVP}}{(CH_3O)_2\overset{\overset{O}{\|}}{P}OCH{=}CCl_2} + CH_3Cl$$

A survey of the literature showed that, within a period of a few months, this reaction has been discovered independently by several investigators both in Germany and in the United States. This situation is a very good illustration of how scientists often think along the same lines simultaneously and come up with identical solutions.

One interesting development by Shell Chemical Co. is a technique to allow the use of the relatively toxic DDVP as a safe household insecticide. When hung indoors, plastic strips impregnated with this insecticide release minute quantities of it into the atmosphere. The vapor kills pests such as flies and mosquitoes. Because the insecticide vapor cannot be seen and yet knocks down flies quite effectively, this product could be called the invisible fly swatter. The amount of insecticide released is of course insufficient to harm people, cats, or dogs.

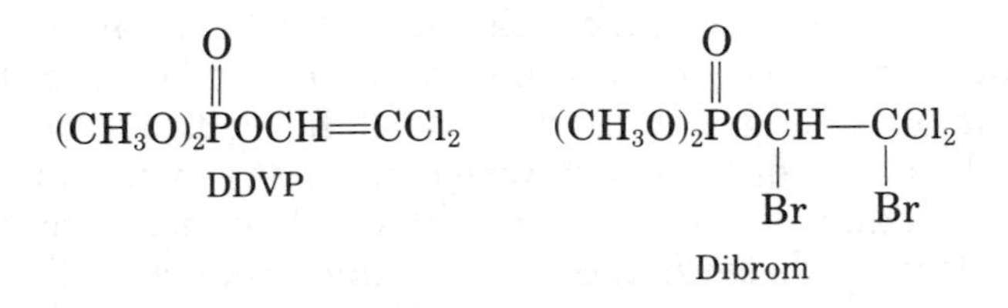

In many cases, one organic phosphorus insecticide is a direct derivative of another. A good example is Dibrom. It can be made by the addition of bromine to the carbon-to-carbon double bond of DVPP. The removal of the double bond results in increasing the dermal LD_{50} (to rabbits)(i.e., reducing the toxicity) from 107 to 1100 mg/kg. In 1982, the usage in the United States of Dibrom was one million pounds and that for DVPP was one hundred thousand pounds.

Another interesting example is that of Orthene, which can be made by acetylating the insecticide Monitor.

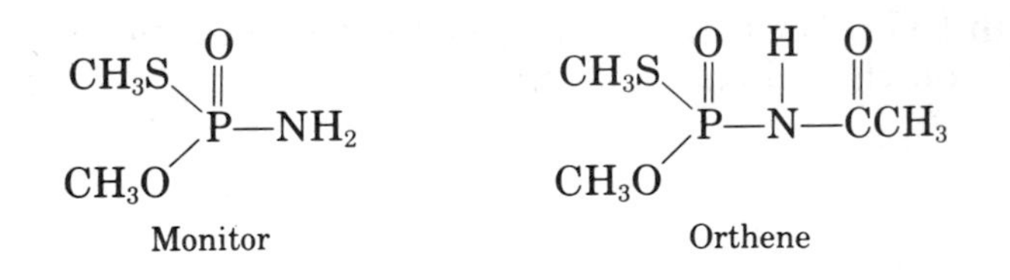

Monitor has a dermal LD_{50} (to rabbits) of 118 mg/kg for the 75% technical grade. Acetylating it to form Orthene greatly reduces the dermal toxicity. The dermal LD_{50} for Orthene is greater than 10,250 mg/kg. The 1982 usage in the United States for Orthene was 2.2 million pounds compared to 1.3 million pounds for Monitor.

Systemic Insecticides

Systemic insecticides, as a class, are absorbed through the roots or leaves of a plant to render the sap toxic to sucking insects such as aphids without harming the plant. Contact insecticides sometimes kill both harmful and useful insects (such as bees) indiscriminately. They even kill insects that are the natural enemies of the harmful ones. Systemic insecticides do not harm the useful predators, which do not suck plant juices. Some of the organic phosphorus systemic insecticides, acaricides, and nematocides are listed in Table 19.III. Many of these compounds also kill by contact action.

One of the first systemic insecticides discovered was schradan, $[(CH_3)_2N]_2P(=O)OP(=O)[N(CH_3)_2]_2$. The compound as synthesized by the chemist is not particularly toxic to insects or to warm-blooded animals. However, once schradan is absorbed into the plant, it is converted into a highly toxic chemical. This conversion may also be effected by incubating the compound in slices of mammalian liver or even in ground-up lettuce leaves. The toxicity is increased 1000-fold by this incubation. Chemists and biologists believe that the toxic compound is an oxidation product of some enzyme system in the plants or animals and that the product may have the structure

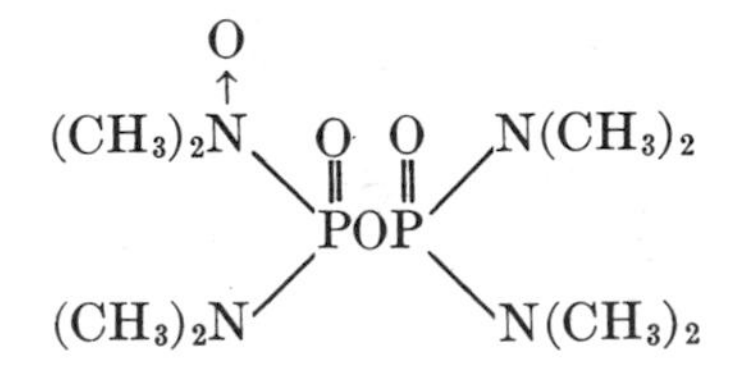

oxidized schradan

If this hypothesis is true, a similar increase in toxicity should be obtained by a chemical reaction. This situation was found to be the case. Schradan was oxidized with sodium hypochlorite to a highly toxic product, which, however, is quite unstable. Therefore, just schradan should be used and the plant system can do the converting to the toxic product. For one reason or another, schradan never became commercially important.

The useful systemic insecticides are those that eventually decompose in the plants and leave no toxic residue. One systemic insecticide, Amiton, when injected into elm trees, effectively controls the beetles that are a major factor in spreading Dutch elm disease. Unfortunately, Amiton's toxicity is so high and its stability is so great that if the treated elm tree leaves were burned in the autumn, enough toxic smoke would be given off to endanger a whole community.

The research on organic phosphorus insecticides has advanced to

TABLE 19.III. Organic Phosphorus Systemic Insecticides, Acaricides, and Nematicides

Common and Trade Name	*Formulas*	*Chemical Names*	*U.S. Consumption, 1982 (Million lb)*	*Acute Dermal LD_{50} (mg/kg)*
phorate, Thimet	$(C_2H_5O)_2P(=S)SCH_2SC_2H_5$	*O,O*-diethyl *S*-ethylthiomethyl phosphorodithioate	4.6	20–30[a]
disulfoton, Disyston	$(C_2H_5O)_2P(=S)SCH_2CH_2SC_2H_5$	*O,O*-diethyl *S*-[2-(ethylthio)ethyl] phosphorodithioate	1.9	6–25[b]
demeton, Systox	$\{(C_2H_5O)_2P(=S)-O-CH_2CH_2SC_2H_5$; $(C_2H_5O)_2P(=O)-S-CH_2CH_2SC_2H_5\}$	mixture of *O,O*-diethyl *S*-(and *O*)-2-(ethylthio)ethyl phosphorothioates	0.1	8.2–14

dimethoate, Cygon	$(CH_3O)_2P(=S)-S-CH_2C(=O)-NH-CH_3$	*O,O*-dimethyl *S*-(methylcarbamoylmethyl) phosphorodithioate	1.3	>1000[a,c]
fenamiphos, Nemacur	$(C_2H_5O)((CH_3)_2CHNH)P(=O)-O-C_6H_3(CH_3)-SCH_3$	ethyl 3-methyl-4-(methylthio)phenyl (1-methylethyl)-phosphoroamidate	0.6	178–225[d]
monocrotophos, Azodrin	$(CH_3O)_2P(=O)-O-C(CH_3)=C(H)-C(=O)NHCH_3$	*O,O*-dimethyl *O*-[2-(methylcarbamoyl)-1-methylvinyl] phosphate	2.4	450
mevinphos, Phosdrin	$(CH_3O)_2P(=O)-OC(CH_3)=C(H)-C(=O)-OCH_3$	methyl 3-hydroxy-α-crotonate, dimethyl phosphate	0.5	33.8[d]
dicrotophos, Bidrin	$(CH_3O)_2P(=O)-OC(CH_3)=C(H)-C(=O)-N(CH_3)_2$	dimethyl *cis*-[2-(dimethylcarbamoyl)-1-(methylvinyl) phosphate	0.1	112–400[d]

Table 19.III *(Continued)*

Common and Trade Names	*Formulas*	*Chemical Names*	*U.S. Consumption, 1982 (Million lb)*	*Acute Dermal LD_{50} (mg/kg)*
phosphamidon, Dimecron	$(CH_3O)_2P(=O)\text{—O—}C(CH_3)=C(Cl)C(=O)\text{—}N(C_2H_5)_2$	dimethyl phosphate, ester with 2-chloro-*N*,*N*-diethyl-3-hydroxycrotonamide	0.1	267[d]
Amiton, Tetram	$(C_2H_5O)_2P(=O)SCH_2CH_2N(C_2H_5)_2 \cdot H_2C_2O_4$	2-(diethoxyphosphinylthio) ethyldiethylammonium hydrogenoxalate	0	—[e]
schradan	$[(CH_3)_2N]_2P(=O)\text{—O—}P(=O)[N(CH_3)_2]_2$	octamethylpyrophosphoramide	0	9[f,g]

[a] Guinea pig.
[b] Rat.
[c] 50% active ingredients in solution.
[d] Rabbit.
[e] Not determined.
[f] Acute oral LD_{50}.
[g] Male rat.

the point where even animal systemics have been developed. Such compounds could be sprayed on or fed to the animals at a controlled dose that would kill insects feeding on their tissues but be harmless to the hosts. An example of such a compound is coumaphos or Coral:

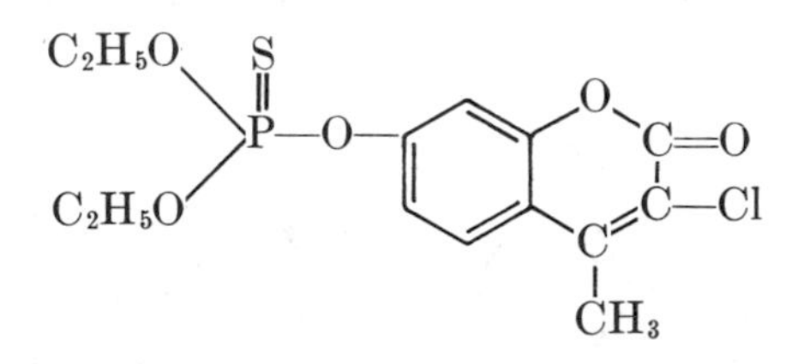

O-(3-chloro-4-methyl-2-oxo-2*H*-1-benzopyran-7-yl)
O,O-diethyl phosphorothioate

The biggest development in this area involves the cattle grub. These parasites come from the eggs of heel flies that have laid their eggs on the hair of the cattle. The eggs hatch into maggots, which tunnel under the skin of the cows and then bore through the animal's body in about 6 months. Finally, the maggot lodges in the back of the cattle and bites a hole through the skin for breathing. This causes a tumorlike inflammation about the size of a human thumb on the back of the animal. When the maggot has matured to about 1 in. long, it squeezes itself through the hole in the hide and drops to the ground. It will then develop into a fly and start the cycle over again by laying eggs on the cattle.

Cows infected with these maggots lose weight and give less milk. The maggots often lodge in the area from which prime steaks come, and hides with holes in them lose their value as premium leather. The systemic insecticides solve all these problems by killing the cattle grubs.

Research on organic phosphorus insecticides is still lively. The search continues for the wide-spectrum insecticide that is toxic only to harmful insects and not to warmblooded animals. The compound should be odorless, tasteless, and inexpensive. Its discovery and development represent a fascinating challenge to research chemists.

The research on organic phosphorus poisons is also a challenge to the biologist, the physiologist, and the entomologist. After all, we still do not have the complete picture of why certain compounds are effective although others are not.

Other Toxic Organic Phosphorus Compounds

A toxic organic phosphorus compound that caused paralysis of many thousands of people in the 1930s was tri-*o*-cresyl phosphate (TOCP) (*5*). The compound was used to replace castor oil for making Fluid Extract of Ginger, U.S.P., intended for use in a beverage. This replacement was done to reduce cost and without consideration of any possible toxicity. Subsequently, TOPC was found to metabolize into a cyclic phosphate metabolite that is responsible for the neurotoxicity:

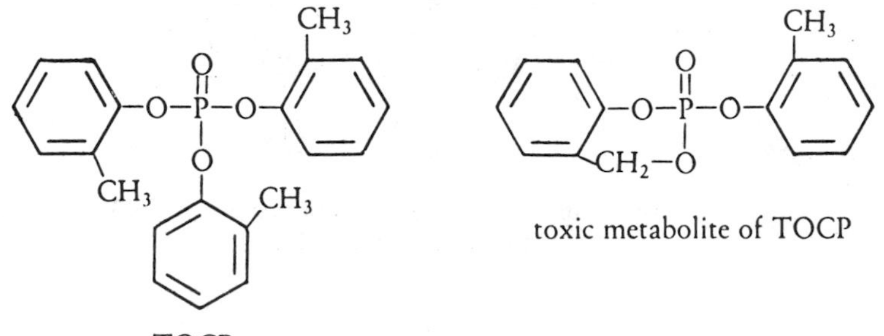

TOCP

toxic metabolite of TOCP

The toxicity is probably the result of the inhibition of the esterases. However, the toxic effect is not reversed by those acetylchlolinesterase reactivating agents, which are normally effective for organic phosphorus nerve gases and related compounds.

Today, large quantities of tricresyl phosphate are used in flame-retardant hydraulic fluids and as plasticizers for plastics (*see* Chapters 10 and 15). Very strict specifications are set on the amount of *o*-cresol allowed in making the tricresyl phosphate. The cresols used are primarily the meta and para isomers:

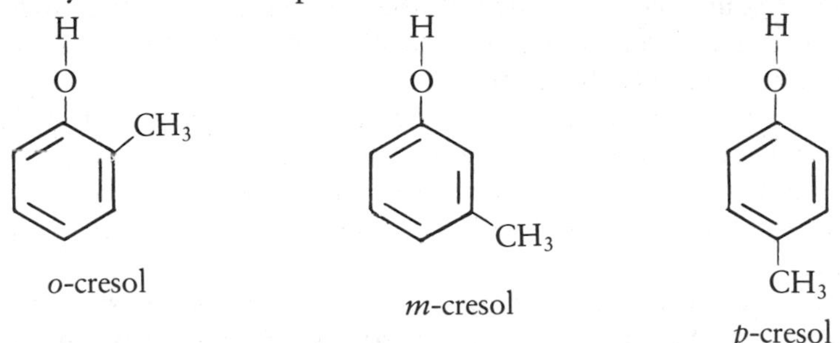

o-cresol

m-cresol

p-cresol

The tricresyl phosphates formed from *m*- and *p*-cresol cannot form the cyclic toxic metabolite.

To go one step further toward making a safer phosphate ester for functional fluids and plasticizers, the chemists have developed *tert*-butylphenyl phosphate. For this compound, the methyl (CH_3–) group in the cresol is replaced by a *tert*-butyl [$(CH_3)_3C$–] group.

Because of the chemical structure of the *tert*-butyl group, this compound does not form a cyclic toxic metabolite. However, it still possesses most of the desirable functional properties of tricresyl phosphate that are required for various industrial applications.

Not all toxic phosphorus compounds owe their toxicity to the inhibition of the chlolinesterase enzyme. Another series of toxic phosphorus compounds is represented by the following:

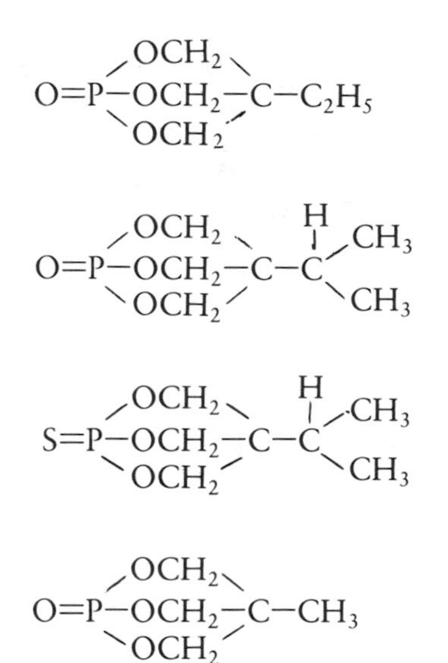

Animals exposed to these compounds usually die within a few minutes after first suffering from convulsive seizures. The signs of poisoning are probably the result of the stimulation of the central nervous system. When Professor John Verkade of Iowa State University prepared these compounds, he had no idea of their toxicity. He made them for use as ligands in his basic research on coordination chemistry. Only years later, when he was cooperating with Professor John Casida, was their toxicity discovered (6).

Literature Cited

1. Schrader, G. *Die Entwirklung neuer insektizider Phosphorsäurester*; Verlag Chemie GmbH: Bergstrasse, Weinheim, West Germany, 1963.
2. Corbridge, D. E. C. *Phosphorus, an Outline of Its Chemistry, Biochemistry, and Technology*, 2nd ed.; Elsevier Scientific: New York, 1980; p 346.
3. *Chemical Economics Handbook*; SRI International: Menlo Park, CA, 1984; Pesticides 573.3005 O.

4. *Farm Chemical Handbook*; Meister Publishing: Willoughby, OH, 1984.
5. Hilderbrand, R. L. *The Role of Phosphonates in Living Systems*; CRC: Boca Raton, FL, 1983: p 149.
6. Casida, J. E.; Eto, M.; Moscioni, A. D.; Engel, J. L.; Milbrath, D. S.; Verkade, J. G. *Toxicol. Appl. Pharmacol.* **1976**, *36*, 261–179.

Chapter Twenty

Mechanism of Toxic Action of Organic Phosphorus Poisons

Toxic Action

Why are some organophosphorus compounds such potent killers of insects or animals? How do they work? What are the antidotes, and how were they discovered? At present, not all the answers to the mechanism of biological action of these compounds are known. However, the biological theory developed thus far fits quite well with the structure and chemistry of some of these poisons. The theoretical basis for their action was developed from hindsight. That is, the toxic action was discovered first, and the theory that accounts for the action was formulated and checked out later.

Nevertheless, the theories, though incomplete, have served as very useful guiding principles in the search for newer and more effective compounds. These theories also serve as guides for the development of antidotes. However, because we do not have the complete mechanistic picture, a bit of luck still plays a big part in each discovery. Such luck usually goes, however, to researchers who are diligent and alert enough to take advantage of unexpected leads.

Biologists generally agree that many poisonous organic phosphorus compounds are toxic because they inhibit the enzyme acetylcholinesterase (AChE). Although the exact mechanism is still a subject for further research, sufficient data indicate that the following simplified description is valid. In a nerve transmission system, a small gap exists between the end of the nerve fiber and the end of the muscle fiber. The gap is called the myoneural junction. It is about 100-Å (0.000001-mm) wide. An impulse for action in the muscle is started by changes in the electrical potential of the nerve ending. This releases a chemical

1002–0/87/0319$06.00/1

called acetylcholine, [$CH_3C(=O)OCH_2CH_2N^+(CH_3)_3$], which diffuses across the myoneural junction and is accepted by the muscle fiber. Acetylcholine changes the electric potential of the muscle fiber and causes the fiber to contract. The contraction is what we observe or feel as muscle action.

After contraction of the muscle fiber, the enzyme AChE immediately catalyzes the hydrolysis of the acetylcholine. The two products of hydrolysis are acetic acid [$CH_3C(=O)OH$] and choline [$HOCH_2CH_2N^+(CH_3)_3$]. These two compounds then diffuse away from the muscle fiber. These compounds will be brought together again by other body enzymes elsewhere.

Acetylcholine must be hydrolyzed and removed from the muscle once it contracts the fiber. When acetylcholine is gone, the muscle fiber returns to its original state and is ready for contraction again when the next controlled impulse from the nerve fiber sends in more acetylcholine. If acetylcholine remained in the muscle fiber longer than usual, the muscle would twitch and remain contracted.

If another nerve impulse sends in more acetylcholine before the acetylcholine from the previous impulse is dissipated, the muscle fiber cannot cope with the situation and becomes paralyzed. Paralyzed muscles are usually uncomfortable but not fatal unless the chest muscles that control breathing are involved; in that case, a person dies quickly from asphyxia. Thus, the enzyme acetylcholinesterase must be available to hydrolyze acetylcholine and move it away from the muscle.

According to present theory, certain organic phosphorus compounds are toxic because they inhibit AChE so that it can no longer hydrolyze acetylcholine. This inhibition is chemical. The following describes how these organic phosphorus compounds work.

The complete chemical and physical structure of the enzyme AChE is still not known. We do know that it is a large protein molecule made up of many amino acids and exists as a coiled chain. The enzyme has two known active sites. One site is called the esteratic site because it attracts ester groups to it. The other site is called the anionic or negative site because it attracts a cation (or an electrically positive section of a compound). The distance between these two sites for mammals is estimated at 4.3–4.7 Å. For insects, the distance is estimated to be 5.0–5.5 Å (*1*). When AChE acts on acetylcholine, the first step is the formation of a complex. In this complex, the anionic or negative site of the AChE locks onto the cationic or positive end of the acetylcholine molecule. The esteratic site locks into the acetyl [$CH_3C(=O)–$] end of the acetylcholine:

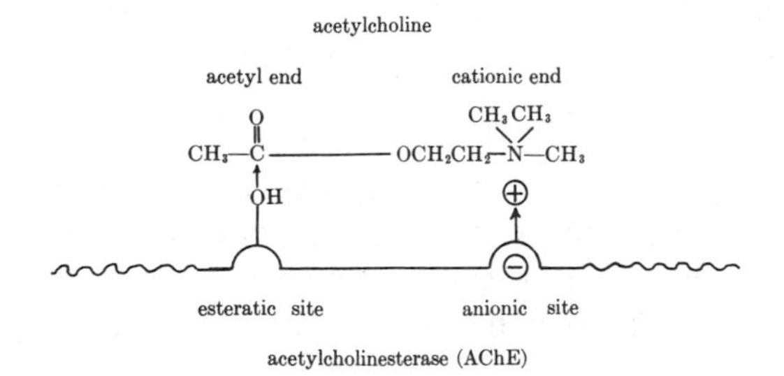

In this intimate condition, the esteratic site can easily react with the acetyl group of the acetylcholine and form acetylated acetylcholinesterase:

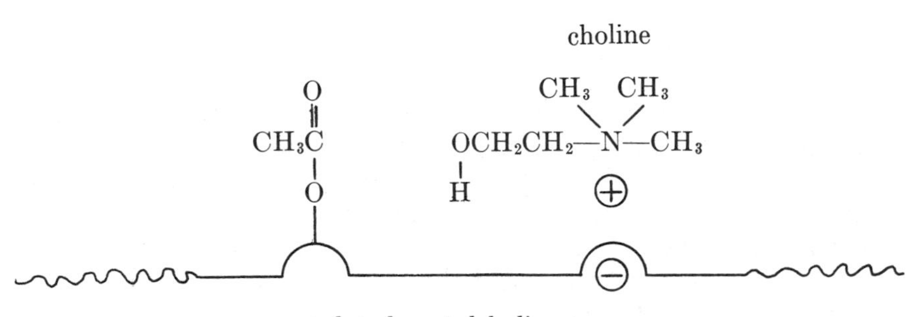

acetylated acetylcholinesterase

This reaction breaks up acetylcholine. Acetylated acetylcholinesterase then hydrolyzes in the water of the surrounding tissue to regenerate (AChE) and liberate acetic acid and choline.

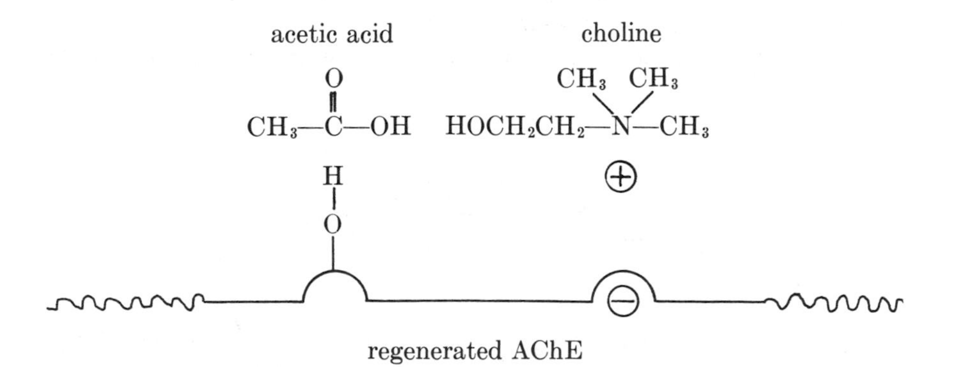

regenerated AChE

Acetic acid and choline then diffuse away from the muscle fiber and elsewhere re-form acetylcholine for use again in the next nerve impulse.

A crude mixture of AChE isolated from a live biological system, such as a homogenate of fly brains, can split acetylcholine in a test tube. If the test tube contains a buffer solution of sodium bicarbonate ($NaHCO_3$), the acetic acid that is liberated reacts with sodium bicarbonate to give off carbon dioxide gas:

$$CH_3\overset{\overset{\displaystyle O}{\|}}{C}{-}OH + NaHCO_3 \rightarrow CH_3\overset{\overset{\displaystyle O}{\|}}{C}{-}ONa + H_2O + CO_2\uparrow$$

[This test carried out away from the biological specimen is called an in vitro test, in contrast to an in vivo test conducted in the live specimen.] If the AChE is inhibited or prevented from splitting acetylcholine to yield acetic acid, no gaseous CO_2 will evolve. If AChE is partially inhibited, less CO_2 is evolved. The biologist uses this in vitro test extensively to measure the degree of AChE inhibition by agents such as organic phosphates.

Two sites in AChE must be locked onto two sites in the acetylcholine before AChE can be acetylated. This finding is proved by the fact that when the anionic site in AChE is occupied by a stronger cation from another molecule, AChE can no longer hydrolyze acetylcholine.

Present theory for organic phosphorus poisoning proposes that an organophosphorus molecule (an ester) forms an intimate complex with AChE. AChE is thus phosphorylated, rather than acetylated. For example, with diethyl *p*-nitrophenyl phosphate, or paraoxon,

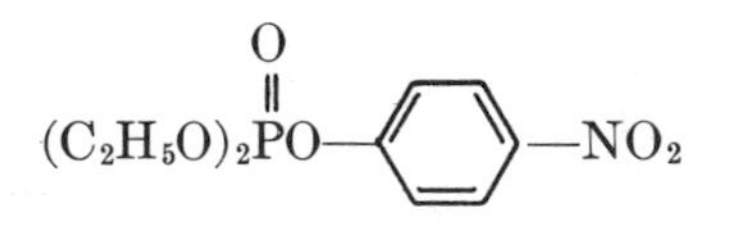

the phosphorylated AChE is the diethyl phosphate ester of AChE. [Paraoxon is used as an example because the widely used parathion

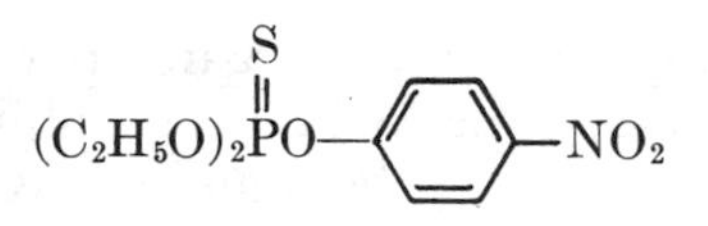

has been proved experimentally to be converted to paraoxon in the body before it becomes toxic.] Most researchers agree that the esteratic site of AChE contains an –OH group from the amino acid serine. The serine –OH is therefore the site for phosphorylation:

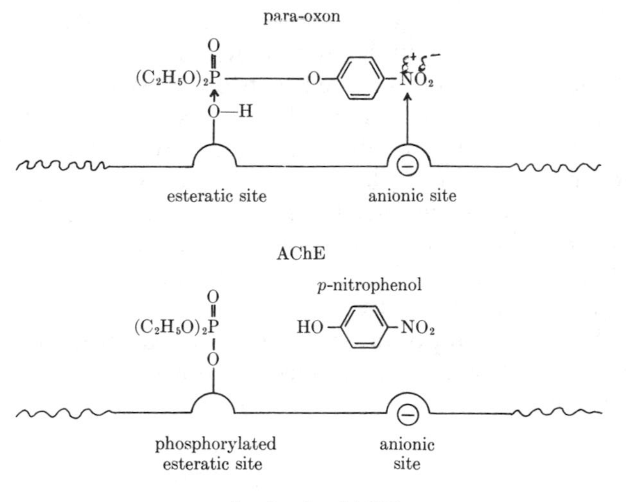

In the diagram, δ^+ and δ^- indicate a partially ionic bond, with the shared electrons drawn toward the atom labeled δ^-. The diethyl phosphate group $[(C_2H_5O)_2P(=O)–]$ attached to the esteratic site of the AChE makes AChE unavailable for forming a complex with incoming acetylcholine. AChE thus cannot split acetylcholine, and we say that the AChE is inhibited. When excess unsplit acetylcholine accumulates on the muscle, the muscle fiber starts to twitch and is finally paralyzed.

The diethyl phosphate group in phosphorylated AChE does not hydrolyze as readily as the acetyl group in acetylated AChE. However, hydrolysis does occur, slowly, to regenerate AChE so that it becomes available again for splitting acetylcholine. For this reason, if you are not killed quickly by an organic phosphorus poison, you will live. In other words, if the AChE is not inhibited by an overdose of organic phosphorus poison, the inhibited AChE may be regenerated in time to prevent death.

In this description of the formation of the complex between AChE and acetylcholine, much was said about the esteratic site complexed with the ester group and the anionic site complexed with the cationic site. In the case of an organic phosphorus compound, such as paraoxon, a phosphate ester group [$(C_2H_5O)_2P({=}O)$–] was available to complex with the esteratic site of the AChE. However, we can say that only a weak cationic or positive site also exists to complex with the anionic or negative site of the AChE.

Some researchers speculate that complexing is also caused by a spatial effect. That is, the molecular structural geometry of the other portion of the paraoxon molecule fits into the AChE chain like a key in a lock. This molecular fit is of the utmost importance. If, in addition, a cationic site in the phosphorus compound also happens to be in the correct position to lock onto the anionic site of AChE, the inhibition of AChE is enhanced. This last point was shown by the comparison of the activity of the compound $(C_2H_5O)_2P({=}O)SCH_2CH_2S^+(CH_2CH_3)_2$ with that of $(C_2H_5O)_2P({=}O)SCH_2CH_2SCH_2CH_3$. Pictorially, these two compounds are structurally similar; however, the first compound containing a cationic (positive) site reacts 1000 times faster than the second compound, which does not possess a cationic site.

Antidotes

What remedies are there for organic phosphorus poisoning (*2, 3*)? The therapeutic remedy devised is based on the theories of toxic action discussed previously. If the poisoning is not too severe and is of short duration, two methods can be applied together. These methods are based on the fact that nerve stimuli send acetylcholine to the muscle fiber and AChE in the muscle fiber, inhibited by phosphorylation, is unable to split acetylcholine.

In one method, the muscle fiber tip is covered with a shot of atropine; this compound blocks the muscle fiber from further acetylcholine invasion, and phosphorylated AChE will gradually lose its phosphate group by slow hydrolysis. Once freed of the phosphate group, AChE then resumes its regular duty of splitting acetylcholine.

However, when a muscle is paralyzed, especially the muscles of breathing, AChE must be regenerated as fast as possible. Here chemists have developed a method using oximes. One very promising oxime is 2-[(hydroxyimino)methyl]methylpyridinium iodide, or 2-PAM:

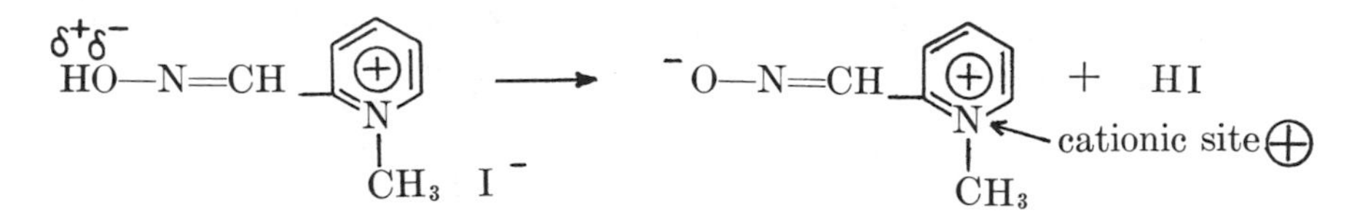

2-PAM has a cationic site that can lock into the anionic site in AChE.

The other end of the molecule carries a negative charge, either partial, as shown on the left, or perhaps a formal negative charge as the equilibrium leading to the zwitterionic form on the right is attained. This negatively charged group seeks out the positive phosphorus atom in the phosphate ester group and displaces it away from the AChE; this action is called nucleophilic displacement.

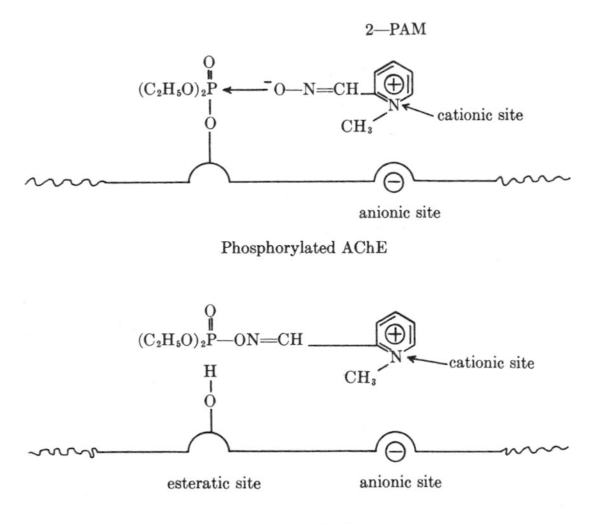

When AChE is regenerated with 2-PAM, the compound

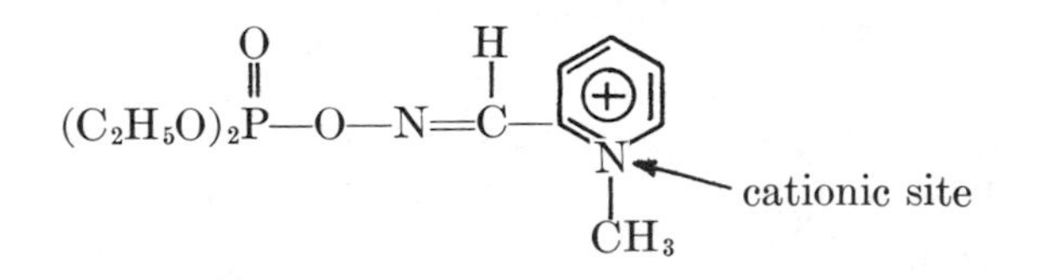

is also produced. This byproduct also has a cationic site and a phosphate ester group. Because such a structure should be ideal for phosphorylating the enzyme AChE, it should also be a poison. In other words, this reaction should be reversible, and in fact it is. In poisoning by the nerve gas sarin

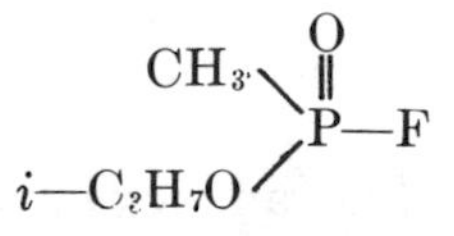

the antidote 2-PAM, in displacing the

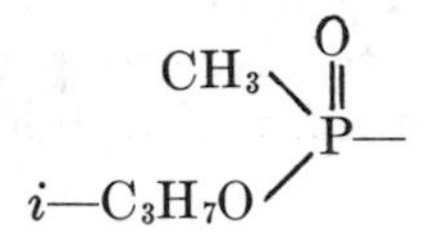

group from AChE, forms a new compound that is more toxic than sarin, but the new compound is less stable to hydrolysis than sarin, so the new compound is destroyed by hydrolysis in the surrounding aqueous medium; this action shifts the equilibrium toward AChE regeneration.

Paraoxon reacts with AChE despite the fact that it has no obvious cationic site for AChE to lock onto. Perhaps the molecule fits well geometrically with the enzyme AChE. This speculation on the stereospecific molecular geometry of two compounds fitting together received some support from the experimental evidence that the isomer of 2-PAM, 3-PAM, is not an antidote. In 2-PAM, the six-corner pyridine ring is numbered with the corner occupied by the N as 1, thus

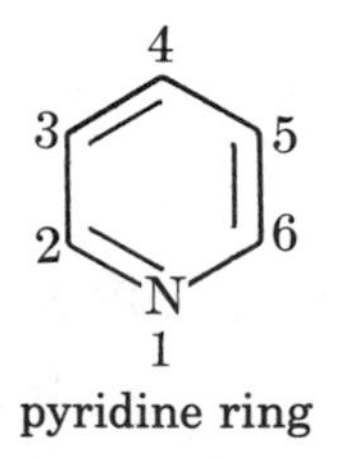

pyridine ring

The chemical name of 2-[(hydroxyimino)methyl]methylpyridinium iodide means that the (hydroxyimino)methyl group, HON=CH–, is attached to the ring on corner 2. When the name is shortened to 2-PAM, the 2 is retained to indicate the location of the attachment. If the attachment of the HON=CH– group were on corner 3, the compound would be 3-[(hydroxyimino)methyl]methylpyridinium iodide with the structure

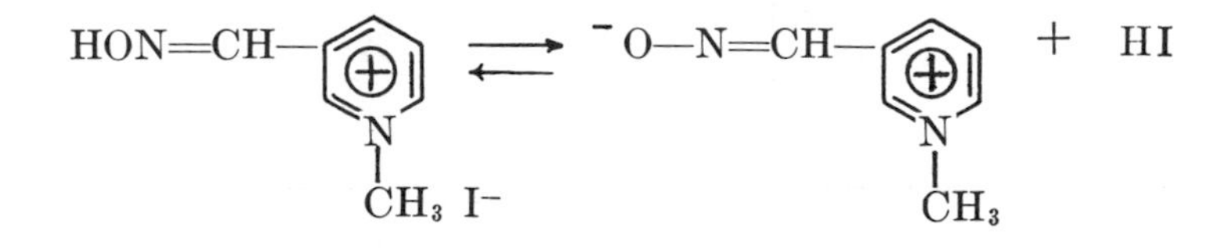

The compound is thus called 3-PAM. The substituent group of 2-PAM and 3-PAM is therefore the same, but its location on the ring is different. 2-PAM and 3-PAM are isomers of each other. Although 3-PAM also has a cationic site, it is not a regenerator for phosphorylated AChE, probably because it does not physically lock with AChE. The presence of the –ON=CH– group in corner 3 of the ring gives 3-PAM the wrong spatial configuration.

In cases of poisoning by phosphorus insecticides or nerve gas, the immediate treatment is an injection of atropine. Although this compound has no therapeutic effect, it blocks the action of the acetylcholine; therefore, the injected antidote such as 2-PAM has sufficient time to reverse the phosphorylation or phosphonylation of cholinesterase. In fact, 2-PAM chloride is included with atropine in the standard U.S. Army injector introduced in 1980 (*4*).

The speed of the treatment is very crucial for survival. The longer the time before atropine and an antidote are used, the more difficult the regeneration of AChE. If phosphorylated AChE is allowed to age, AChE regeneration becomes more difficult. On aging, one of the organic groups in the phosphate group is hydrolyzed. For example, in the inhibition of AChE by diisopropyl fluorophosphate (DFP), the phosphorylated AChE is

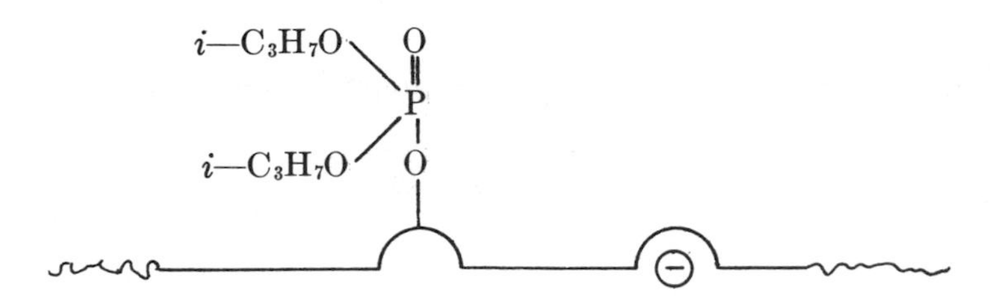

This compound is a neutral phosphate ester. When the ester is aged, an isopropyl group is hydrolyzed and the phosphorylated AChE becomes

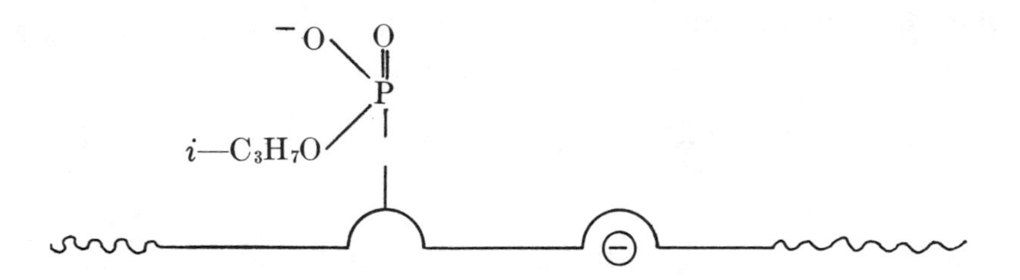

Reactivation or regeneration of AChE depends on attack of the phosphorus atom by an anion, either an HO^- from water or a

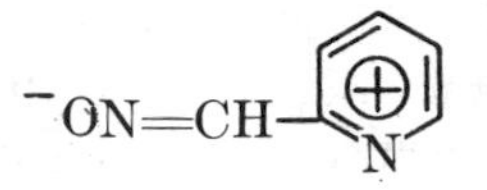

from 2-PAM. Aging removes one isopropyl group from the phosphorylated AChE to leave

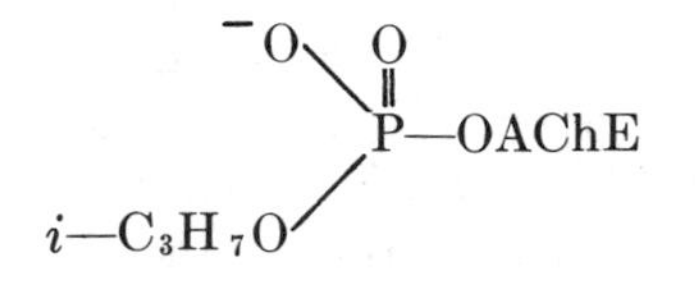

also an anion. This negatively charged anion will repel any oncoming negatively charged anions. Thus, the effectiveness of HO^- or

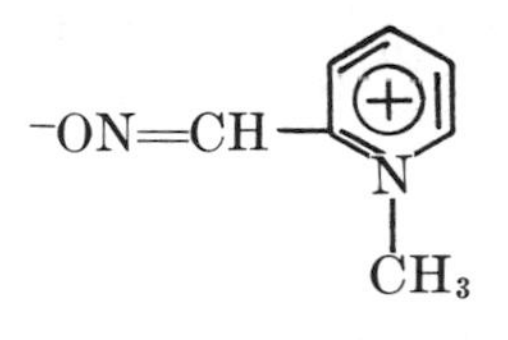

to hydrolyze the monoisopropyl phosphate group

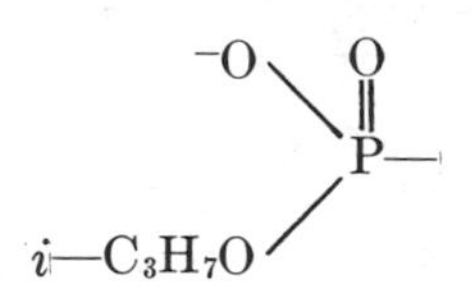

to regenerate AChE becomes very difficult. As a result, the patient dies.

In the case of the nerve gas Soman (*see* Chapter 18), the aging process takes only 2–3 min. After that time period, the poisoning is irreversible and the person dies.

Second generations of oxime antidotes or cholinesterase reactivators are now available. One compound is called obidoxime, bis{4-[(hydroxyimino)methyl]methylpyridinium} ether dichloride (*3*):

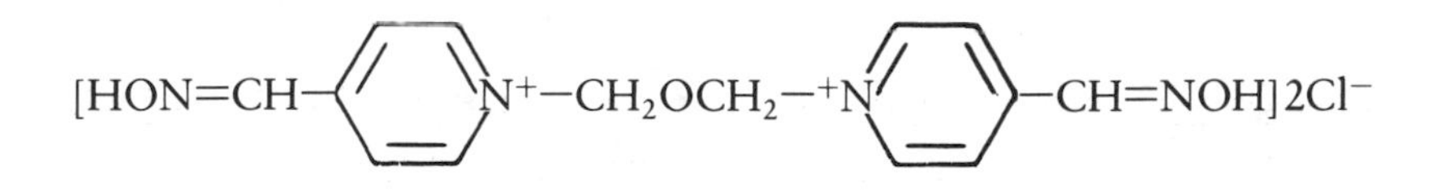

Apparently, the structure of these compounds also fits into that of cholinesterase.

Two other second generation oxime antidotes are designated as HI-6 and HS-6:

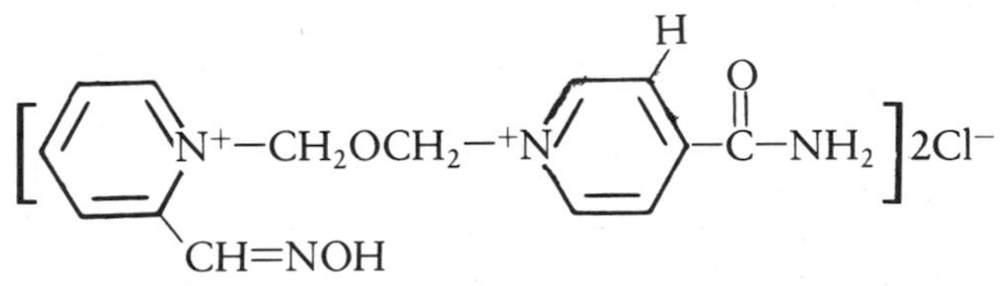

HI-6

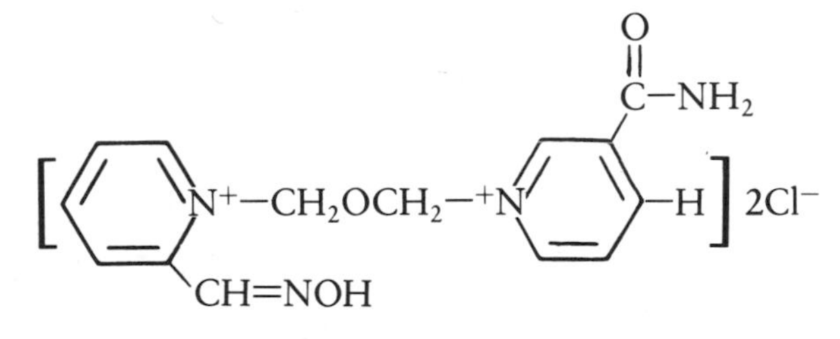

HS-6

Because we have a theory for the poisoning action of organic phosphorus compounds and also a theory for counteracting this action, we ought to be able to design compounds of specific structures for the counteraction. As indicated previously, this designing is now being done to a limited extent. However, present theories are only rough approximations of what is truly occurring. We still do not know enough about the physiology of mammals and insects. For that matter, we know very little about any one insect. Furthermore, the

AChEs are different in humans, animals, and insects. Even in insects the enzymes differ from species to species.

Mammals contain other enzymes besides AChE that can hydrolyze all phosphate esters and that are not inhibited by them. Also, enzymes exist that are quite similar to AChE and that, for the lack of a better name, are called pseudocholinesterases. These enzymes hydrolyze practically everything.

Studies have shown that the oxime antidotes developed thus far are not equally effective against each nerve gas (*see* Table 20.I).

Table 20.I. Effectiveness of Pyridinium Oximes against Standard Nerve Gases

Oxime	*Sarin*	*Tabun*	*Soman*	*VX*
2-PAM chloride	+	+	0	+
Obidoxime	+ + +	+ +	0	+ + +
HI-6	+ + + +	0	+ +	+ + + +
HS-6	+ + +	0	+	+ + +

NOTE: The number of plus signs indicates the degree of effectiveness; 0 represents no effect.
SOURCE: Reference 4.

Literature Cited

1. Fest, C.; Schmidt, K.-J. In *Chemistry of Pesticides*; Büchel, K. H., Ed.; Wiley: New York, 1983; p 98.
2. O'Brian, R. D. In *Insecticide Biochemistry and Physiology*; Wilkinson, C. F., Ed.; Plenum: New York, 1976; pp 271–320.
3. Fest, C.; Schmidt, K.-J. *The Chemistry of Organophosphorus Pesticides*; Springer-Verlag: New York, 1973; pp 251–261.
4. *Chem. Br.* **1984,** *8*, 7.

Chapter Twenty-one

Organic Phosphorus Plant Growth Control Agents, Herbicides, and Fungicides

THE STORY (*1*) WAS TOLD that back in 1893, in the Azores, a small fire occurred in a greenhouse containing some pineapple plants. The owners had expected that these plants would be damaged by the flames and smoke from the fire. To their surprise, all the pineapple plants burst into bloom. The reason for this surprising phenomenon is that, in the fire, not all the organic material was burned to carbon dioxide and water. Some of the organic material was decomposed by the heat to release ethylene ($CH_2{=}CH_2$). A more recent case involved a greenhouse located near a chemical plant. The owner found that his plants in the greenhouse bloomed prematurely. He figured that the cause must be pollution from the neighboring chemical plant. Upon investigation, however, the real culprit was found to be ethylene. Ethylene came from the incomplete combustion in the wood stove used for heating the greenhouse.

Ethylene is a plant growth hormone. It not only promotes flowering in the bromeliad or pineapple family of plants but also causes seeds to sprout and fruits to ripen and fall off. Many fruits generate their own ethylene biologically. For example, ripening apples give off ethylene. In fact, the ethylene generated by apples may be used for the ripening of other fruits. So, the common practice of hastening the ripening of other fruits at home by sealing them in a plastic bag with a ripening apple has a scientific basis. Similarly, a flower bud will open faster if put in a plastic bag with a ripe apple.

Research has shown that plants make ethylene from an unusual amino acid, 1-aminocylopropane-1-carboxylic acid, or ACC. On the

1002–0/87/0331$06.00/1

basis of present knowledge, ACC produces ethylene through the overall reaction

$$\underset{\text{ACC}}{\text{(H}_2\text{C)}_2\text{C(NH}_3^+\text{)C(=O)O}^-} \longrightarrow \underset{\text{ethylene}}{CH_2{=}CH_2} + CO_2 + HCN + H_2$$

Chemists are not satisfied with the vagaries of nature to generate ethylene. They want a compound that will give them some control over the time and place where the ethylene should be released. The result is the development of (β-chloroethyl)phosphonic acid. This compound has the common name of ethephon. It is applied to the plants by spraying as a dilute water solution. Once absorbed in a plant, the compound gradually breaks down to release ethylene:

$$ClCH_2CH_2P(=O)(O^-)(OH) + OH^- \longrightarrow CH_2{=}CH_2 + Cl^- + O{=}P(O^-)(OH)(OH)$$

The reactions for the preparation of ethephon are

$$PCl_3 + 3\overset{O}{CH_2CH_2} \longrightarrow \underset{\text{tris(β-chloroethyl) phosphite}}{P(OCH_2CH_2Cl)_3}$$

$$P(OCH_2CH_2Cl)_3 \xrightarrow{\Delta} \underset{\text{bis(β-chloroethyl) (β-chloroethyl)phosphonate}}{ClCH_2CH_2\overset{O}{\overset{\|}{P}}(OCH_2CH_2Cl)_2}$$

$$ClCH_2CH_2\overset{O}{\overset{\|}{P}}(OCH_2CH_2Cl)_2 + 2H_2O \longrightarrow \underset{\text{ethephon}}{ClCH_2CH_2\overset{O}{\overset{\|}{P}}(OH)_2} + 2ClCH_2CH_2OH$$

When ethephon is sprayed on pineapple plants to the extent of about 0.5 lb/acre, it induces all the plants to flower at about the same time; this treatment results in the synchronous development and ripening of the fruit and easier mechanical harvesting. The mechanical harvesting of tomatoes is also facilitated by spraying the plants with ethephon solution. All the tomatoes thus ripen at about the same time. In this case, the amount of ethephon needed is 0.75–1.25 lb/acre. Ethephon is also used to increase the latex flow in rubber trees. The result of an ethephon application has been an increase of two- to fourfold in latex flow. However, the treatment on the rubber plants is not permanent. The increased flow gradually falls off after 10–15 days. After 2 months, the flow rate returns to that of the untreated tree. Retreatment has no effect. Also, rubber trees that have been treated with ethephon do not live as long as untreated trees (2).

Ethephon also induces the formation of more female flowers in some squash and cucumber species. Cucumber plants at the two-leaf stage, when sprayed with a solution containing 150 ppm of ethephon, will produce more female flowers. The increase in female flowers, of course, means more fruit formation per cucumber plant.

Ethephon is not the only organic phosphorus compound that has a biological effect on some plants. Another biologically active phosphorus compound is bis(phosphonomethyl)glycine (*3*). It is also known by the common name of glyphosine and has a structure

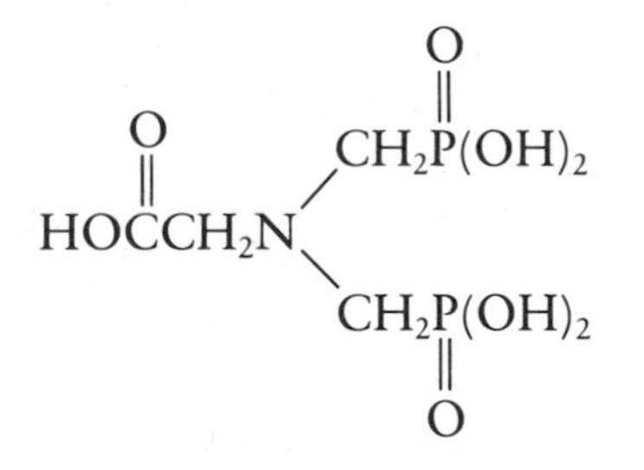

When a solution of this compound is sprayed on sugar cane plants to the extent of about 4.5 lb/acre, the yield of sucrose is increased by 10–20%. This increase represents about 1 ton of sugar per acre. However, the timing of the spraying is important. The spraying must be done 4–8 weeks before harvest. Glyphosine is believed to increase the stability of sucrose in the plant. Glyphosine kills the tip growth of the plant and thus slows down the process of the conversion of the sucrose to plant fibers. This biochemical phenomenon also permits a longer harvest season. We were told that the behavior of glyphosine as a plant hormone was discovered accidentally. The original goal for synthesizing the compound was to find a new detergent builder and a chelating and water-softening agent.

In Chapter 7, we discussed the compound aminotris(methylenephosphonic acid), known as NMPA, and sold commercially as Dequest 2000:

aminotris(methylenephosphonic acid)

NMPA is an excellent chelating and water-softening agent. It is the phosphorus analogue of a very well known chelating and water-softening agent, the nitrilotriacetic acid, or NTA.

NTA

The chemical structure shows that glyphosine is analogous to NTA in that two of the —C(=O)OH groups in NTA are replaced by two $—P(=O)(OH)_2$ groups. As expected, the compound is an excellent chelating agent. In evaluation of the compound for other functions, however, the compound was discovered to possess plant hormone activity. As a matter of fact, the compound is also reported to have herbicidal activity (*4*).

A very closely related compound, *N*-(phosphonomethyl)glycine (NPMG), or glyphosate, is a nonselective herbicide. It has the structure

$$HO\overset{\overset{\displaystyle O}{\|}}{C}CH_2\overset{\overset{\displaystyle H}{|}}{N}CH_2\overset{\overset{\displaystyle O}{\|}}{P}(OH)_2$$

NPMG is an effective herbicide when used at a rate of 1–5 lb/acre. A formulation containing as the active compound, isopropylammonium NPMG, was introduced commercially in the early 1970s as a weed killer (*5*). It has the trade name Round-Up. Another compound, trimethylsulfonium NPMG, has been reported in the patent literature as a herbicide (*6*), as has the compound iminourea NPMG (*7*).

NPMG is a nonselective, systemic, postemergence herbicide. It kills even when only about 20% of the plant is wetted by the spray. The reason for this efficacy is that the active compound is absorbed through the foliage and translocated throughout the plant, including the root system. Because NPMG is a nonselective herbicide, it kills crop plants as well. For killing the tall weeds selectively grown above the lower growing crop plants, a rope-wick applicator was invented by Dole of the U.S. Department of Agriculture (*8*).

This device (Figure 21.1) consists of a series of short nylon ropes with one end connected to a reservoir of the herbicide solution. The other end of the rope dangles down low enough to touch and wet the tall weeds but not low enough to reach the low-growing crops, such as soybeans. The herbicide solution passes down the rope by both capillary action and gravity flow. This procedure is a relatively economical method of application. It has an advantage over the spraying method in that it is selective. No damage is done to other plants by spray-drift. In 1980, 20 million acres of soybean and cotton were

Figure 21.1. Rope-wick applicator. (Reproduced from reference 24. Copyright 1981 American Chemical Society.)

treated this way, and the use of this method of application is expected to grow when the method is approved for use in other crops.

The mode of action of NPMG on a plant is not definitively known. One hypothesis (*9*, *10*) is that after the compound is absorbed and translocated in the plant, NPMG inhibits the biological pathway through which the plant makes aromatic amino acids, such as phenylalanine. This amino acid is required by plants to produce needed proteins. Thus, the killing process is rather slow, taking 1 week to 10 days. Jaworski (*9*) reported in 1972 that the effect of glyphosate on the plants he studied could be reversed by adding phenylalanine to the nutrient medium. The best results were usually achieved by adding a combination of three aromatic amino acids: phenylalanine, tyrosine, and tryptophan.

Jaworski hypothesized that the action of NPMG could be due to its inhibition of the enzyme chorismate mutase or prephenate dehydratase, which is required for the process that converts prephenic acid eventually to phenylalanine. He also suggested the possibility exists that this inhibition could have resulted from the chelating by NPMG of the metal in the enzyme. Subsequent work by Steinrucken and Amrhein (*10*) led them to believe that the target of NPMG inhibition is 5-enolpyruvylshikimic acid-3-phosphatase, which affects the enzymatic step to form chorismic acid.

The structure of NPMG and its active derivatives appears to be responsible for their high activity as herbicides. Toy and Uhing (*11*) earlier prepared a related compound, the phosphinic analogue of NPMG, (*N*-phosphinomethyl)glycine:

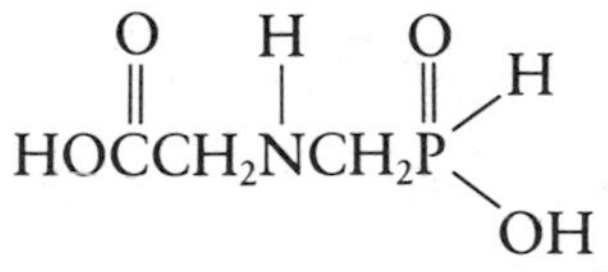

$$HO\overset{O}{\overset{\|}{C}}CH_2\overset{H}{\overset{|}{N}}CH_2\overset{O}{\overset{\|}{P}}\begin{matrix} H \\ OH \end{matrix}$$

The difference between this compound and NPMG is a matter of one oxygen. Yet, this compound was found to be ineffective as a herbicide when evaluated at a practical usage level. Toy and Uhing were also first to prepare NPMG itself as a member of a class of compounds of substituted (aminomethyl)phosphinic acids and substituted (aminomethyl)phosphonic acids. These compounds were prepared for applications that included their use as chelants. They all have the required chemical structures to form the five-membered chelating rings with the proper metal cations (*see* Chapter 7 on chelating). In the actual evaluation on representative members of these classes of compounds, they were found to indeed be chelants, although not of sufficient activity for commercialization. Unfortunately for them, Toy and Uhing did not evaluate NPMG as a herbicide.

Another compound structurally related to *N*-(phosphonomethyl)-glycine is phosphonoiminodiacetic acid (*12*). In 1949, this compound was reported to be an effective chelating agent. Since then, phosphonoiminodiacetic acid has also been found to possess herbicidal acitivity, although it is not nearly as effective as NPMG. Glyphosine, the plant growth regulator, is also structurally related to NPMG. As mentioned earlier, glyphosine is reported to have herbicidal activity when applied at a high concentration.

We prepared several related substituted (aminomethyl)phosphonic acids such as ethylenediaminetetra(methylphosphonic acid)

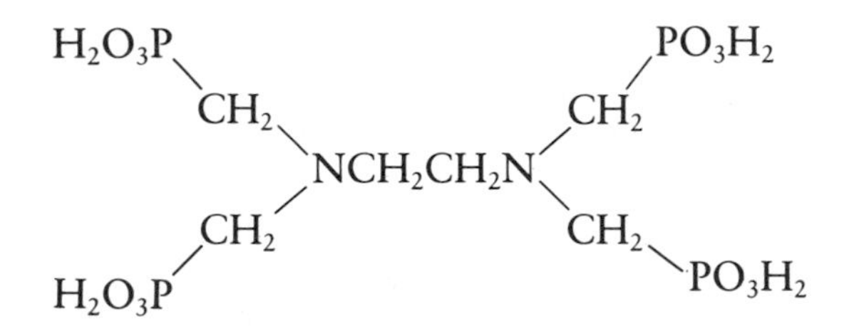

and aminotris(methylenephosphonic acid)

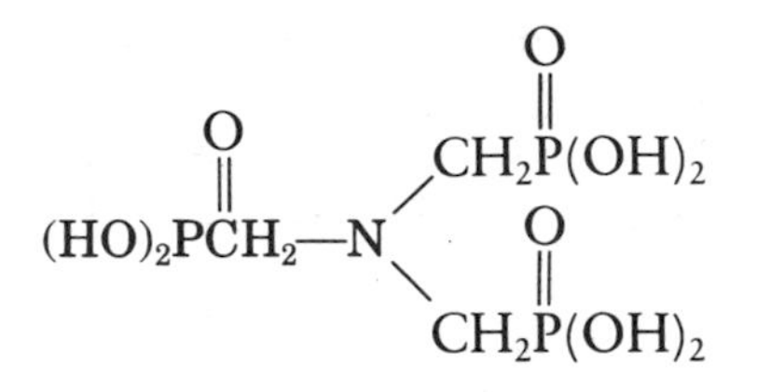

as a class of compounds with chelating properties. As a matter of fact, since the publication of our work, other workers (*13*, *14*) have evaluated NPMG itself and found that it is indeed a good chelating agent. These compounds were prepared by oxidation of the corresponding phosphinic compound with mercuric chloride. This preparation is a convenient laboratory procedure, and it has been used successfully by other workers in this field (*15*). Another method (*16*) reported for the synthesis of NPMG is based on the use of the chelant (phosphonomethyl)iminodiacetic acid. (Phosphonomethyl)iminodiacetic acid is prepared according to the following equation:

$$H_3PO_3 + \underset{\text{iminodiacetic acid}}{(HOOCCH_2)_2NH} + CH_2O \xrightarrow{H^+} \underset{\text{phosphonomethyliminodiacetic acid}}{(HOOCCH_2)_2NCH_2\overset{\overset{\displaystyle O}{\|}}{P}(OH)_2} + H_2O$$

The disodium salt, when oxidized with oxygen, is converted to the disodium salt of NPMG. This sodium salt can easily be converted to the acid by acidification.

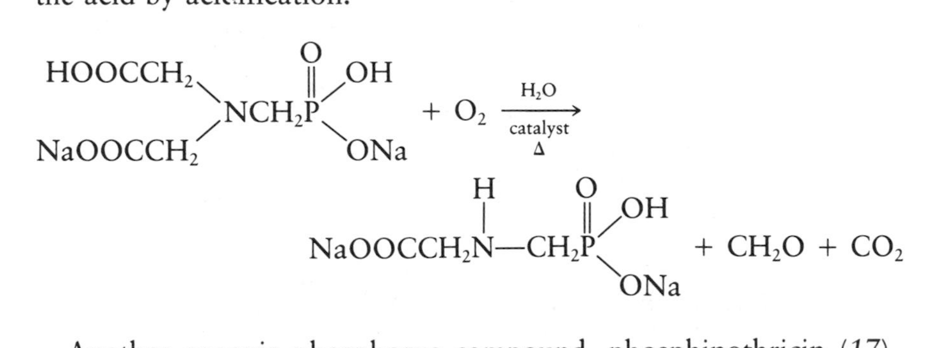

Another organic phosphorus compound, phosphinothricin (*17*), also has nonselective postemergence herbicidal activity. Its action on plants is believed to be through inhibition of the enzyme glutamine synthetase, the initial enzyme required for the plant to assimilate ammonia. This inhibition results in the accumulation of a toxic quantity of ammonia in the cell (*18*). However, this herbicide was not discovered through rational biological design. As is typical of most of the commercially available herbicides and insecticides, phosphinothricin was discovered through biological screening and evaluation. The compound was discovered as a microbial metabolite of *Streptomyces viridochromoquenes* (*17*) and also as a metabolite of *Streptomyces hydroscopicus* (*19*). Its structure is

$$CH_3-\overset{\overset{\displaystyle O}{\|}}{\underset{\underset{\displaystyle OH}{|}}{P}}-CH_2CH_2-\overset{\overset{\displaystyle NH_2}{|}}{CH}-\overset{\overset{\displaystyle O}{\|}}{C}-\overset{\overset{\displaystyle H}{|}}{N}-\overset{\overset{\displaystyle CH_3}{|}}{CH}-\overset{\overset{\displaystyle O}{\|}}{C}-\overset{\overset{\displaystyle H}{|}}{N}-\overset{\overset{\displaystyle CH_3}{|}}{CH}-COOH$$

The compound was initially recognized for its antibiotic properties. However, the researchers working on it also recognized its herbicidal potential (*20*).

At present, a simpler synthetic analogue is under commercial development as a nonselective herbicide. It is called Glufosinate (*23*):

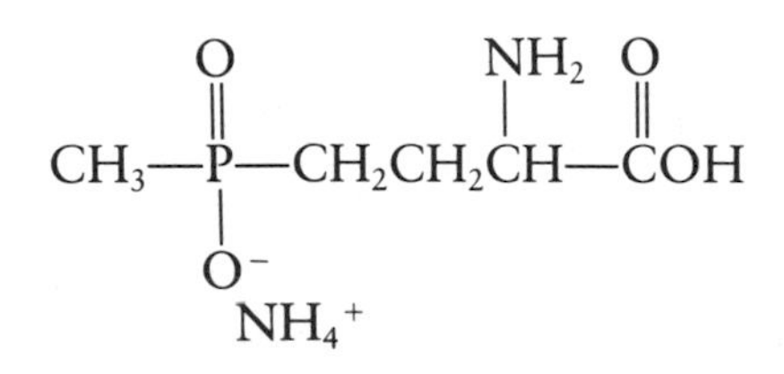

Speculation exists that the interference of this compound with glutamine synthetase is a result of its structure, which mimics that of glutamic acid:

$$\mathrm{HO\overset{\overset{O}{\|}}{C}CH_2CH_2\overset{\overset{NH_2}{|}}{C}H{-}\overset{\overset{O}{\|}}{C}OH}$$

That is, the phosphinic group, $CH_3P({=}O)(OH)$–, of Glufosinate mimics one of the carboxylic groups, HOOC–, of glutamic acid.

One method of synthesis reported in the literature is based on the following reactions:

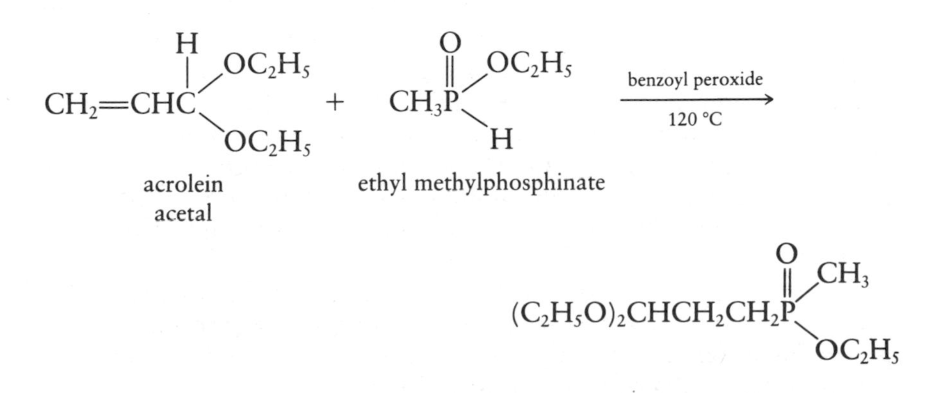

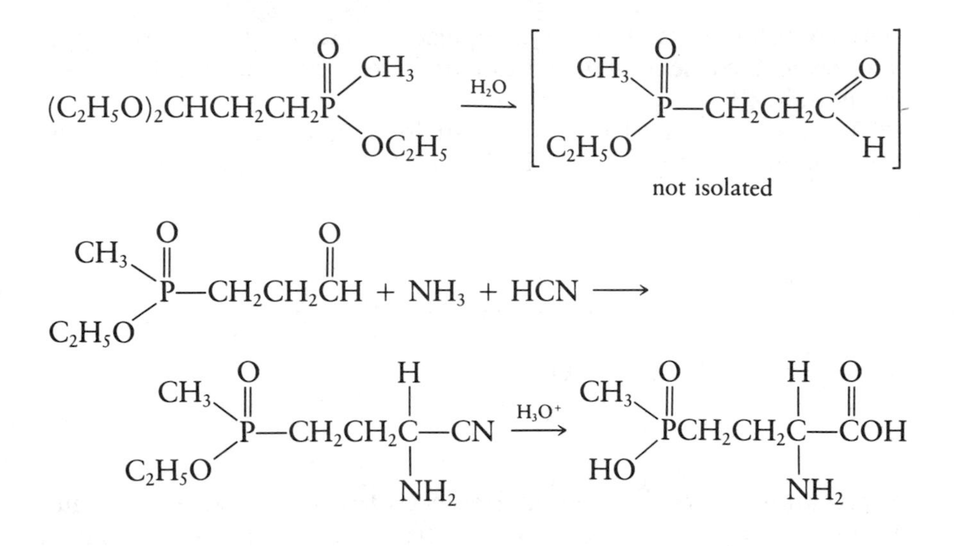

A phosphorus compound with specific defoliant action on some plants is *S,S,S*-tributyl trithiophosphate. It is a good defoliant for cotton (*see* Chapter 15). *S,S,S*-Tributyl trithiophosphate is prepared by the action of phosphorus oxychloride on butyl mercaptan:

$$\underset{\text{phosphorus oxychloride}}{POCl_3} + \underset{\text{butyl mercaptan}}{3CH_3(CH_2)_2CH_2SH} \longrightarrow \underset{S,S,S\text{-tributyl trithiophosphate}}{OP[SCH_2(CH_2)_2CH_3]_3} + 3HCl$$

As a defoliant, *S,S,S*-tributyl trithiophosphate removes leaves from the cotton plants without killing them. Green leaves on a cotton plant have a tendency to stain the white cotton bolls during machine harvesting. Stained cotton commands a lower price. If the plants are defoliated by a spray of a water emulsion of a compound such as *S,S,S*-tributyl trithiophosphate, the potential for staining is eliminated. A related compound, *S,S,S*-tributyl trithiophosphite, prepared by the reaction of phosphorus trichloride with butyl mercaptan, also acts as a cotton defoliant. One disadvantage of both of these compounds is that they have the obnoxious odor of butyl mercaptan. Also, reports that these compounds have a delayed neurotoxic effect exist.

Some phosphorus compounds are effective as fungicides (*21*); this property is especially true of a class of compounds known as phosphites. Representatives of this class are the sodium and, particularly, the aluminum salts of *O*-ethyl phosphite:

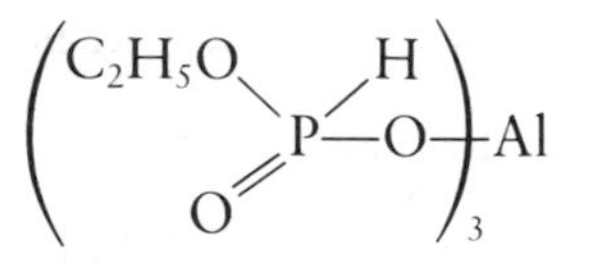

phosetylaluminum

This aluminum salt has a common name of phosetylaluminum. It controls various plant diseases caused by oomycetes (*21*), a class of plant fungus. The active moiety of the compound is the *O*-ethyl phosphite. The aluminum salt was chosen over other metal salts because it is better tolerated by crop plants. This salt is used to control diseases caused by such fungi as downy mildew on grapevines. It also has good activity against phomopsisriticola (excoriosis) and Guignardiabidwellii (black rot) on grapevines. It is also effective on some fungi that attack peanuts and apples. The spectrum of activity of *O*-ethyl phosphite is determined in part by the host plant.

The salts of *O*-ethyl phosphite are systemic fungicides in that they are transported from the treated leaves to the roots and to the new shoots that develop after foliar treatment. Thus, the fungicide protects the whole plant from diseases caused by oomycetes. The main metabolite of phosetylaluminum detected in the plant tissue is phosphorus acid (H_3PO_3), and the original compound is present in barely detectable amounts.

Another interesting herbicide is Betasan. Its chemical name is *O*,-*O*-bis(1-methylethyl) *S*-[2-[(phenylsulfonyl)amino]ethylphosphoro-

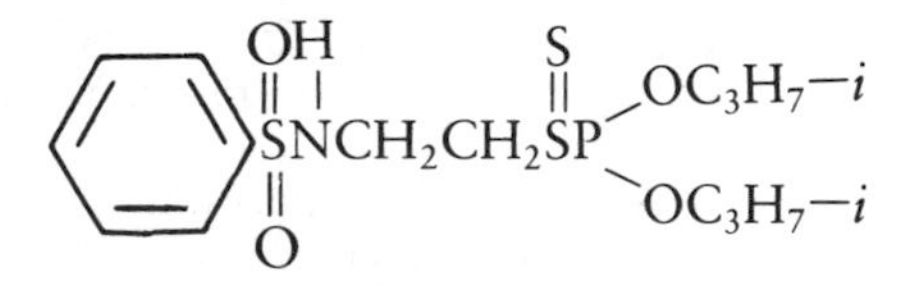

Betasan was synthesized during a systematic research program to find an effective insecticide. Although evaluation showed that it was not an effective insecticide, further evaluation by a botanist showed it to be a very good herbicide. It effectively and selectively kills crab grass as it emerges from the seed while not harming the desired grasses and other ornamental lawn or garden plants.

Most of the phosphorus herbicides described so far require the use of 0.5–4 lb/acre. The synthesis chemist is challenged to find compounds potent at concentrations as low as 0.5 lb/acre. Recently, patents (22) have appeared with claims for phosphorus herbicides that are effective at less than 0.1 lb/acre.

An example of this class of compound is

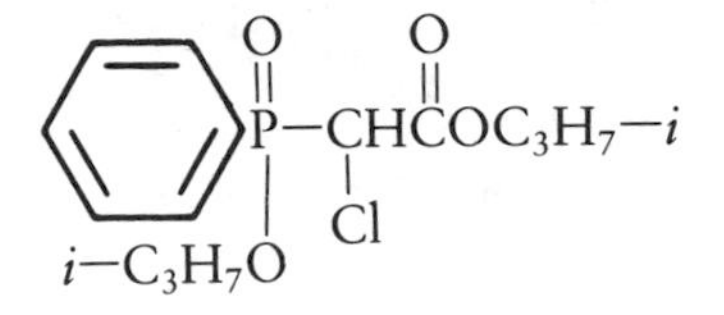

This compound is effective as a preemergence herbicide. Normally, the compound will prevent the emergence of weed plants, yet it will not injure useful plants such as peanuts, cotton, soybean, and sugar beets.

Literature Cited

1. Dagani, R. *Chem. Eng. News* **1984,** *Feb. 13*, 21.
2. Abeles, F. B. *Ethylene in Plant Biology*; Academic: New York, 1973; p 79.
3. Hamm, P. C. U.S. Patent 3 556 762, 1971 (to Monsanto).
4. Irani, R. R. U.S. Patent 3 455 675, 1969 (to Monsanto).
5. Franz, J. E. U.S. Patent 3 799 758, 1974 (to Monsanto).
6. Large, G. U.S. Patent 4 315 765, 1982 (to Stauffer Chemical).
7. Bakel, I. U.S. Patent 4 397 676, 1983 (to Geshuri Laboratories).
8. Sanders, H. J. *Chem. Eng. News* **1981,** *Aug. 3*, 34.
9. Jaworski, E. G. *J. Agric. Food Chem.* **1972,** *20* (6), 1195–1198.
10. Steinrucken, H. C.; Amrhein, N. *Biochem. Biophys. Res. Commun.* **1980,** *94* (4), 1207–1212.
11. Toy, A. D. F.; Uhing, E. H. U.S. Patent 3 160 632, 1964 (to Stauffer Chemical).
12. Schwarzenbach, G.; Ackermann, H.; Ruhstuhl, P. *Helv. Chim. Acta* **1949,** *32*, 1175.
13. Madsen, H. E. L.; Christensen, H. H.; Gottlieb-Petersen, C. *Acta Chem. Scand., Ser. A* **1978,** *32* (*1*), 79–83.
14. Motekaitis, R. J.; Martell, A. E. *J. Coord. Chem.* **1985,** *14,* 139–149.
15. Motekaitis, R. J.; Murase, I.; Martell, A. E. *J. Inorg. Nucl. Chem.* **1971,** 3353–3365.
16. Franz, J. E. U.S. Patent 4 147 719, 1979 (to Monsanto Chemical).
17. Bayer, E.; Gugel, K. H.; Hägele, K.; Hagenmaier, H.; Jessipow, S.; Konig, W. A.; Zahner, H. *Helv. Chim. Acta* **1972,** *55*, 224.
18. Kocher, H. *Aspects Appl. Biol.* **1983,** *4*, 227–234.
19. Kondo, Y.; Shomura, T.; Ogawa, Y.; Tsuruoka, T.; Watanabe, H.; Totsukawa, K.; Suzuki, T.; Motiyama, C.; Yoshida, J.; Inouye, S.; Niida, T. *Meiji Seika Kenkyu Nempo* **1973,** *13*, 34.
20. Polish Patent 198 145, 1977.
21. Staub, T. H.; Hubele. *Chemie der Pflanzenschutz- und Schadingsbekämpfungsmittel*; Wegler, R., Ed.; Springer-Verlag: Berlin, 1981; Vol. 6, pp 407–410.
22. Sauers, R. F. U.S. Patent 4 255 521, Sept. 30, 1980 (to E. I. du Pont de Nemours).
23. Rupp, W.; Finke, M.; Bieringer, H.; Langeüddke, P.; Kleiner, H.-J. U.S. Patent 4 168 963, 1979 (to Hoechst Aktiengesellschaft).
24. *Chem. Eng. News* **1981,** *59* (*Aug. 3*), 34.

Glossary

A adenine
ampere
Å angstrom
ABS acrylonitrile–butadiene–styrene
ACC 1-aminocyclopropane-1-carboxylic acid
AChE acetylcholinesterase
ADA American Dental Association
ADP adenosine diphosphate
Al aluminum
AMP adenosine monophosphate
Ar aromatic
atm atmosphere
ATP adenosine triphosphate
B boron
Ba barium
BOD biological oxygen demand
c cent
C carbon
cytosine
°C degree Celsius
Ca calcium
cal calorie
cAMP cyclic adenosine monophosphate
Cl chloride
cm centimeter
cm^3 cubic centimeter
CoA coenzyme A
cP centipoise
Cr chromium
Cu copper
DAP diammonium phosphate
DDVP dimethyl dichlorovinyl phosphate
DEPA bis(2-ethylhexyl) phosphate
DFP diisopropyl phosphorofluoridate
DIOP diphosphine

DIPAMP (*R*,*R*)-1,2-bis(phenylanisylphosphine)ethane
DNA deoxyribonucleic acid
DPD dicalcium phosphate dihydrate
DSP disodium phosphate
EDTA ethylenediaminetetraacetic acid
emf electromotive force
EPA Environmental Protection Agency
equiv equivalent
F fluoride
°F degree Fahrenheit
Fe iron
ft^2 square feet
g gram
G guanine
gal. gallon
Ge germanium
GRS government rubber–styrene
h hour
H hydrogen
HEDPA 1-hydroxyethylidene-1,1-diphosphonic acid
HEW Department of Health, Education, and Welfare
hU 5,6-dihydrouridine
I inosine
IMP insoluble sodium metaphosphate
in. inch
K potassium
kcal kilocalorie
kg kilogram
L liter
La lanthanum
lb pound
LD_{50} minimum lethal dose required to kill 50% of the subjects
Li lithium
μm micron
$m_2{}^2G$ 2,2-dimethylguanosine
m′G 1-methylguanosine
m′I 1-methylinosine
MAP monoammonium phosphate
MFP sodium monofluorophosphate
mg milligram
Mg magnesium
mil 0.0001 in.
min minute
mm millimeter

mol mole
mRNA messenger ribonucleic acid
$\eta^{25}{}_{D}$ index of refraction
Na sodium
NAD nicotinamide adenine dinucleotide
NAD_{ox} oxidized nicotinamide adenine dinucleotide
NAD_{red} reduced nicotinamide adenine dinucleotide
Nd neodymium
Ni nickel
NPD aminotri(methylphosphonic acid)
NPMG *N*-(phosphonomethyl)glycine
NTA nitrilotriacetic acid
NTPA nitrilotris(methylenephosphonic acid)
O oxygen
OPAP dioctylphenyl phosphate
oz ounce
P phosphorus
2-PAM 2-[(hydroxyimino)methyl]methylpyridinium iodide
3-PAM 3-[(hydroxyimino)methyl]methylpyridinium iodide
Pb lead
Pd palladium
pH $-\log [H^+]$
P_i inorganic phosphorus
ppb parts per billion
ppm parts per million
psi pseudouridine
PVC poly(vinyl chloride)
R alkyl group
RDA radioactive dentin abrasion
RNA ribonucleic acid
rRNA ribosomal ribonucleic acid
S sulfur
SALP sodium aluminum phosphate
SAPP sodium acid pyrophosphate
SAS sodium aluminum sulfate
Sb antimony
Si silicon
Sn tin
SNTA sodium nitrilotriacetate
STMP cyclic sodium trimetaphosphate
STPP sodium tripolyphosphate
T ribothymidine
thymine
TDBPP tris(dibromopropyl) phosphate

TEPP tetraethyl pyrophosphate
THPC tetrakis(hydroxymethyl)phosphonium chloride
THPOH tetrakis(hydroxymethyl)phosphonium hydroxide
TKPP tetrapotassium pyrophosphate
TOCP tri-*o*-cresyl phosphate
TOPO trioctylphosphine oxide
tRNA transfer ribonucleic acid
TSP trisodium phosphate
TSPP tetrasodium pyrophosphate
U uracil
uranium
V vanadium
Zn zinc
Zr zirconium

Index

A

B

C

D

E

I

K

L

M

N

R

S

T

U

V

W

Y

Z

Copy editing, production, and indexing by Deborah H. Steiner
Text design by Pamela Lewis
Managing Editor: Janet S. Dodd

Typeset by Techna Type, Inc., York, PA,
and Hot Type Ltd., Washington, DC
Printed and bound by Maple Press Company, York, PA